THE INTERACTION OF OCEAN WAVES AND WIND

The Interaction of Ocean Waves and Wind is the first book to describe in detail the two-way interaction between wind and ocean waves and shows how ocean waves affect weather forecasting on time scales of 5 to 90 days.

Winds generate ocean waves, but at the same time airflow is modified due to the loss of energy and momentum to the waves; thus, momentum loss from the atmosphere to the ocean depends on the state of the waves. This volume discusses ocean-wave evolution according to the energy balance equation, which describes the rate of wave-spectrum change by advection and by physical processes such as wind input, dissipation and nonlinear interactions. An extensive overview of nonlinear transfer is given, and as a by-product the role of four-wave interactions in the generation of extreme events, such as freak waves, is discussed. Effects on ocean circulation are described. Coupled ocean-wave–atmosphere modelling gives improved weather and wave forecasts.

This volume will interest ocean-wave modellers, physicists and applied mathematicians, and engineers interested in shipping and coastal protection.

Peter Janssen is head of the Ocean Waves Section at the European Centre for Medium-Range Weather Forecasts (ECMWF), Reading, UK.

Cover: water colour by Danielle Mérelle.

T0192552

THE INTERACTION OF OCEAN WAVES AND WIND

PETER JANSSEN

European Centre for Medium-Range Weather Forecasts

CAMBRIDGE
UNIVERSITY PRESS

CAMBRIDGE UNIVERSITY PRESS
Cambridge, New York, Melbourne, Madrid, Cape Town, Singapore, São Paulo, Delhi

Cambridge University Press
The Edinburgh Building, Cambridge CB2 8RU, UK

Published in the United States of America by Cambridge University Press, New York

www.cambridge.org
Information on this title: www.cambridge.org/9780521121040

First published 2004
This digitally printed version 2009

A catalogue record for this publication is available from the British Library

Library of Congress Cataloguing in Publication data

ISBN 0 521 46540 0 hardback

Janssen, Peter, 1951–
The interaction of ocean waves and wind/Peter Janssen.
p.   cm.
Includes bibliographical references and index.
ISBN 0 521 46540 0
1. Ocean–atmosphere interaction.   I. Title.
GC190.2.J36 2004
551.46′3 – dc22   2004045181

ISBN 978-0-521-46540-3 hardback
ISBN 978-0-521-12104-0 paperback

Contents

Preface

This is a book about ocean waves, their evolution and their interaction with the environment. It presents a summary and unification of my knowledge of wave growth, nonlinear interactions and dissipation of surface gravity waves, and this knowledge is applied to the problem of the two-way interaction of wind and waves, with consequences for atmosphere and ocean circulation.

The material of this book is, apart from my own contributions, based on a number of sources, ranging from the works of Whitham and Phillips to the most recent authoritative overview in the field of ocean waves, namely the work written by the WAM group, *Dynamics and Modelling of Ocean Waves*. Nevertheless, the present book is limited in its scope because it will hardly address interesting issues such as the assimilation of observations, the interpretation of satellite measurements from, for example, the radar altimeter, the scatterometer and the synthetic-aperture radar, nor will it address shallow-water effects. These are important issues but I felt that the reader would be served more adequately by concentrating on a limited number of subjects, emphasizing the role of ocean waves in practical applications such as wave forecasting and illuminating their role in the air–sea momentum exchange.

I started working on this book some 8 years ago. It would never have been finished had it not been for the continuous support of my wife Danielle Mérelle. Her confidence in my ability to complete this work far exceeded my own. I thank my parents, Aloysius Janssen and Rosa Burggrave, for supporting me to follow a university education. I am indebted to my Ph.D. advisor Martin Weenink and L. J. F. Broer for their introduction into the field of nonlinear physics. Also, it is a pleasure to acknowledge the contributions of P. G. Saffman and G. B. Whitham to my education in ocean waves. Things really started to happen when I joined the WAve Model (WAM) group. Most of the members of the WAM group thought that this was a unique opportunity for collaboration, and we had the time of our lives. I would like to thank Gerbrand Komen, Klaus and Susanne Hasselmann, Mark Donelan and Luigi Cavaleri for all the fruitful discussions and the collaborations. Furthermore, I would like to thank Luciana Bertotti, Heinz Günther, Anne Guillaume, Piero Lionello and Liana Zambresky for sharing the burden of the development of a beautiful piece of software, and for all the fun we had.

Last, but not least, I would like to thank Pedro Viterbo and Jim Doyle for disentangling all the intricacies involved in the actual coupling of an atmospheric model and an ocean-wave prediction system. The former and present members of ECMWF's ocean-wave team, Jean Bidlot, Björn Hansen, Saleh Abdalla, Hans Hersbach and Øyvind Saetra are thanked for their dedicated efforts to further develop the WAM model software, while support by Lennart Bengtsson, David Burridge, Anthony Hollingsworth, Adrian Simmons and, in particular, Martin Miller is much appreciated.

Saleh Abdalla, Jean Bidlot, Luigi Cavaleri and Miguel Onorato are thanked for critically reviewing parts of the manuscript. The fine artwork by Anabel Bowen is really appreciated while Carsten Maass is thanked for his advice with LaTeX.

1
Introduction

The subject of ocean waves and their generation by wind has fascinated me greatly since I started to work in the Department of Oceanography at the Royal Netherlands Meteorological Institute (KNMI) at the end of 1979. The wind-induced growth of water waves on a pond or a canal is a daily experience for those who live in the lowlands, yet it appeared that this process was hardly understood. Gerbrand Komen, who arrived 2 years earlier at KNMI and who introduced me to this field, pointed out that the most prominent theory explaining wave growth by wind was the Miles (1957) theory which relied on a resonant interaction between wind and waves. Since I did my Ph.D. in plasma physics, I noticed immediately an analogy with the problem of the interaction of plasma waves and electrons; this problem has been studied extensively both experimentally and theoretically. The plasma waves problem has its own history. It was Landau (1946), who discovered that depending on the slope of the particle distribution function at the location where the phase velocity of the plasma wave equals the particle velocity, the plasma wave would either grow or damp. Because of momentum and energy conservation this would result in a modification of the particle-velocity distribution. For a spectrum of growing plasma waves with random phase, this problem was addressed in the beginning of the 1960s by Vedenov et al. (1961) and by Drummond and Pines (1962). The principal result these authors found was that because of the growth of the plasma waves the velocity distribution would change in such a way that for large times its slope vanishes in the resonant region, thereby removing the cause of the instability. Thus, a new state emerges consisting of a mixture of stable, finite-amplitude plasma waves and a modified particle-velocity distribution.

Based on this analogy, I realized that the approach by Miles (1957), which relied on linear theory, could not be complete, because energy and momentum were not conserved. Taking nonlinear effects into account would enable me to determine how much momentum transfer there is from the wind to the waves, which would give rise to a wave-induced stress on the airflow. This resulted then in a slowing down

1

of the airflow, and hence in a modified wind profile. Considering, for simplicity, the two-dimensional problem only (hence wave propagation in one direction), I performed the necessary calculations which were similar in spirit to those of the plasma problem. They indeed confirmed my expectation that in the presence of growing water waves the wind profile would change. The role of the particle-velocity distribution in this problem was played by the vorticity of the mean flow; hence, in the absence of all kinds of other effects (e.g. turbulence) a new state would emerge consisting of stable, finite-amplitude water waves and a mean flow of which the gradient of the mean vorticity would vanish in the resonant region. It should be noted that a number of years earlier, Fabrikant (1976) reached a similar conclusion while Miles (1965) also addressed certain aspects of this problem. This theory has become known as the quasi-linear theory of wind-wave generation.

A number of colleagues at KNMI pointed out to me, however, that my treatment was far from complete if it was to be of practical value. And, indeed, I had neglected lots of complicating factors such as nonlinear wave–wave interactions, dissipation due to white capping, flow separation, air turbulence, water turbulence, etc. For example, it is hard to imagine that in the presence of air turbulence the mean airflow would have a linear dependence on height (corresponding to the vanishing of the gradient of its vorticity), since the turbulent eddies would try to maintain a logarithmic profile. Thus, in general, it is expected that there will be competition between the effect of ocean waves, through the wave-induced stress, and turbulence; presumably, the steeper the waves, the larger the wave effect will be. Nevertheless, it was evident that knowledge of the momentum transfer from air to sea required knowledge of the evolution of ocean waves, which apart from wind input is deter-mined by nonlinear wave–wave interactions and dissipation due to white capping. In short, in order to show the practical value of the idea of the wave effect on the airflow, the running of a wave model was required.

At the beginning of the 1980s a spectral ocean-wave model, including wave–wave interactions, was not considered to be a viable option. The reason for this was that there was not enough computer power available to determine the nonlinear transfer in a short enough time to be of practical value for wave forecasting. This picture changed with the introduction of the first supercomputers and with the work of Hasselmann and Hasselmann (1985) who proposed an efficient parametrization of the nonlinear transfer. Combined with the promise of the wealth of data on the ocean surface from remote-sensing instruments on board new satellites such as ERS-1, ERS-2 and Topex-Poseidon, this provided sufficient stimulus to start a group of mainly European wave modellers who called themselves the WAve Model (WAM) group. Apart from a keen interest in advancing our knowledge regarding the physics of ocean waves and assimilation of wave observations, the main goal was to develop a spectral wave model based on the so-called 'energy balance equation' which included the physics of the generation of ocean waves by wind, dissipation due to white capping and, of course, nonlinear interactions. I joined the WAM group in 1985 because of my interest in wave prediction and, in the back of my mind, with the hope that perhaps I could now study the consequences of the slowing down of the airflow in the presence of ocean waves.

The interests and backgrounds of the members of the WAM group varied greatly. It brought together experimentalists, theorists, wave forecasters and people with a commercial interest. Nevertheless, owing to the great enthusiasm of the group, the tremendous efforts by Susanne Hasselmann to develop a first version of the WAM model, and, not least, the computer facilities generously provided by the European Centre for Medium-Range Weather Forecasts (ECMWF), developments progressed rapidly. After a number of studies on the limited area of the North Sea and the north-east Atlantic with promising results, a global version of the WAM model was running quasi-operationally at ECMWF by March 1987. Surface windfields were obtained from the ECMWF atmospheric model. The reason for the choice of this date was that by mid March a large experimental campaign, measuring two-dimensional wave spectra, started in the Labrador Sea (Labrador Sea Extreme Waves Experiment (LEWEX)). Results of the comparison between observed and modelled spectra were later reported at the final LEWEX meeting by Zambresky (1991). By August 1987, a first version of an altimeter wave-height data assimilation system had already been tested by Piero Lionello, while a number of verification studies on wave-model performance were well underway by the end of 1987. Zambresky (1989) compared 1 year of WAM model results with conventional buoy observations, while Janssen *et al.* (1989) and Bauer *et al.* (1992) compared results with altimeter wave-height data from the Seasat mission and Romeiser (1993) compared with Geosat altimeter data. Meanwhile the WAM model, which originally was a deep-water model with some simple shallow-water effects, was generalized extensively to include bottom and current refraction effects, while the problem of swell dissipation being too strong (as was evident from the comparison studies with altimeter data) was alleviated by modifying the dissipation source term. Finally, extensive efforts were devoted to beautifying the wave-model code and making it more efficient, and in July 1992 the WAM model became operational at ECMWF. By the end of 1994, the WAM model was distributed to more than 75 institutes, reflecting the success of the WAM group. A more detailed, scientific account of all this may be found in Komen *et al.* (1994).

In the meantime, while taking part in the WAM group, I tried to assess the relevance of my findings on the slowing down of airflow by ocean waves. First of all, observational evidence suggested that the drag coefficient C_D increases with wind speed U_{10}. Here the drag coefficient C_D follows from the kinematic stress τ and the wind speed at 10 m height according to $C_D = \tau / U_{10}^2$. The increase of

C_D with U_{10} for airflow over ocean waves is in contrast with the classical results of airflow over a smooth, flat plate. For such a surface, the slowing down of the airflow is caused by viscous dissipation. As a result, since for larger windspeed, and hence larger Reynolds number, the effect of viscosity becomes less important, the drag coefficient decreases with wind speed. Apparently, in the presence of ocean waves there are additional ways to transfer air momentum; an obvious candidate for such a process is the generation of surface waves by wind. This was realized by Charnock (1955) who suggested that the roughness length of airflow over ocean waves should therefore depend on two parameters, namely acceleration of gravity g and friction velocity $u_* = \tau^{1/2}$. Dimensional considerations then gave rise to the celebrated Charnock relation for the roughness length, and, although in the mid 1950s there was hardly any observational evidence, a realistic estimate for the Charnock parameter was given as well. In Charnock's analysis, it was tacitly assumed that the sea state was completely determined by the local friction velocity u_*. However, observations of the windsea state obtained during the Joint North Sea Wave Project (JONSWAP, in 1973) suggested that the shape of the ocean-wave spectrum depends on the stage of development of the sea state or the so-called 'wave age'. In the early stages of development, called 'young' windsea, the wave spectrum showed a very sharp peak while the high-frequency waves were steep. On the other hand, when the sea state approaches equilibrium, the wind waves were less steep and the spectral peak was less pronounced. This led Stewart (1974) to suggest that the Charnock parameter is not really a constant, but should depend on the stage of development of wind waves.

Thus, the work of Charnock and Stewart suggested that wind-generated gravity waves, which receive energy and momentum from the airflow, should contribute to the slowing down of the airflow. In other words, ocean waves and their associated momentum flux may be important in controlling the shape of the wind profile over the oceans. However, the common belief in the field was that air turbulence was dominant in shaping the wind profile while the effect of surface gravity waves was considered to be small (Phillips, 1977). On the other hand, Snyder *et al.* (1981) found that the momentum transfer from wind to waves might be considerable, therefore the related wave-induced stress may be a substantial fraction of the total stress in the surface layer. This turned out to be the case, particularly for 'young' windseas, which are steep. The consequence is that the momentum transfer from air to ocean and therefore the drag coefficient at 10 m height depend on the sea state. The first experimental evidence for this was found by Donelan (1982), and it was confirmed by Smith *et al.* (1992) during the Humidity Exchange of the Sea (HEXOS) experiment.

It therefore seemed natural to combine results of the quasi-linear theory of wind-wave generation with knowledge on the evolution of wind waves, in order to be able to determine the sea-state dependence of air–sea momentum transfer. Of course, it should be realized that the quasi-linear theory is strictly speaking not valid because, for example, effects of air turbulence on the wave-induced motion are disregarded, and also effects of flow separation are ignored. Nevertheless, I thought it worthwhile to study whether it was posssible to obtain, in the context of this theory, realistic estimates of the air–sea momentum transfer. This turned out to be the case. However, results were found to depend in a sensitive manner on the state of the high-frequency waves because these are the fastest-growing waves and therefore carry most of the wave-induced stress. The close relation between aerodynamic drag and the sea state implied that an accurate knowledge of momentum transfer required a reliable determination of the high-frequency part of the spectrum. It turned out that this could be provided by the WAM model.

The consequence was that a reliable knowledge of momentum transfer required the running of a wave model because of the two-way interaction between wind and waves. I therefore started wondering whether the sea-state dependence of the drag would be relevant in other areas of geophysics such as in storm-surge modelling, weather prediction, the atmospheric climate and gas transfer. Although observations (Donelan, 1982) and theory (Janssen, 1989) did suggest an enhancement of drag by a factor of 2 for young windsea, which is quite significant, it appears that the relevance of this wave effect can only be assessed after performing some numerical experiments. One of the reasons for this is that when a change is being made in one part of a complicated system, (unexpected) compensations may occur that are induced by other parts of the system. Consider, as an example, the impact of the sea state on the evolution of a depression. When the wind starts blowing, the young sea state will give an increased roughness which on the one hand may result in an enhanced filling up of the pressure low; on the other hand, however, the enhanced roughness may lead to an increased heat flux which, through vortex stretching, results in a deeper depression. The final outcome can, therefore, only be determined in the context of a coupled ocean-wave–atmosphere model.

To date, a number of studies have shown the relevance of the sea-state-dependent momentum transfer for storm-surge modelling (Mastenbroek *et al.*, 1993), weather prediction (Doyle, 1995; Janssen *et al.*, 2002), the atmospheric climate (Janssen and Viterbo, 1996) and ocean circulation (Burgers *et al.*, 1995). These studies suggest that the modelling of momentum transfer (and also of heat and moisture) can only be done adequately in the context of a coupled model. Ideally, one would therefore imagine one grand model of our geosphere, consisting of an atmospheric- and an ocean-circulation model, where the necessary interface between ocean and atmosphere is provided by an ocean-wave model.

This book is devoted to the problem of the two-way interaction of wind and waves and the possible consequences for air–sea interaction. I therefore start with an introduction to the subject of ocean waves. First, important concepts and tools such as dynamical equations, the dispersion relation, the role of the group velocity and the Hamiltonian and the Lagrangian for ocean waves are introduced. This is followed by an emphasis on the need for a statistical description of ocean waves by means of the wave spectrum. The evolution equation for the wave spectrum, called the 'energy balance equation', is derived from Whitham's averaged Lagrangian approach. The energy balance equation describes the rate of change of the wave spectrum due to advection and refraction on the one hand and, on the other hand, the rate of change due to physical processes such as wind input, nonlinear interactions and dissipation by white capping. After a brief discussion of advection and refraction I will give a thorough discussion of the energy transfer from wind to ocean waves, the consequent slowing down of the airflow and of nonlinear interactions. This is followed by a brief discussion of the least-understood aspect of wave dynamics, namely dissipation due to white capping.

Next, the role of the various source terms in shaping the wave spectrum is studied, resulting in an understanding of the evolution of the windsea spectrum. At the same time the sea-state dependence of the air–sea momentum transfer is considered and its sensitive dependence on the high-frequency part of the wave spectrum is emphasized.

Because air–sea interaction depends in a sensitive way on the quality of the sea state, the present status of ocean-wave forecasting needs to be addressed. This is done by presenting a validation of ECMWF wave forecast and analysis results against conventional buoy data and against altimeter wave-height data obtained from the ERS-2 satellite.

Having established the role of an ocean waves in the field of air–sea interaction, it is suggested that the standard model of the geosphere, which usually consists of an atmospheric- and an ocean-circulation model, should be extended by means of an ocean-wave model that provides the necessary interface between the two. The role of ocean waves in air–sea interaction is then illustrated by studying the impact of the sea-state-dependent momentum transfer on storm surges, and by showing that ocean waves also affect the evolution of weather systems such as depressions. Finally, ocean waves are also shown to affect, in a systematic manner, the atmospheric climate on a seasonal time scale.

2

The energy balance of deep-water ocean waves

In this chapter we shall try to derive, from first principles, the basic evolution equation for ocean-wave modelling which has become known as the energy balance equation. The starting point is the Navier–Stokes equations for air and water. The problem of wind-generated ocean waves is, however, a formidable one, and several approximations and assumptions are required to arrive at the desired result. Fortunately, there are two small parameters in the problem, namely the steepness of the waves and the ratio of air density to water density. As a result of the relatively small air density, the momentum and energy transfer from air to water is relatively small so that, because of wind input, it will take many wave periods to have an appreciable change of wave energy. In addition, the steepness of the waves is expected to be relatively small. In fact, the assumption of small wave steepness may be justified a posteriori. Hence, because of these two small parameters one may distinguish two scales in the time–space domain, namely a short scale related to the period and wavelength of the ocean waves and a much longer time and length scale related to changes due to small effects of nonlinearity and the wind-induced growth of waves.

Using perturbation methods, an approximate evolution equation for the amplitude and the phase of the deep-water gravity waves may be obtained. Formally, in lowest order one then deals with free surface gravity waves while higher-order terms represent the effects of wind input, nonlinear (four-) wave interactions and dissipation. In this manner the problem of wind-generated surface gravity waves (shown schematically in Fig. 2.1) may be solved.

After Fourier transformation, a set of ordinary differential equations for amplitude and phase of the waves is obtained which may be solved on a computer. This approach is followed in meteorology. The reason for its success is that the integration period (between 5 and 10 days) is comparable to the period of the long atmospheric waves. For water waves this approach is not feasible, however, because of the disparity between a typical wavelength of ocean waves (in the range of 1 to

7

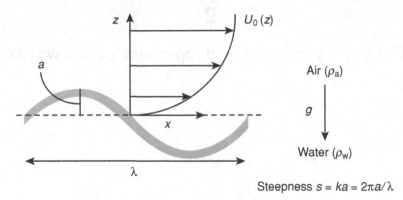

Fig. 2.1. Schematic of the problem in two dimensions.

1000 m) and the size of a typical ocean basin (of the order of 10 000 km). A way of circumventing this problem is to employ a multiple-scale approach. Since there are two scales in the problem at hand, and since the solution for the free gravity waves is known, we only have to consider the evolution of the wave field on the long time and space scale, thus making the wave forecasting problem on a global scale a tractable one.

Furthermore, in practice there is no need for detailed information regarding the phase of the ocean waves. In fact, there are no observations of the phase of ocean waves on a global scale. Usually, we can content ourselves with knowledge about the distribution of wave energy over wavenumber $\mathbf{k}$. In other words, only knowledge of the wave spectrum $F(\mathbf{k})$ is required. A statistical description of the sea state, giving the wave spectrum averaged over a finite area, seems therefore the most promising way to proceed. From the slow time evolution of the wave field it follows that the wave spectrum F is a slowly varying function of time as well. Its evolution equation, called the energy balance equation, is the final result of this chapter. We conclude the chapter by giving a brief overview of our knowledge on observations of wave evolution. This will be accompanied by the introduction of a number of relevant physical parameters, all derived from the wavenumber spectrum F, that are frequently used in the remainder of this work.

2.1 Preliminaries

Referring to Fig. 2.1 for the geometry, our starting point is the usual evolution equation for an incompressible, two-layer fluid consisting of air and water. Consider a fluid with density ρ that flows with a velocity $\mathbf{u}$. In general, density and velocity depend on position $\mathbf{x} = (x, y, z)$ and time t. A right-handed coordinate system is chosen in such a way that the coordinate z points upwards while the acceleration of gravity **g** points in the negative z-direction. The rate of change of the velocity is caused by the Coriolis force, by the pressure, p, gradient, by acceleration of gravity and by the divergence of the stress tensor τ. Denoting the interface between air and water by $\eta(\mathbf{x}, t)$, we then have

$$\nabla \cdot \mathbf{u} = 0,$$

$$\left(\frac{\partial}{\partial t} + \mathbf{u} \cdot \nabla\right)\mathbf{u} + \mathbf{f} \times \mathbf{u} = -\frac{1}{\rho}\nabla p + \mathbf{g} + \nabla \cdot \tau, \tag{2.1}$$

where

$$\rho = \begin{cases} \rho_a, & z > \eta, \\ \rho_w, & z < \eta, \end{cases}$$

and the subscripts a and w refer to air and water respectively. For surface gravity waves, the Coriolis acceleration may be ignored because the frequency of the waves is much higher than the Coriolis parameter **f**. Velocities and forces, such as the normal and tangential stresses, are continuous at the interface. A particle on either side of the surface, described by $z = \eta(\mathbf{x}, t)$ will move in a time Δt from $(\mathbf{x}, z = \eta)$ to $(\mathbf{x} + \Delta\mathbf{x}, z + \Delta z = \eta(\mathbf{x} + \Delta\mathbf{x}, t + \Delta t))$ with $\Delta\mathbf{x} = \mathbf{u}\Delta t$ and $\Delta z = w \Delta t$. Thus, by Taylor expansion of $z + \Delta z$ and by taking the limit $\Delta t \to 0$ one obtains the kinematic boundary condition

$$\frac{\partial \eta}{\partial t} + \mathbf{u} \cdot \nabla \eta = w. \tag{2.2}$$

Here, **u** is the horizontal velocity at the interface while w is its vertical velocity. In order to complete the set of equations, one has to express the stress tensor τ in terms of properties of the mean flow. The stress contains the viscous stress and in addition may contain contributions from unresolved turbulent fluctuations (the Reynolds stress).

Finally, boundary conditions have to be specified. In deep water one imposes the condition that for $z \to \pm\infty$ the wave motion should vanish. However, for waves in water of finite depth the normal component of the water velocity should vanish at the bottom.

In order to derive the energy balance equation we shall discuss the properties of pure gravity waves. Thus the following approximations are being made.

- Neglect viscosity and stresses. This gives the Euler equations. Continuity of the stress at the interface of air and water is no longer required. The parallel velocity at the interface may now be discontinuous.
- We disregard the air motion altogether because $\rho_a/\rho_w \ll 1$. In our discussion on wave growth, the effects of finite air–water density ratio are, of course, retained.

- We assume that the water velocity is irrotational. This is a reasonable assumption for water waves. In the framework of the Euler equations, it can, in fact, be shown that the vorticity remains zero when it is zero initially.

The condition of zero vorticity is automatically satisfied for velocity fields that are derived from a velocity potential ϕ. Hence,

$$\mathbf{u} = \nabla\phi \qquad (2.3)$$

and since the flow is divergence free the velocity potential satisfies Laplace's equation inside the fluid,

$$\nabla^2\phi + \frac{\partial^2\phi}{\partial z^2} = 0 \qquad (2.4)$$

with two conditions at the surface $z = \eta(x, y, t)$,

$$z = \eta, \begin{cases} \dfrac{\partial\eta}{\partial t} + \nabla\phi \cdot \nabla\eta = \dfrac{\partial\phi}{\partial z}, \\[2mm] \dfrac{\partial\phi}{\partial t} + \dfrac{1}{2}(\nabla\phi)^2 + \dfrac{1}{2}(\dfrac{\partial\phi}{\partial z})^2 + g\eta = 0 \text{ (Bernoulli)}, \end{cases} \qquad (2.5)$$

and a condition at the bottom $z = -D$, which is assumed to be flat,

$$z = -D, \ \nabla\phi = 0. \qquad (2.6)$$

We remark that the Bernoulli equation in Eq. (2.5) follows immediately from the Euler equations with zero vorticity, combined with the boundary condition of zero pressure at the surface.

The set of equations (2.4)–(2.6) determines the evolution of free gravity waves. At first sight this appears to be a relatively simple problem, because the relevant differential equation is Laplace's equation, which may be solved in a straightforward manner. The important point to note is, however, that Laplace's equation needs to be solved in a domain that is not known beforehand, but which is part of the problem. This is what makes the problem of free surface waves such a difficult, but also such an interesting, problem as the nonlinearity enters our problem through the boundary conditions at the surface $z = \eta(\mathbf{x}, t)$.

In order to make progress we need to introduce two additional tools that will facilitate the further development of the theory of surface gravity waves. The system of equations (2.4)–(2.6) has the elegant property that it conserves the total energy which is a necessary requirement for the existence of a Hamiltonian and a Lagrangian. The Hamiltonian for water waves, first discovered by Zakharov (1968), is useful in deriving the nonlinear wave–wave interactions in a systematic way, while the Lagrangian, first discovered by Luke (1967), plays a key role in obtaining the energy balance equation.

It is well-known that Eqs. (2.4)–(2.6) conserve the total energy E of the fluid,

$$E = \frac{1}{2}\rho g \int d\mathbf{x} \, \eta^2 + \frac{1}{2}\rho \int d\mathbf{x} \int_{-D}^{\eta} dz \left((\nabla\phi)^2 + (\frac{\partial\phi}{\partial z})^2 \right). \tag{2.7}$$

Here, the first term is the potential energy of the fluid while the second term is its kinetic energy.

By choosing appropriate canonical variables, Zakharov (1968), Broer (1974) and Miles (1977) independently found that E may be used as a Hamiltonian. The proper canonical variables are

$$\eta, \text{ and } \psi(\mathbf{x}, t) = \phi(\mathbf{x}, z = \eta, t). \tag{2.8}$$

The boundary conditions at the interface are then equivalent to Hamilton's equations

$$\frac{\partial\eta}{\partial t} = \frac{\delta E}{\delta\psi}, \quad \frac{\partial\psi}{\partial t} = -\frac{\delta E}{\delta\eta}, \tag{2.9}$$

where $\delta E/\delta\psi$ and $\delta E/\delta\eta$ are functional derivatives.

The formulation of the water-wave problem in terms of a Hamiltonian has certain advantages. If one is able to solve the potential equation

$$\nabla^2\phi + \frac{\partial^2\phi}{\partial z^2} = 0$$

with boundary conditions

$$\phi(\mathbf{x}, z = \eta) = \psi, \quad \frac{\partial}{\partial z}\phi(\mathbf{x}, z = -D) = 0,$$

thereby expressing ϕ in terms of the canonical variables η and ψ, then the energy E can be evaluated in terms of these canonical variables and the evolution in time of η and ψ follows at once from the Hamilton equations (2.9).

There is also a Lagrangian formulation of the water-wave problem. Luke (1967) found that the variational principle

$$\delta \int d\mathbf{x} \, dt \, \mathcal{L} = 0 \tag{2.10}$$

with

$$\mathcal{L} = -\rho \int_{-D}^{\eta} dz \left\{ \frac{\partial\phi}{\partial t} + \frac{1}{2}(\nabla\phi)^2 + \frac{1}{2}(\frac{\partial\phi}{\partial z})^2 + gz \right\}$$

gives Laplace's equation and the appropriate boundary conditions.

One would expect that the Lagrangian and Hamiltonian description of surface waves is equivalent. Indeed, Miles (1977) was able to derive the Hamilton equations (2.9) from Luke's variational principle.

Intermezzo   Readers not familiar with Hamiltonians and Lagrangians are advised to study the following brief account on the fundamentals of classical mechanics. We first discuss Hamilton's equations. Consider a particle with momentum p and position q in a potential well V. The total energy of the particle, with mass m, is then given by the sum of kinetic energy T and potential energy V, or

$$E = \frac{1}{2}\frac{p^2}{m} + V(q). \tag{2.11}$$

Regard p and q as canonical variables. Then, with $\dot{q} = \partial q/\partial t$, etc., Hamilton's equations become

$$\dot{q} = \frac{\partial E}{\partial p} = \frac{p}{m}, \tag{2.12a}$$

$$\dot{p} = -\frac{\partial E}{\partial q} = -\frac{\partial V}{\partial q}. \tag{2.12b}$$

Eliminating momentum p we obtain with velocity $v = \dot{q}$,

$$m\dot{v} = -\frac{\partial V}{\partial q} = \text{Force}, \tag{2.13}$$

which we recognize as Newton's law where the force is derived from the potential V.

In classical mechanics, the Hamiltonian formulation follows from the principle of 'least' action. In order to see this, consider the Lagrangian

$$\mathcal{L} = T - V = \frac{1}{2}m\dot{q}^2 - V(q) = \mathcal{L}(q, \dot{q}). \tag{2.14}$$

Newton's law then follows from the condition that the action $\mathcal{A}$ be extremal, where

$$\mathcal{A} = \int_{t_1}^{t_2} dt\, \mathcal{L}(q, \dot{q}). \tag{2.15}$$

The action is now extremal if $\delta\mathcal{A} = 0$, which is equivalent to the requirement that over an arbitrarily chosen time interval (t_1, t_2) the difference between kinetic and potential energy is minimized. Here,

$$\delta\mathcal{A} = \int_{t_1}^{t_2} dt\, [\mathcal{L}(q + \delta q, \dot{q} + \delta\dot{q}) - \mathcal{L}(q, \dot{q})].$$

Taylor expansion of the first term, and disregarding terms of higher order in δq and $\delta\dot{q}$, gives

$$\delta\mathcal{A} = \int_{t_1}^{t_2} dt\, \left[\delta q\frac{\partial}{\partial q}\mathcal{L} + \delta\dot{q}\frac{\partial}{\partial\dot{q}}\mathcal{L}\right]$$

$$= \int_{t_1}^{t_2} dt\, \left[\frac{\partial}{\partial q}\mathcal{L} - \frac{\partial}{\partial t}\frac{\partial}{\partial\dot{q}}\mathcal{L}\right]\delta q,$$

where the last equality follows from a partial integration of the second integral. As the vanishing of the first variation of the action should hold for arbitrary δq one finds that the action is extremal if q satisfies the Euler–Lagrange equations

$$\mathcal{L}_q - \frac{\partial}{\partial t}\mathcal{L}_{\dot{q}} = 0 \Leftrightarrow m\ddot{q} = -\frac{\partial V}{\partial q}. \tag{2.16}$$

Defining the momentum p as

$$p \equiv \mathcal{L}_{\dot{q}}$$

one may eliminate $\dot{q}$ in favour of p, $\dot{q} = \dot{q}(p)$. Then, regarding from now on p and q as independent variables, the Hamiltonian $H = H(p, q)$ is given by

$$H(p, q) = \dot{q}\mathcal{L}_{\dot{q}} - \mathcal{L} = \frac{1}{2}\frac{p^2}{m} + V(q) \tag{2.17}$$

and Hamilton's equations (2.12a) and (2.12b) now follow by differentiating H with respect to q and p.

All this is, however, less straightforward to do for a continuum such as the one we are dealing with. Nevertheless, Miles (1977) was able to derive from Luke's variational principle the Hamilton equations (2.9).

2.2 Linear theory

We have now paid sufficient attention to the basics and it is high time to derive the dispersion relation for surface gravity waves. In linear theory all nonlinear terms are disregarded because of the assumption of small wave steepness and the evolution equations (2.4)–(2.6) for potential flow become

$$\nabla^2 \phi + \frac{\partial^2 \phi}{\partial z^2} = 0, \tag{2.18a}$$

$$z = 0, \begin{cases} \dfrac{\partial \eta}{\partial t} = \dfrac{\partial \phi}{\partial z}, \\[2mm] \dfrac{\partial \phi}{\partial t} + g\eta = 0 \text{ (Bernoulli)}, \end{cases} \tag{2.18b}$$

$$z = -D, \ \frac{\partial \phi}{\partial z} = 0. \tag{2.18c}$$

We are interested in gravity waves that propagate along the surface and that have maximum amplitude at the surface. The elementary sinusoidal solutions with amplitude a take the form

$$\eta = a\,e^{i\theta}, \phi = Z(z)\,e^{i\theta}, \tag{2.19}$$

where the phase θ is given as

$$\theta = \mathbf{k} \cdot \mathbf{x} - \omega t,$$

with **k** the wavenumber and ω the angular frequency. Wavenumber and angular frequency are related to the wavelength λ and the frequency f of the wave according to $k = 2\pi/\lambda$ and $\omega = 2\pi f$. From Laplace's equation the chosen form of ϕ is a solution provided that Z satisfies the ordinary differential equation

$$Z'' - k^2 Z = 0, \ k = |\mathbf{k}| = \sqrt{k_x^2 + k_y^2}. \tag{2.20}$$

The boundary condition on $z = -D$ requires $Z'(-D) = 0$. For water of constant depth D the problem in Eq. (2.20) may be solved in terms of exponential functions, hence

$$Z \sim \cosh k(z + D).$$

Using the second equation of Eq. (2.18b) we find

$$\eta = a \, \mathrm{e}^{\mathrm{i}\theta},$$

$$\phi = -\mathrm{i}\eta \frac{g}{\omega} \cosh k(z + D) / \cosh kD. \tag{2.21}$$

However, the first equation of Eq. (2.18b) has still to be satisfied. This is only possible when the angular frequency obeys the linear dispersion relation

$$\omega^2 = gk \tanh kD. \tag{2.22}$$

This is the dispersion relation of surface gravity waves on still water. It is straightforward to extend this result to the case of gravity waves on a current $\mathbf{U}_0$. In the dispersion relation the angular frequency ω is then replaced by the Doppler-shifted frequency $\omega - \mathbf{k} \cdot \mathbf{U}_0$. For a given wavenumber, Eq. (2.22) has in general two solutions, which represent the case of waves propagating to the right and waves propagating to the left. In addition, it is important to distinguish between deep-water and shallow-water waves.

In deep water we have $D \to \infty$ and therefore Eq. (2.22) becomes

$$\omega^2 = gk. \tag{2.23}$$

The phase speed of the waves, defined as $c = \omega/k$, is then given as

$$c = \sqrt{g/k} = g/\omega. \tag{2.24}$$

Therefore, the high-frequency waves have the lowest phase speed. The energy of the waves is, as will be seen shortly, advected by the group velocity $\partial\omega/\partial k$. In deep water the group velocity v_g becomes

$$v_g = \frac{\partial\omega}{\partial k} = \frac{1}{2}\frac{g}{\omega}, \tag{2.25}$$

hence, the group velocity is exactly half the phase speed. Furthermore, given the solution (2.21), it is straightforward to obtain the energy of the waves. Using Eq. (2.7) we find for the wave energy density $\mathcal{E}$,

$$\mathcal{E} = 2\rho g |a|^2. \tag{2.26}$$

In the case of shallow water we take the limit of small depth and therefore Eq. (2.22) becomes

$$\omega = \pm k\sqrt{gD}. \tag{2.27}$$

The phase speed of shallow-water waves is given by

$$c = \frac{\omega}{k} = \sqrt{gD}, \tag{2.28}$$

which is independent of wavenumber, hence there is no dispersion. As a consequence, the group velocity equals the phase speed,

$$v_g = \frac{\partial \omega}{\partial k} = \sqrt{gD}. \tag{2.29}$$

The energy of shallow-water waves is given by the same expression as the one for deep-water waves, viz. Eq. (2.26). The most extreme examples of shallow-water waves are called tsunamis. These are generated by earthquakes in, for example, the Gulf of Alaska. The resulting surface elevation, although of small amplitude, has a large extent, thus the relevant wavelength may be of the order of a few tenths of a kilometre. These are truly shallow-water waves as the average depth of the North Pacific is of the order of 5 km. As a consequence, the phase speed of these long waves may become quite large, of the order of 800 km/h.

Comparing deep- and shallow-water waves we note that there is an important difference between the two cases. Deep-water waves are highly dispersive whereas truly shallow-water waves are not because they have the same phase speed. Later, it will be seen that this has important consequences for the nonlinear evolution of surface gravity waves. In fact, when dispersive waves are interacting with each other the interaction time will be finite because each wave propagates with a different phase speed. If there is, on the other hand, no dispersion then waves will stick together for a long time with the result that even for small steepness the effect of nonlinearity may become very strong. Consequently, this may give rise to a considerable steepening of the surface elevation, which, in the case of no dispersion, results in shock waves, while for weak dispersion, solitary wave solutions occur. Although the theory of solitary waves and shock waves is a fascinating subject (an elegant account of this is given by Whitham (1974)), I shall not treat this topic here. The reason is that I would like to develop the theory of random, weakly nonlinear waves which simply cannot deal with the strong nonlinear case. Therefore, only dispersive waves are considered, which implies in practice that our results are only valid for waves with $kD = \mathcal{O}(1)$ or larger, where $\mathcal{O}$ denotes order of magnitude.

2.3 Wave groups

In the previous section we discussed some of the properties of a single wave. In practice we know, however, that waves come in groups (see Fig. 2.2). Everyone who has done some sunbathing on the beach and has listened to the breaking ocean waves knows that the seventh wave is the biggest. Even songs are devoted to this subject ('Love is the seventh wave' by Sting on the album entitled '*The Dream of the Blue Turtles*').

If the wave groups are sufficiently long, we can give a reasonably accurate description of their evolution by using a plane-wave solution with slowly varying phase and amplitude. Thus, similarly to geometrical optics, wave groups may be described by

$$\eta = a(\mathbf{x}, t)\, e^{i\theta(\mathbf{x},t)} + \text{c.c.}, \tag{2.30}$$

where c.c. denotes the complex conjugate and both amplitude a and phase θ are slowly varying functions of space and time. Here, 'slow' has a relative meaning; it refers to the basic length and time scale imposed by the wave, namely its wavelength and period. Thus, we require

$$\frac{1}{a}\nabla a \ll k,\; \frac{1}{a}\frac{\partial a}{\partial t} \ll \omega,\; \text{etc.} \tag{2.31}$$

Since also the phase is slowly varying we may define a local angular frequency and wavenumber according to

$$\omega = -\frac{\partial \theta}{\partial t},\; \mathbf{k} = \nabla\theta. \tag{2.32}$$

Assuming that the phase function θ is at least twice differentiable (e.g. $\partial^2\theta/\partial x\partial t = \partial^2\theta/\partial t\partial x$), Eq. (2.32) implies the following consistency relation, known as the

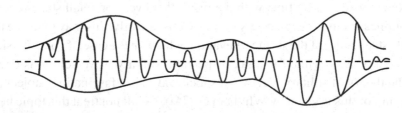

Fig. 2.2. Waves come in groups.

equation of conservation of the number of wave crests:

$$\frac{\partial \mathbf{k}}{\partial t} + \nabla \omega = 0. \tag{2.33}$$

This equation tells us that, if the frequency of the wave depends on position $\mathbf{x}$ (because of, for example, a slowly varying current and/or depth), the wavenumber changes in time. Equation (2.33) therefore provides one of the key elements in the energy balance equation, namely refraction.

The evolution of amplitude a and frequency ω is not arbitrary either. To obtain these evolution equations one could, in principle, substitute the Ansatz (2.30), together with a similar Ansatz for the potential ϕ, into the basic equations (2.4)–(2.6). Then by means of a perturbation analysis the appropriate evolution equations for amplitude a and the dispersion relation for ω may be obtained. This perturbation analysis is, however, not a straightforward one because in higher-order equations so-called 'secular terms' will arise which make the perturbation series invalid after a finite time. A uniformly valid perturbation series is obtained by introducing multiple scales in space and time. Removal of secular terms then will give rise to the slow time and space evolution equation for amplitude and angular frequency.

We shall not follow this approach here. In its stead we prefer to give a derivation which starts from the Lagrangian (2.10). This approach, introduced by Whitham (1974), is much more instructive. It gives a better insight into the underlying structure of wave evolution and it applies to any wave system that has a Lagrangian.

To that end, we simply substitute the expansion

$$\eta = a\,e^{i\theta} + a_2\,e^{2i\theta} + \cdots + \text{c.c.} \tag{2.34}$$

and the corresponding series for the potential into the Lagrangian density (Eq. (2.10)) and we average the Lagrangian over the rapidly varying phase θ. The resulting average Lagrangian

$$\langle \mathcal{L} \rangle = \frac{1}{2\pi} \int_0^{2\pi} d\theta\, \mathcal{L} \tag{2.35}$$

depends on the unknown amplitudes of the potential series and the amplitudes $a, a_2, a_3, \ldots$ of the series for the surface elevation. In addition, $\langle \mathcal{L} \rangle$ depends on angular frequency ω and wavenumber $\mathbf{k}$. The appropriate evolution equations follow from the variational principle

$$\delta \int d\mathbf{x}\, dt\, \langle \mathcal{L} \rangle = 0. \tag{2.36}$$

A considerable simplification of the Lagrangian density may be achieved by eliminating the higher-order amplitudes $a_2, a_3, \ldots$ and the amplitudes of the potential series through the variational principle (Eq. (2.36)). For example, variation with respect to the amplitudes $a_2, a_3, \ldots$ gives

$$\langle \mathcal{L} \rangle_{a_i} = 0, \, i \geq 2$$

and this relation enables one to express $a_2, a_3, \ldots$ in terms of a, while making use of a similar relation for the amplitudes of the potential series.

Neglecting wave-induced currents[1] and discarding terms that involve derivatives of ω and $\mathbf{k}$ with respect to time and position, the average Lagrangian $\langle \mathcal{L} \rangle$ becomes

$$\langle \mathcal{L} \rangle = \frac{1}{2}\mathcal{E} \left\{ \frac{(\omega - \mathbf{k} \cdot \mathbf{U}_0)^2}{gkT} - 1 \right\} - \frac{1}{2} \frac{k^2 \mathcal{E}^2}{\rho g} \left\{ \frac{9T^4 - 10T^2 + 9}{8T^4} \right\} + \mathcal{O}(\mathcal{E}^3),$$

(2.37)

where $\mathcal{E} = 2\rho g \, |a|^2$ is the wave energy density, $\mathbf{U}_0$ is the mean water current and $T = \tanh(kD)$. In passing, it should be noted that we shall discuss the role of the terms involving derivatives of ω and $\mathbf{k}$ in the chapter on four-wave interactions in Section 4.3 discussing the nonlinear Schrödinger equation. Here, we simply assume that their contribution to wave evolution can be ignored.

The key result now is that we have obtained an average Lagrangian $\mathcal{L}$ (from now on we omit the angular brackets) that only depends on ω, $\mathbf{k}$ and amplitude a:

$$\mathcal{L} = \mathcal{L}(\omega, \mathbf{k}, a),$$

(2.38)

where $\omega = -\partial\theta/\partial t$ and $\mathbf{k} = \nabla\theta$. Hence, the evolution equations of a wave group follow from the variational principle

$$\delta \int d\mathbf{x} \, dt \, \mathcal{L}(\omega, \mathbf{k}, a) = 0.$$

(2.39)

Variation with respect to the amplitude a then gives the dispersion relation

$$\frac{\partial}{\partial a}\mathcal{L} = 0,$$

(2.40a)

while variation with respect to the phase θ (note that θ appears in $\mathcal{L}$ only through derivatives) gives the evolution equation for the amplitude

$$\frac{\partial}{\partial t}\mathcal{L}_\omega - \nabla \cdot \mathcal{L}_\mathbf{k} = 0,$$

(2.40b)

while a third equation follows from consistency (see Eq. (2.33))!

$$\frac{\partial}{\partial t}\mathbf{k} + \nabla\omega = 0.$$

(2.40c)

---

[1] Wave-induced currents are of considerable interest in ocean circulation, see Section 5.4.1 for a discussion.

The set of equations (2.40a)–(2.40c) describes the evolution of a slowly varying wave group. We remark that the above set is quite general as it is valid for any wave system that has a Lagrangian!

Before we return to our problem of surface gravity waves, we introduce a transport velocity

$$\mathbf{u} = -\mathcal{L}_{\mathbf{k}}/\mathcal{L}_{\omega} \tag{2.41}$$

so that Eq. (2.40b) becomes

$$\frac{\partial}{\partial t}\mathcal{N} + \nabla \cdot (\mathbf{u}\mathcal{N}) = 0, \tag{2.42}$$

where we have introduced the notation $\mathcal{N} = \mathcal{L}_{\omega}$ for the action density. This equation describes the evolution of the action density $\mathcal{N}$ and it will turn out that in linear theory the transport velocity $\mathbf{u}$ is just equal to the group velocity of the waves.

We now apply our results to the Lagrangian (2.37) in the linear approximation, i.e. we disregard terms nonlinear in $\mathcal{E}$. The Lagrangian (2.37) may then be written in the following convenient form:

$$\mathcal{L} = \frac{1}{2}D(\omega, \mathbf{k})\mathcal{E}, \tag{2.43}$$

where

$$D(\omega, \mathbf{k}) = (\omega - \mathbf{k} \cdot \mathbf{U}_0)^2/gkT - 1.$$

The dispersion relation then immediately follows from Eq. (2.40a), or

$$D(\omega, \mathbf{k}) = 0, \tag{2.44}$$

hence, with the introduction of the so-called 'intrinsic frequency' $\sigma = \sqrt{gkT}$, we have

$$\omega = \mathbf{k} \cdot \mathbf{U}_0 \pm \sigma. \tag{2.45}$$

Thus, in the presence of a current the angular frequency of the waves has a Doppler shift $\mathbf{k} \cdot \mathbf{U}_0$. We remark that the current $\mathbf{U}_0$ and the depth D are allowed to be slowly varying functions of space and time. Also, note that the vanishing of D implies that in the extremum $\mathcal{L} = 0$. Since it can be shown that the Lagrangian is just equal to the difference between kinetic and potential energies, this means the usual result that, for linear waves, kinetic and potential energies are equal.

Finally, differentiating $\mathcal{L}$ of Eq. (2.43) with respect to ω the action density $\mathcal{N}$ becomes

$$\mathcal{N} = \frac{\mathcal{E}}{\sigma} \tag{2.46}$$

which obeys the balance equation

$$\frac{\partial}{\partial t}\mathcal{N} + \nabla \cdot (\mathbf{v_g}\mathcal{N}) = 0, \tag{2.47}$$

where $\mathbf{v_g}$ is the group velocity $\partial\omega/\partial\mathbf{k}$. Here, it is remarked that the group velocity follows directly from the relation (2.41), or,

$$\mathbf{v_g} = -D_\mathbf{k}/D_\omega.$$

The importance of the action balance equation cannot be overemphasized. Equation (2.47) has the form of a conservation law in which the rate of change in time of a density, i.e. $\mathcal{N}$, is determined by a flux of that density, i.e., $\mathbf{v_g}\mathcal{N}$. In fact, if one has zero flux at the boundaries of the ocean basin, one finds that the integral

$$N_{\text{tot}} = \int_{\mathcal{D}} \mathrm{dx}\, \mathcal{N}$$

over the domain $\mathcal{D}$ is conserved. We emphasize that in the case of slowly varying bottom and currents, it is not the wave energy $E = \int \mathrm{dx}\, \mathcal{E}$ that is conserved, but the total action N_{tot}!

This conclusion may come as a surprise because we started from evolution equations which conserve energy (see Eq. (2.7)). However, in this context we would like to refer to the well-known example of a pendulum in which its length is slowly varied in time. In that case, energy E and frequency ω change when the pendulum length is varied but the so-called 'adiabatic' invariant

$$A = \frac{E}{\omega}$$

is constant. An illuminating discussion on adiabatic invariants may be found in Whitham (1974).

Therefore, in slowly varying circumstances the wave energy $\mathcal{E}$ is not conserved, but the total energy of the *system* which includes a contribution from the current is certainly conserved. Conservation of the total energy of the system follows from invariance in time of the Lagrangian (Whitham, 1974). Once more denoting the energy of the system by E one finds

$$E = \omega\mathcal{L}_\omega - \mathcal{L}, \tag{2.48}$$

and, using Eqs. (2.43) and (2.44), this becomes

$$E = \omega\mathcal{N} = \frac{1}{2}\omega D_\omega\mathcal{E}. \tag{2.49}$$

Clearly for finite current the energy of the system E differs from the wave energy $\mathcal{E} = \sigma\mathcal{N}$. The energy flux equals $-\omega\mathcal{L}_\mathbf{k}$, and in linear theory this equals $\mathbf{v_g}E$.

Hence, the energy E of the system obeys the conservation equation

$$\frac{\partial}{\partial t}E + \nabla \cdot (\mathbf{v_g}E) = 0. \tag{2.50}$$

This may be verified directly by using the evolution equation for action density (Eq. (2.47)), the dispersion relation (2.44) and the equation for the number of wave crests (Eq. (2.40c)). For zero energy flux at the boundaries of the ocean basin one then finds that the integral

$$E_{\text{tot}} = \int_{\mathcal{D}} \mathbf{dx}\, E \tag{2.51}$$

is conserved. Another conserved quantity is wave momentum P, because the Lagrangian (2.43) is also invariant in space. The appropriate expression for the wave momentum becomes

$$P = k\mathcal{L}_\omega = k\mathcal{N}, \tag{2.52}$$

and the corresponding conservation relation is

$$\frac{\partial}{\partial t}P + \nabla \cdot (\mathbf{v_g}P) = 0. \tag{2.53}$$

Therefore, for zero momentum flux at the boundaries, momentum P is conserved as well.

We close our discussion on some properties of wave groups with the following remarks. Comparing Eqs. (2.46) and (2.52), a well-known relation between wave momentum and wave energy is obtained, namely

$$P = \mathcal{E}/c_0 \tag{2.54}$$

where c_0 is the phase speed of the waves referring to the intrinsic frequency, $c_0 = \sigma/k$. Remarkably, in the linear approximation with $\mathcal{L} = 0$ a similar-looking relation exists between wave momentum and the total energy. Using Eqs. (2.49) and (2.52) one finds

$$P = E/c \tag{2.55}$$

where now c is the phase speed of the waves, $c = \omega/k$. Nevertheless, it is emphasized that the relation between wave energy and wave momentum, Eq. (2.54), is more fundamental because it holds for any nonlinear Lagrangian, while Eq. (2.55) is only relevant in the linear approximation.

Furthermore, Eqs. (2.46) and (2.52) stress the fundamental role that is played by the action density $\mathcal{N}$, being similar to the role of the particle distribution function. If the action density is known, wave momentum is obtained as k times the action density while wave energy follows as σ times the action density. This suggests a parallel with quantum mechanics where similar relations apply for momentum and energy of particles (with the constant of proportionality being Planck's constant $\hbar$). In fact, Tsytovich (1970) developed the nonlinear theory of wave–wave interactions along these lines.

Finally, since wave momentum will play a crucial role in the discussion of wind–wave interaction, we give an explicit expression for P in terms of the amplitude of the waves. Using Eq. (2.54) and the expression for wave energy given below Eq. (2.37) one finds for deep-water gravity waves

$$P = 2\rho\sigma \, |a|^2 \tag{2.56}$$

This relation even holds when effects of capillarity are taken into account, but it is not valid for shallow-water waves.

2.4 The energy balance equation

The purpose of this section is to outline a derivation of the basic evolution equation for an ensemble of random, weakly nonlinear water waves. This equation is called the 'action balance equation', but one frequently refers to it as the energy balance equation.

In the previous section we saw how free wave packets evolve on varying currents in ocean basins with variable depth. There are, however, many other causes of the change in wave packets with time. For example, waves grow because of the energy and momentum input by wind and they lose energy because of white capping. In addition, finite-steepness waves may interact nonlinearly with other waves in such a way that energy and momentum are conserved. As long as the perturbations are small they can be added and the action balance equation becomes

$$\frac{\partial}{\partial t}\mathcal{N} + \nabla \cdot (\mathbf{v}_g \mathcal{N}) = S = S_{\text{in}} + S_{\text{nl}} + S_{\text{ds}}, \tag{2.57}$$

where the source terms on the right-hand side represent effects of wind input S_{in}, nonlinear interactions S_{nl} and dissipation due to white capping S_{ds}. Even interactions between physical processes are allowed as long as the time scale of such a process is much longer than the 'typical' frequency of the waves, in other words the slowly varying assumption must hold.

In the following chapters we shall show how to derive the wind-input term and the nonlinear interactions while we use simple scaling arguments to choose the dissipation term. In this section we discuss in some detail the properties of the left-hand side of the action balance equation, which is called the adiabatic part. However, before we start this discussion we need to introduce the concept of the wave spectrum. As already pointed out, solving the deterministic action balance equation (2.57) is not practical because, apart from the initial amplitudes, knowledge of the phase of the waves is required as well. In order to avoid this problem we concentrate on a statistical description of the ocean surface.

We therefore introduce the homogeneous and stationary theory of a random wave field. In such a theory, wave components are assumed to be independent and have random phase. As a consequence, the probability distribution of the ocean surface elevation is approximately Gaussian. The (near) Gaussian property of the ocean surface follows in principle from the central limit theorem which tells us that if the waves have random and independent phase then the probability distribution is Gaussian. The waves are to a good approximation independent because they have propagated into a given area of the ocean from different distant regions. Even if initially one would start with a highly correlated state then, because of dispersion, waves become separated, thereby decreasing the correlation. In fact, for dispersive waves the loss of correlation is exponentially fast. On the other hand, finite-steepness waves may give rise to correlations between the different wave components because of (resonant) wave–wave interactions. However, the effect is small for small steepness. Therefore, in practice one nearly always finds that, for dispersive ocean waves, the Gaussian property holds in good approximation.

In the remainder of this book we content ourselves with knowledge about average quantities such as the moments

$$\langle \eta(\mathbf{x}_1) \rangle, \langle \eta(\mathbf{x}_1)\eta(\mathbf{x}_2) \rangle, \text{ etc,}$$

where the brackets denote an ensemble average and $\mathbf{x}_1$ and $\mathbf{x}_2$ denote two positions on the ocean surface. In most practical situations it turns out that we then have sufficient information about the ocean surface. Since it is assumed that the mean of the surface elevation vanishes, only the so-called 'two-point correlation function'

$$\langle \eta(\mathbf{x}_1)\eta(\mathbf{x}_2) \rangle \tag{2.58}$$

needs to be considered. Because of the assumed small wave steepness, all higher-order correlations may be expressed in terms of the two-point correlation function. In addition, it is assumed that on the scale of the wavelength the wave field is homogeneous, i.e. the two-point correlation function depends on the distance $\boldsymbol{\xi} = \mathbf{x}_1 - \mathbf{x}_2$ only. We therefore have to study the properties of the following two-point correlation function:

$$R(\boldsymbol{\xi}) = \langle \eta(\mathbf{x} + \boldsymbol{\xi})\eta(\mathbf{x}) \rangle. \tag{2.59}$$

The (frozen) wavenumber spectrum $F(\mathbf{k})$ is now defined as the Fourier transform of the correlation function R:

$$F(\mathbf{k}) = \frac{1}{(2\pi)^2} \int d\boldsymbol{\xi} \, e^{i\mathbf{k}\cdot\boldsymbol{\xi}} R(\boldsymbol{\xi}). \tag{2.60}$$

It is fairly straightforward to establish a relation between the wavenumber spectrum and the complex amplitudes of the surface elevation. Realizing that there are two possible wave modes (a positive-frequency mode and a negative-frequency mode) the general solution of the linear wave problem discussed in Section 2.2 may be written as

$$\eta(\mathbf{x}, t) = \int_{-\infty}^{\infty} d\mathbf{k} \, \hat{\eta}_+(\mathbf{k}) \, e^{i(\mathbf{k}\cdot\mathbf{x}-\omega_+t)} + \int_{-\infty}^{\infty} d\mathbf{k} \, \hat{\eta}_-(\mathbf{k}) \, e^{i(\mathbf{k}\cdot\mathbf{x}-\omega_-t)} \qquad (2.61)$$

where according to Eq. (2.45) $\omega_\pm = \mathbf{k} \cdot \mathbf{U}_0 \pm \sigma$. Since η is supposed to be real we have

$$\hat{\eta}_-(\mathbf{k}) = \hat{\eta}_+^*(-\mathbf{k})$$

and therefore

$$\eta(\mathbf{x}, t) = \int_{-\infty}^{\infty} d\mathbf{k} \, \hat{\eta}(\mathbf{k}) \, e^{i(\mathbf{k}\cdot\mathbf{x}-\omega t)} + \text{c.c.}, \qquad (2.62)$$

where we have omitted the subscript $+$ on $\hat{\eta}$ and $\omega = \omega_+$. Substituting Eq. (2.62) into Eq. (2.59) and requiring a homogeneous, stationary two-point correlation function, i.e. one that only depends on the distance $\boldsymbol{\xi}$, and is independent of time t, the complex amplitude $\hat{\eta}$ should satisfy

$$\langle \hat{\eta}(\mathbf{k}_1)\hat{\eta}(\mathbf{k}_2) \rangle = 0,$$
$$\langle \hat{\eta}(\mathbf{k}_1)\hat{\eta}^*(\mathbf{k}_2) \rangle = |\hat{\eta}(\mathbf{k}_1)|^2 \, \delta(\mathbf{k}_1 - \mathbf{k}_2), \qquad (2.63)$$

where $\mathbf{k}_1$ and $\mathbf{k}_2$ are arbitrary wavenumber vectors. Because of Eq. (2.63) the two-point correlation becomes

$$R(\boldsymbol{\xi}) = \int_{-\infty}^{\infty} d\mathbf{k} \, |\hat{\eta}(\mathbf{k})|^2 \, e^{i(\mathbf{k}\cdot\boldsymbol{\xi})} + \text{c.c.} \qquad (2.64)$$

In view of Eq. (2.60) the wavenumber spectrum is therefore given by

$$F(\mathbf{k}) = 2 \, |\hat{\eta}(\mathbf{k})|^2. \qquad (2.65)$$

Setting $\boldsymbol{\xi}$ to zero in Eq. (2.64) and using Eq. (2.65) we have

$$\langle \eta^2 \rangle = R(0) = \int_{-\infty}^{\infty} d\mathbf{k} \, F(\mathbf{k}) \qquad (2.66)$$

and, as expected, the integral over the wavenumber spectrum equals the wave variance $\langle \eta^2 \rangle$. Realizing that, for propagating linear waves, potential and kinetic energy are equal (because the Lagrangian vanishes), we find that the ensemble average of the wave energy $\langle E \rangle$ is related to the wavenumber spectrum in the following manner:

$$\langle E \rangle = \rho g \langle \eta^2 \rangle = \rho g \int_{-\infty}^{\infty} d\mathbf{k} \, F(\mathbf{k}); \qquad (2.67)$$

thus the wavenumber spectrum indeed gives the distribution of wave energy over wavenumber.

The wavenumber spectrum we have introduced is independent of space and time. Formally, we have taken a Fourier transform over the whole ocean domain. In order to allow for spatial dependence of the wavenumber spectrum we simply adopt the procedure of taking the Fourier transform over a domain with such an extent that the two-point correlation function may still be regarded as homogeneous. On the other hand, the domain should be large enough that it contains a sufficient number of ocean waves, say of the order of 100, in order that the spectrum gives a valid representation of the sea state. With a typical wavelength of about 100 m the extent of such a domain is therefore of the order of 10 km.

We are now finally in a position to derive the action balance equation for a continuous spectrum from the action balance equation (Eq. (2.57)) for a single wave group. By analogy with the discrete case, we introduce the action density spectrum $N(\mathbf{k})$ as

$$N(\mathbf{k}) = \frac{g F(\mathbf{k})}{\sigma} \qquad (2.68)$$

where, as before, $\sigma = \sqrt{gk \tanh(kD)}$, and we leave out the constant water density factor. It is tempting now to use the action balance equation Eq. (2.57) for the discrete case to obtain the action balance equation for the continuous case. There is one pitfall, however. The action density spectrum $N(\mathbf{k}, \mathbf{x}, t)$ has as independent variables wavenumber $\mathbf{k}$, position $\mathbf{x}$ and time t, while in the discrete case the wavenumber is a local variable that depends on position and time.

The most convenient way to proceed is therefore to establish the following connection between the continuous action density spectrum and the discrete analogy $\mathcal{E}/\sigma$ of Eq. (2.46). We are interested in the action density contained in modes with wavenumbers in an interval $\Delta \mathbf{k}$ around $\mathbf{k}$. Introduce, therefore, a function $\epsilon(\mathbf{k})$ such that

$$\epsilon(\mathbf{k}) = \begin{cases} 1, & \text{for } -\frac{1}{2}\Delta \mathbf{k} < \mathbf{k} < \frac{1}{2}\Delta \mathbf{k}, \\ 0, & \text{otherwise.} \end{cases} \qquad (2.69)$$

Thus, $\epsilon(\mathbf{k})$ is a filter that selects modes in the interval $\Delta \mathbf{k}$ around wavenumber $\mathbf{k}$. Denoting the action density of a wave group with wavenumber $\mathbf{k}'$ by $\mathcal{N}_{\mathbf{k}'}$, the appropriate connection between the discrete and continuous case is

$$N(\mathbf{k})\Delta\mathbf{k} = \sum_{\mathbf{k}'} \mathcal{N}_{\mathbf{k}'}\epsilon(\mathbf{k}' - \mathbf{k}), \tag{2.70}$$

where $\mathcal{N}_{\mathbf{k}'} = 2g|a_{\mathbf{k}'}|^2/\sigma$. Hence, the sum in Eq. (2.70) is over all wave groups with wavenumber in the interval $\Delta\mathbf{k}$ around $\mathbf{k}$. It is emphasized that the 'local' wavenumber $\mathbf{k}'$ depends on position and time, and therefore the number of wave groups in a particular interval may vary from time to time.

The evolution equation for $N(\mathbf{k})$ may now readily be obtained by evaluating

$$\left.\frac{\partial N}{\partial t}\right|_{\mathbf{x},\mathbf{k}},$$

i.e. the rate of change of N in time, keeping $\mathbf{x}$ and $\mathbf{k}$ fixed, by using Eq. (2.70) and the action balance equation Eq. (2.57). The result is

$$\frac{\partial N}{\partial t} + \nabla_{\mathbf{x}} \cdot (\nabla_{\mathbf{k}}\Omega\, N) - \nabla_{\mathbf{k}} \cdot (\nabla_{\mathbf{x}}\Omega\, N) = S. \tag{2.71}$$

Here, Ω represents the dispersion relation:

$$\Omega = \mathbf{k} \cdot \mathbf{U} + \sigma, \sigma = \sqrt{gk\tanh(kD)}. \tag{2.72}$$

Equation (2.71) tells us that the rate of change in time of the action density spectrum is determined by advection with the group velocity $\mathbf{v}_g = \nabla_{\mathbf{k}}\Omega$, by refraction (the third term stems from the time and space dependence of the local wavenumber of a wave group) and by physical processes such as the generation of ocean waves by wind, nonlinear interactions, and dissipation by white capping. These two processes are all contained in the source term S, which is just the ensemble mean of the right-hand side of Eq. (2.57).

Equation (2.71) is called the action balance equation and is the basic evolution equation for the continuous wave spectrum. Nowadays it is regarded as the starting point of modern wave models. The idea of a spectral transport equation was first suggested by Gelci *et al.* (1957), while the derivation as given here closely follows the work of Willebrand (1975). In fact, Willebrand's work is more general in the sense that the starting point is a nonlinear wave system, rather than the simple linear approach we have followed here. As a consequence, nonlinear corrections to the group velocity appear. These nonlinear corrections are thought to be small, however, at least in deep water.

At this point it is good to emphasize the relevance of the action balance equation by putting matters into a historical perspective. Interest in wave prediction started during the Second World War because of the practical need for knowledge of the sea state during landing operations. The first operational predictions were based on the work of Sverdrup and Munk (1947), who introduced a parametrical description of the sea state and who used empirical windsea and swell laws. Examples of these descriptions will be given later in this chapter. Manual techniques based on this approach have been used by operational forecasters for many years and these techniques turned out to be a convenient means of obtaining a short-term forecast for a given wind field (Groen and Dorrestein, 1976). In the meantime, an important advance was the introduction of the concept of a wave spectrum (Pierson *et al.*, 1955), while the corresponding dynamical evolution equation was proposed by Gelci *et al.* (1957). The source term at that time was purely empirical. This changed after the new theories of wave generation by Phillips (1957) and Miles (1957) had been published and the source function for the nonlinear transfer had been derived (Hasselmann, 1962). As a result, it was concluded that the source function consists of three terms representing the input of wind, the nonlinear transfer and the dissipation by white capping (and in shallow waters by bottom friction). This form of the energy balance equation is still in use today, particularly in the context of wave prediction models.

The reason that the effects of wind, nonlinear transfer, and dissipation can simply be added is that compared with a typical wave period the corresponding rates of change of the spectrum are small. In other words, these effects are small, and can therefore be obtained by means of perturbation expansions as discussed in the following chapters. Because of the smallness of these terms, products of the effects of wind and nonlinear transfer, for example, can safely be ignored. However, before we close this chapter, it would be useful to discuss some properties of the action balance equation in the absence of sources and sinks. This is then followed by section 2.6 on empirical growth laws for wave height and the frequency spectrum in fetch- and duration-limited cases. At the same time this presents us with an opportunity to introduce a number of important quantities relevant to wave prediction.

2.5 Kinematic part of the energy balance equation

In this section we shall briefly discuss some properties of the kinematic part of the energy balance equation; in particular, shoaling and refraction – by bottom topography and ocean currents – are investigated in the context of a statistical description of gravity waves. But first we generalize the form of the energy balance equation allowing us to write down this fundamental law of wave prediction in any coordinate system, in particular for spherical coordinates.

Let x_1 and x_2 be the spatial coordinates and k_1, k_2 the wave coordinates, and let

$$\mathbf{z} = (x_1, x_2, k_1, k_2) \tag{2.73}$$

be their combined four-dimensional vector. The most elegant formulation of the 'energy' balance equation is in terms of the action density spectrum N which is the energy spectrum divided by the so-called 'intrinsic frequency' σ. The action density plays the same role as the particle density in quantum mechanics. Hence, there is an analogy between wave groups and particles, because wave groups with action N have energy σN and momentum kN. Thus, the most fundamental form of the transport equation for the action density spectrum $N(\mathbf{k}, \mathbf{x}, t)$ without the source term can be written in the flux form

$$\frac{\partial}{\partial t}N + \frac{\partial}{\partial z_i}(\dot{z}_i N) = 0, \tag{2.74}$$

where $\dot{z}$ denotes the propagation velocity of a wave group in the four-dimensional phase space of $\mathbf{x}$ and $\mathbf{k}$ and i denotes an index which runs from 1 to 4. This equation holds for any field $\dot{z}$, and also for velocity fields which are not divergence free in four-dimensional phase space. In the special case when $\mathbf{x}$ and $\mathbf{k}$ represent a canonical vector pair – this is the case, for example, when they are the usual Cartesian coordinates – the propagation equations for a wave group (also known as the Hamilton–Jacobi propagation equations) read:

$$\dot{x}_i = \frac{\partial}{\partial k_i}\Omega, \tag{2.75a}$$

$$\dot{k}_i = -\frac{\partial}{\partial x_i}\Omega, \tag{2.75b}$$

where Ω denotes the dispersion relation in Eq. (2.72).

The Hamilton–Jacobi equations have some intriguing consequences because the field $\dot{z}$ for a continuous ensemble of wave groups is divergence free in four-dimensional phase space,

$$\frac{\partial}{\partial z_i}\dot{z}_i = 0. \tag{2.76}$$

First, the transport equation for the action density may be expressed in the advection form

$$\frac{\mathrm{d}}{\mathrm{d}t}N = \frac{\partial N}{\partial t} + \dot{z}_i\frac{\partial}{\partial z_i}N = 0. \tag{2.77}$$

Thus, along a path in four-dimensional phase space defined by the Hamilton–Jacobi equations (2.75a) and (2.75b), the action density N is conserved. This property only holds for canonical coordinates for which the flow divergence vanishes (Liouville's theorem – first applied by Dorrestein (1960) to wave spectra).

Second, the analogy between Hamilton's formalism of particles with Hamiltonian H and wave groups obeying the Hamilton–Jacobi equations should be pointed out. Indeed, wave groups may be regarded as particles in this respect and the Hamiltonian H and angular frequency Ω play similar roles. Because of this similarity, Ω

is expected to be conserved as well (under the restriction that Ω does not depend on time). This can be verified by direct calculation of the rate of change of Ω following the path of a wave group in phase space,

$$\frac{\mathrm{d}}{\mathrm{d}t}\Omega = \dot{z}_i \frac{\partial}{\partial z_i}\Omega = \dot{x}_i \frac{\partial}{\partial x_i}\Omega + \dot{k}_i \frac{\partial}{\partial k_i}\Omega = 0. \tag{2.78}$$

The vanishing of $\mathrm{d}\Omega/\mathrm{d}t$ follows at once upon using the Hamilton–Jacobi equations (2.75a) and (2.75b). Note that the restriction of no time dependence of Ω is essential for the validity of Eq. (2.78), just as the Hamiltonian H is only conserved when it does not depend explicitly on time t. Thus, Eq. (2.78) gives the important message that angular frequency is conserved when following a wave group and this property will play an important role in our discussion of refraction.

We now turn to the form of the action density balance equation in the flux form, Eq. (2.74). This formulation is more general and has certain advantages. When one transforms from one set of coordinates to another there is no guarantee that the flow remains divergence free and therefore the flux form of the action balance equation is the preferred starting point. The transformation of the standard Cartesian geometry transport equation to spherical geometry is then fairly straightforward (see also Groves and Melcer, 1961; WAMDI, 1988; Komen *et al.*, 1994). Let us therefore consider the spectral action density $\hat{N}(\omega, \theta, \phi, \lambda, t)$ with respect to angular frequency ω and direction θ (measured clockwise relative to true north) as a function of latitude ϕ and longitude λ. The reason for the choice of angular frequency as the independent variable (instead of, for example, the wavenumber k) is that for a fixed topography and current the frequency Ω is conserved when following a wave group, therefore the transport equation simplifies. In general, the conservation equation for $\hat{N}$ thus reads

$$\frac{\partial}{\partial t}\hat{N} + \frac{\partial}{\partial \phi}(\dot{\phi}\hat{N}) + \frac{\partial}{\partial \lambda}(\dot{\lambda}\hat{N}) + \frac{\partial}{\partial \omega}(\dot{\omega}\hat{N}) + \frac{\partial}{\partial \theta}(\dot{\theta}\hat{N}) = 0, \tag{2.79}$$

and since $\dot{\omega} = \partial\Omega/\partial t$ the term involving the derivative with respect to ω drops out in the case of time-independent current and bottom topography. Finally, the action density $\hat{N}$ is related to the normal spectral density N with respect to a local Cartesian frame (x, y) through $\hat{N}\,\mathrm{d}\omega\,\mathrm{d}\theta\,\mathrm{d}\phi\,\mathrm{d}\lambda = N\,\mathrm{d}\omega\,\mathrm{d}\theta\,\mathrm{d}x\,\mathrm{d}y$, or

$$\hat{N} = NR^2 \cos\phi, \tag{2.80}$$

where R is the radius of the Earth. Substitution of Eq. (2.80) into Eq. (2.79) yields the transport equation

$$\frac{\partial}{\partial t}N + (\cos\phi)^{-1}\frac{\partial}{\partial \phi}(\dot{\phi}\cos\phi N) + \frac{\partial}{\partial \lambda}(\dot{\lambda}N) + \frac{\partial}{\partial \omega}(\dot{\omega}N) + \frac{\partial}{\partial \theta}(\dot{\theta}N) = 0, \tag{2.81}$$

where, with v_g the magnitude of the group velocity,

$$\dot{\phi} = (v_g \cos\theta + U_0)/R, \tag{2.82a}$$

$$\dot{\lambda} = (v_g \sin\theta + V_0)/(R\cos\phi), \tag{2.82b}$$

$$\dot{\theta} = v_g \sin\theta \tan\phi/R + (\dot{\mathbf{k}} \times \mathbf{k})/k^2, \tag{2.82c}$$

$$\dot{\omega} = \partial\Omega/\partial t \tag{2.82d}$$

represent the rates of change of the position and propagation direction of a wave packet. Here, U_0 and V_0 are the components of the current in northerly and easterly directions respectively. Equation (2.81) is the basic transport equation that is used in numerical wave prediction. The remainder of this section is devoted to a discussion of some of the properties of Eq. (2.81). We first discuss some peculiarities of Eq. (2.81) for the infinite-depth case in the absence of currents and we then discuss the special cases of shoaling and refraction due to bottom topography and currents.

Great circle propagation on the globe

A wave group propagates along a great circle on the globe. The proof of this is rather tedious (see Komen *et al.*, 1994, pp. 210–211). This property is related to the presence of refraction on the globe, even in the absence of depth and current refraction ($\dot{\mathbf{k}} = 0$). From Eq. (2.82c) we see that there is then a rate of change of direction according to

$$\dot{\theta} = v_g \sin\theta \tan\phi/R. \tag{2.83}$$

This refraction is entirely due to the change in time of the local northward-pointing vector, and is therefore apparent because it is only related to the choice of coordinate system. Nevertheless it is important to realize that this effect exists. Because of this property, southward-propagating swells generated in the Gulf of Alaska will, after a few weeks, arrive in the Indian Ocean!

Shoaling

Consider now finite-depth effects in the absence of currents. Shoaling of waves already occurs for wave propagation parallel to the direction of the depth gradient. In this case there is no depth refraction because $\dot{\mathbf{k}} \times \mathbf{k} = 0$. In addition, we take the wave direction θ to be zero (northerly propagation) so that the longitude is constant ($\dot{\lambda} = 0$) and $\dot{\theta} = 0$. For time-independent topography (hence $\partial\Omega/\partial t = 0$) the transport equation becomes

$$\frac{\partial}{\partial t}N + (\cos\phi)^{-1}\frac{\partial}{\partial\phi}(\dot{\phi}\cos\phi N) = 0, \tag{2.84}$$

where $\dot{\phi} = v_g \cos\theta/R = v_g/R$ and the group speed only depends on latitude ϕ.

Restricting our attention to steady waves we immediately find conservation of the action density flux in the latitude direction, or,

$$\frac{v_g \cos \phi}{R} N = \text{constant}.$$

If, in addition, it is assumed that the variation of depth with latitude occurs on a much shorter scale than the variation of $\cos \phi$, the last factor may be considered to be constant for present purposes. It is then found that the action density is inversely proportional to the group speed v_g, or when the group speed decreases the action density will increase and vice versa. The normal wisdom now is that the group speed decreases for decreasing depth. Therefore, when waves are approaching shallow waters, conservation of flux requires an increase of the action density. However, this is not entirely correct. Using the dispersion relation Eq. (2.72) and starting in deep water, hence $kD \to \infty$, one finds for kD values of up to around 1.5 the group velocity is an increasing function of depth D; however, for smaller kD values the group velocity indeed decreases with depth. As a consequence, when waves approach shallow waters the action density will first decrease with depth up to $kD \approx 1.5$. If the depth continues to decrease this is then followed by an increase of the action density. The latter aspect of this phenomenon is called shoaling. Although this is outside the scope of a statistical, weakly nonlinear description of dispersive waves, it should be mentioned that the most dramatic consequences of shoaling may be seen when tidal waves, generated by earthquakes, approach the coast and result in tsunamis.

Refraction

The second example of finite-depth effects that we discuss is refraction. As a general principle it may be stated then that wave rays (the path of a wave group in **x**-space) will bend towards shallower water, resulting in, for example, focussing phenomena and caustics. In this way, a sea mountain plays a similar role for gravity waves as a lens does for light waves. Furthermore, according to this general principle, close to the coast the waves will always propagate towards the coast, even when far away from the coast they propagate parallel to it. All these examples of this general principle may be explained in terms of refraction. The refraction is given by the rate of change of wave direction, $\dot{\theta}$. Writing Eq. (2.82c) in a more explicit form we have

$$\dot{\theta} = \left(\sin \theta \frac{\partial}{\partial \phi} \Omega - \frac{\cos \theta}{\cos \phi} \frac{\partial}{\partial \lambda} \Omega \right) / (kR). \tag{2.85}$$

Consider then again, as an example, wave propagation to the north ($\theta = 0$), parallel to the coast. Suppose now that depth only depends on longitude in such a way that it decreases towards the coast. Thus, the rate of change of wave direction is then positive as

$$\dot{\theta} = -\frac{1}{kR\cos\phi}\frac{\partial}{\partial\lambda}\Omega > 0,$$

since $\partial\Omega/\partial\lambda < 0$. Therefore, the wave ray will bend towards the coast.

Current effects

Currents may give rise to similar effects as found with depth refraction. However, the most dramatic effects may be found when the waves propagate against the current. For sufficiently large current and high frequency, wave propagation is prohibited and wave breaking and wave reflection occur. The most prominent example of this is found in the Agulhas current, east of South Africa.

The property of wave reflection really follows from the dispersion relation which, for one-dimensional propagation against a current U_0, reads

$$\Omega = \sqrt{gk} - kU_0,$$

where we took the deep-water limit. Then, the group velocity $\partial\Omega/\partial k$ vanishes for $k = g/4U_0^2$ and wave propagation is no longer possible any more.

2.6 Empirical laws for wave growth

In this section a brief overview is given of a number of important empirical facts about the evolution of ocean waves. We will mainly concentrate on the properties of wind waves, and we will discuss extensively so-called 'fetch laws' for wave variance, peak frequency and the high-frequency energy level. This is then followed by a presentation on the parametrization of the frequency spectrum and angular distribution of windsea. We close this section with a discussion of the sea-state dependence of the air–sea momentum transfer. Note that here windsea is loosely defined as those components of the two-dimensional wave spectrum that are directly subject to the influence of the local wind. In contrast, swell is defined as that part of the wave spectrum that has been generated by nonlocal winds and that may have propagated from remote areas towards the point of interest. Hence, the sea state consists of a combination of locally and nonlocally generated components which may interact in a nonlinear way.

Before we discuss examples of empirical growth laws it is noted that there is a direct connection between what has been developed in the previous sections and experimental practice. We have obtained the evolution equation for the action density $N(\mathbf{k})$ which, according to Eq. (2.68), is related to the wavenumber spectrum

$F(\mathbf{k})$ through

$$F(\mathbf{k}) = \sigma(k)N(\mathbf{k})/g. \tag{2.86}$$

Here, the wavenumber spectrum is normalized with the wave variance $\langle \eta^2 \rangle$, where η is the surface elevation,

$$m_0 = \int d\mathbf{k}\, F(\mathbf{k}) = \langle \eta^2 \rangle, \tag{2.87}$$

with m_0 the so-called 'zero-order moment' of the spectrum. Obviously, the integral over the wave spectrum has the dimension of a length squared and therefore a measure for wave height, called the 'significant wave height' H_S, can be introduced according to

$$H_S = 4\sqrt{m_0}. \tag{2.88}$$

The reason for this definition is historical. In the early days the sea state was observed visually. This resulted in a measure of wave height $H_{1/3}$ defined as the average height of the highest $1/3$ waves and it can be shown that for narrow spectra $H_S \approx H_{1/3}$ (Phillips, 1977). The significant wave height is but one example of an integrated parameter. There is a long list of such parameters, and they can all be determined once the wavenumber spectrum is known. Examples are: several forms of the mean period, the mean wave direction and the width of the spectrum.

In-situ observations of the two-dimensional spectrum are rare; it requires rather sophisticated instrumentation to observe the spatial correlation function. However, during dedicated field campaigns wavenumber spectra have been observed successfully (see Donelan *et al.*, 1985). In contrast, many observations of the wavenumber spectrum have been obtained through remote-sensing techniques. The long-wave part of the wavenumber spectrum may be observed from space by means of a so-called 'synthetic aperture radar' (SAR) on board of European Space Agency (ESA) satellites ERS-1, ERS-2 and ENVISAT (see Komen *et al.*, 1994).

It is much easier to obtain the frequency spectrum because this just requires the analysis of time series at a certain location. The two-dimensional frequency spectrum is defined as

$$F_2(\omega, \theta)\,d\omega\,d\theta = F(\mathbf{k})\,d\mathbf{k} = F(k, \theta)\,k\,dk\,d\theta,$$

hence

$$F_2(\omega, \theta) = \frac{k}{v_g} F(k, \theta) \tag{2.89}$$

where $v_g = \partial\omega/\partial k$ is the group velocity. Regarding the directional distribution of waves, conventional buoys provide only limited information. It is more common to observe the one-dimensional spectrum defined as

$$F_1(\omega) = \int d\theta \, F_2(\omega, \theta). \tag{2.90}$$

The frequency spectrum is obtained by means of a straightforward Fourier transformation of the time series for the surface elevation η. As far as notation is concerned we will use the same symbol for the various forms of the spectrum, namely F; the distinction should be clear from their arguments, $F(\mathbf{k})$, $F(\omega, \theta)$ and $F(\omega)$.

Having given the connection between theory and practice of waves it is now high time to give a brief summary of what is empirically known about wave evolution. This knowledge is mainly derived from a number of field campaigns, the most important one of which is no doubt the JOint North Sea WAve Project, abbreviated to JONSWAP. This field campaign was devoted to gaining an understanding of the evolution of the wave spectrum. The measurements were obtained along an array located to the west of the isle of Sylt in the German bight of the North Sea. In the summer time easterly winds occur frequently and therefore this location provided an unique opportunity to study fetch-limited wave evolution, where the fetch is defined as the distance to shore.

Empirical data on ocean waves have been collected for at least a century, and, particularly in the early days, most observations on the sea state were collected on ships by means of visual inspection. These observations were summarized in terms of empirical scaling laws but had a defect in that they were not dimensionally consistent. For example, Stevenson (1874) suggested the following relation between wave height H in metres and fetch F in nautical miles:

$$H = 0.45\sqrt{F},$$

which is independent of wind speed. Others such as Larisch-Mönnich (1925) and Cornish (1934) suggested that wave height would depend linearly on wind speed, e.g.

$$H = 0.48W,$$

where W is the wind speed at about 8 m height. However, the above scaling laws are not dimensionally correct as can be easily verified. To my knowledge, the first dimensionally correct attempt to find an expression for the saturation wave height was given by Rossby and Montgomery (1935) and then followed by Sverdrup *et al.* (1946), while empirical scaling laws for fetch- and duration-limited cases were obtained by Sverdrup and Munk (1947); in essence these scaling laws are still in use today!

As already mentioned, interest in wave prediction grew rapidly during the Second World War because of the need for knowledge of the sea state during landing operations. These predictions were based on the empirical growth laws of Sverdrup and Munk (1947). All this was formalized by Kitaigorodskii (1962, 1970) who proposed similarity laws based on dimensional considerations and some rudimentary knowledge on the evolution of gravity waves by wind. These similarity laws have been extremely useful in analysing wave observations.

In principle, a large number of variables may control wave growth. For example, in the idealized situation of duration-limited waves (when a uniform and steady wind has blown over an unlimited ocean for time t after a sudden onset and starting from a flat, calm sea) the following variables may be relevant: angular frequency ω, acceleration of gravity g, viscosity, surface tension, air and water density, Coriolis parameter f and a wind-speed scale U. Regarding the choice of wind-speed scale there is still an ongoing debate. In practice it is most convenient to choose as wind-speed scale the wind speed at a height of 10 m as this is nowadays the most common height at which the wind velocity is measured. However, this choice of wind-speed scale introduces a height scale, namely 10 m, which has no relevance to the problem of the generation of waves by wind; therefore, another wind-speed scale, namely one based on the friction velocity u_*, may be more appropriate, as this scale is related to the momentum transfer from air to water. We will take up a more complete discussion of this point in Chapter 3. In this section we will start by using U_{10} as wind-speed scale, because most data sets have been analysed in terms of this wind-speed scale. However, in the discussion on the shape of the high-frequency spectra, the friction-velocity u_* will emerge in a quite natural way.

Under the assumptions of the nature of wave motion and the mechanism of wave growth, the energy-containing part of the spectrum is mainly determined by the variables ω, U, g and t. Dimensional considerations then give the following similarity laws for the frequency spectrum $F(\omega)$, the total variance m_0 and the peak frequency ω_p of the spectrum:

$$
\begin{aligned}
g^3 F/U^5 &= f(U\omega/g, gt/U), \\
g^2 m_0/U^4 &= f(gt/U), \\
U\omega_p/g &= f(gt/U).
\end{aligned}
\tag{2.91}
$$

Here, f denotes a general function, and to be definite this function differs for different dimensionless parameters on the left-hand side of Eq. (2.91).

Conditions of duration-limited growth are difficult to fulfil in practice (for an exception see Sanders (1976)), and, from the point of view of the analysis of experimental data, two other idealized cases are more important. One is the case of fully developed waves when a uniform and steady wind has blown over an unlimited ocean long enough for the wave field to become independent of time; this situation may occur in the Trade Winds area. The other, more frequently occurring case, is the so-called 'fetch-limited case', when a uniform and steady wind has blown from a straight coastline long enough for the wave field at distance (fetch) X from the upwind coast to become independent of time. For fetch-limited growth the similarity relations become

$$g^3 F/U^5 = f(U\omega/g, gX/U^2),$$
$$g^2 m_0/U^4 = f(gX/U^2), \qquad (2.92)$$
$$U\omega_p/g = f(gX/U^2).$$

The case of fully developed windsea now follows from either Eq. (2.91) or Eq. (2.92) by taking the limit of infinite duration or fetch. The result is

$$g^3 F/U^5 = f(U\omega/g),$$
$$g^2 m_0/U^4 = \text{constant}, \qquad (2.93)$$
$$U\omega_p/g = \text{constant}.$$

In particular, the equation for the wave variance is important because it shows the sensitive dependence of wave prediction on the wind-speed scale, stressing the need for high-quality winds for accurate wave prediction. Furthermore, the equation on the equilibrium value of the peak frequency suggests (recalling that for deep-water gravity waves the phase speed is g/ω) that in equilibrium the peak phase speed is proportional to the wind-speed scale. In practice, the constant in the last equation of Eq. (2.93) is found to be 0.83 so that for fully developed windsea the peak phase speed is about 20 per cent larger than the 10-m wind speed. Hence, ocean waves propagate considerably faster than the 10-m wind speed. This unexpected result can only be explained when the role of the nonlinear interactions (see Chapter 3) is understood. But note that the use of the 10-m wind speed as the wind-speed scale is to some extent arbitrary.

2.6.1 Wind-speed scaling

A good summary of the state of the art of empirical scaling laws is given by Kahma and Calkoen (1992) (see also Komen *et al.*, 1994). They analysed results from a number a field campaigns and devoted a lot of attention to cleaning these data sets. As an example, Fig. 2.3 shows growth laws for the dimensionless energy and peak frequency obtained from the Bothnian Sea data and the Lake Ontario data. Although the cleaning operation removed a number of outliers, even then Kahma and Calkoen noted that there were considerable differences between growth laws from different campaigns. They could explain these differences by pointing out that there may be another parameter relevant in wind-wave generation, namely one related to stratification effects. When cold air is blowing over a warmer ocean

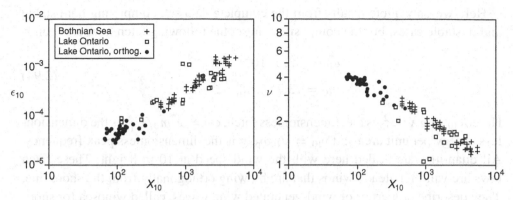

Fig. 2.3. Dimensionless energy ϵ_{10} and peak frequency ν from the Bothnian Sea data and the Lake Ontario data in unstable stratification (from Komen *et al.*, 1994).

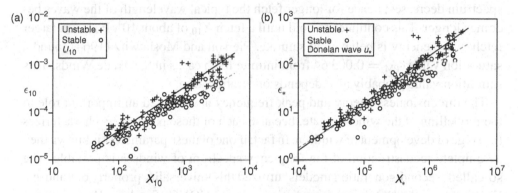

Fig. 2.4. Dimensionless energy in unstable and stable stratification: (a) scaling wind U_{10}; (b) scaling wind u_* using a wave-dependent roughness z_0 (from Komen *et al.*, 1994).

surface, buoyancy effects will give rise to a more effective mixing of momentum in the surface layer, hence more momentum and energy are transferred to the ocean waves, giving larger wave growth and therefore higher wave energy compared with the case of neutral stratification. In addition, in those circumstances there is an increase of gustiness which may also enhance wave growth; this is discussed in the next chapter.

By grouping the observations in two classes, namely stable and unstable stratification, Kahma and Calkoen (1992) found two distinctly different growth laws. When using U_{10} as wind-speed scale the difference in wave energy between the unstable and stable stratification groups is as much as 70 per cent, as illustrated in Fig. 2.4a; however, this difference is considerably reduced when using friction-velocity scaling, as shown in Fig. 2.4b. This point will be discussed in more detail in the Intermezzo of Section 2.6.2.

Here we only quote results from the complete data set, comprising both stable and unstable cases. For the composite data set the following fetch laws are found:

$$\epsilon_{10} = 5.2 \times 10^{-7} \times X_{10}^{+0.9},$$

$$\nu_{10} = 13.7 \times X_{10}^{-0.27}. \tag{2.94}$$

Here $X_{10} = gX/U_{10}^2$ is the dimensionless fetch, $\epsilon_{10} = g^2 m_0/U_{10}^4$ is the dimensionless energy per unit area and $\nu_{10} = U_{10}\omega_p/g$ is the dimensionless peak frequency. All quantities are scaled here with the wind speed at 10 m height. These fetch laws are valid for steady winds that are blowing orthogonally from the shore line. They describe properties of wind-generated wind waves, called windsea for short. These laws are in agreement with the expectation that the longer the fetch the more energy the waves will have, and at the same time the peak frequency of the wave spectrum decreases; hence for longer fetch the typical wavelength of the waves becomes longer. This continues to hold until a fetch X_{10} of about 10^4. For even larger fetch, wave energy is believed to saturate. Pierson and Moskowitz (1964) found a saturation value $\epsilon_{PM} = 0.003\,64$ from infinite fetch cases in the Trade Winds. This saturation value probably also depends on stratification.

The dimensionless energy and peak frequency have played an important role in the modelling of the windsea state, because each of these parameters characterizes the stage of development of windsea. In fact, if one of these parameters is known then a complete reconstruction of the frequency spectrum of windsea is possible. The so-called 'second-generation models' utilized this universality property of windsea. For example, the Dutch wave prediction scheme GONO (Sanders, 1976; Janssen *et al.*, 1984) used what was essentially the dimensionless energy as the stage of development parameter, while the German wave model HYPA (Hasselmann *et al.*, 1976) used the dimensionless peak frequency to characterize the windsea state. But it should be clear that these approaches are equivalent because by elimination of the fetch in Eq. (2.94) it follows that there is a direct relation between the two. This relation is given below but, rather than using the dimensionless peak frequency, we use its inverse which is termed the 'wave age' χ_{10}. Hence, we define the wave age as

$$\chi_{10} = c_p/U_{10}, \tag{2.95}$$

where $c_p = g/\omega_p$ is the phase speed of the peak of the spectrum. Now, young windseas have a wave age which at sea is typically of $\mathcal{O}(1/3)$ while old windsea has a wave age of $\mathcal{O}(1)$. Elimination of the fetch in Eq. (2.94) and making use of the definition in Eq. (2.95) gives the following relation for the dimensionless energy:

$$\epsilon_{10} = 0.0032\,\chi_{10}^{+10/3}. \tag{2.96}$$

Parameters such as the wave age have a certain advantage because they are in fact a generalization of the concept of fetch. That is, by using the wave age, fetch- and duration-limited windseas are treated on an equal footing.

In order that the concept of wave age works, one should show that the spectral shape is universal. For large fetches this issue was addressed by Pierson and Moskowitz (1964) who showed that the scaling laws of Eq. (2.93) make sense. For example, the scaling relation for the wave spectrum was found to work reasonably well. They inferred from their observed wave spectra a universal shape that has become known as the Pierson–Moskowitz spectrum. The high-frequency part of the Pierson–Moskowitz spectrum follows from Phillips (1958). He had argued that the high-frequency part of the spectrum is governed by wave breaking and had found, on dimensional grounds, that the spectrum follows an ω^{-5} law. The relevant constant of proportionality is usually denoted by α_p. This idea on the high-frequency tail of the spectrum was in those days regarded as a well-established result and therefore the spectra were fitted to Phillips' power law at high frequencies. The Pierson–Moskowitz spectrum reads

$$F_{PM}(\omega, \omega_p, \alpha_p) = \alpha_p g^2 \omega^{-5} \exp\left\{-\frac{5}{4}\left(\frac{\omega_p}{\omega}\right)^4\right\}, \tag{2.97}$$

where the Phillips parameter α_p was found to be about 0.0083.

For short fetches an important contribution was provided by the JONSWAP group (Hasselmann *et al.*, 1973). It was found that in the initial stages of wave growth, spectra have a sharper peak than according to the Pierson–Moskowitz spectrum; in other words there is a considerable overshoot (Barnett, 1968). The development of the spectrum in an offshore wind is illustrated by Fig. 2.5 (from Hasselmann *et al.*, 1973). The spectral peak moves towards lower frequencies as fetch increases from 9.5 to 80 km, leaving behind a quasi-saturated high-frequency tail. The spectral peak rises well above the extrapolated tail so that the energy at a particular frequency increases to a maximum as fetch increases, and reduces again as the peak frequency reduces further. The peak enhancement plays an important role in wave evolution, because in the early stages of wave growth the waves near the peak are substantially steeper than for saturated conditions. Therefore, for young wind waves nonlinear effects play an important role.

The fully developed Pierson–Moskowitz spectrum contained two parameters, namely the peak frequency and the Phillips parameter α_p, which describes the spectral level on the high-frequency tail. Hasselmann *et al.* (1973) introduced three new parameters to describe the height γ and width σ_a, σ_b of the enhanced peak. The parameters σ_a and σ_b refer to the width of the asymmetrical peak below and above the peak frequency respectively. An important discovery was that the Phillips parameter α_p depends on fetch and is therefore not a constant as expected from

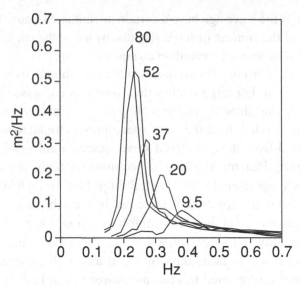

Fig. 2.5. Evolution of wave spectra with fetch for offshore winds (1100 to 1200 h, 15 September, 1968). The spectra are labelled with the fetch in kilometres (from Hasselmann *et al.*, 1973).

Phillips (1958). For these short fetches no consistent dependence of γ, σ_a and σ_b on nondimensional fetch was found, although, to be consistent with the results of Pierson and Moskowitz (1964) for infinite fetch, one would expect that the overshoot disappears, i.e. $\gamma \to 1$ for $X_{10} \to \infty$.

The JONSWAP group therefore proposed the following parametrization of the frequency spectrum of fetch-limited, wind-generated wind waves:

$$F(\omega) = F_{PM}(\omega, \omega_p, \alpha_p) \times \gamma^{\Gamma(\omega, \omega_p, \sigma)}, \qquad (2.98)$$

where F_{PM} is the Pierson–Moskowitz spectrum (2.97), and

$$\Gamma(\omega, \omega_p, \sigma) = \exp\left\{-(\omega - \omega_p)^2 / 2\sigma^2 \omega_p^2\right\}. \qquad (2.99)$$

Here, the peak frequency ω_p follows from the JONSWAP fetch law

$$\nu_{10} = 22 X_{10}^{-0.33}, \qquad (2.100)$$

while the Phillips parameter follows from

$$\alpha_p = 0.076 X_{10}^{-0.22}. \qquad (2.101)$$

Finally, the remaining parameters are constants:

$$\gamma = 3.3, \quad \sigma_a = 0.07, \quad \sigma_b = 0.09. \qquad (2.102)$$

This result marks one of the most important advances in the field of ocean waves. Nevertheless, even nowadays, there is still some discussion on some of its details and interpretation.

A point of concern is that the JONSWAP parametrization for the frequency spectrum imposes a restriction on the choice of fetch laws. In fact, the parametrization as given in Eqs. (2.98)–(2.102) implies a fetch law for the dimensionless wave variance ϵ_{10}. In order to see this, one simply determines the wave variance from Eq. (2.98) by integration over angular frequency. Scaling angular frequency by the peak frequency ω_p, i.e. introducing $x = \omega/\omega_p$, and some rearrangement then gives, in terms of dimensionless parameters,

$$\frac{\nu_{10}^4 \epsilon_{10}}{\alpha_p} = \lambda, \qquad (2.103)$$

where the so-called 'shape parameter' λ represents the dimensionless integral over the JONSWAP spectrum,

$$\lambda = \int_0^\infty dx \, f(x), \qquad (2.104)$$

and $f(x)$ is the JONSWAP spectrum in dimensionless form,

$$f(x) = x^{-5} e^{-5/4x^{-4}} \gamma^\Gamma. \qquad (2.105)$$

Since the spectral parameters γ and σ are independent of fetch it immediately follows that the shape parameter is also independent of fetch. For the choice of parameters given in Eq. (2.102) one finds

$$\lambda = 0.3. \qquad (2.106)$$

Direct observations from JONSWAP give as a mean value $\lambda = 0.25$. As a consequence, the consistency condition of Eq. (2.103) implies, for given fetch laws for ν_{10} and α_p, a fetch law for the dimensionless variance ϵ_{10}. The wave-variance fetch law thus found is not in agreement with the empirical fetch law for the wave variance obtained by fitting observed dimensionless variance against dimensionless fetch. Based on the cleaned data set of JONSWAP, Kahma and Calkoen (1992) find an empirical fetch law

$$\epsilon_{10} = 2.1 \times 10^{-7} \times X_{10}, \qquad (2.107)$$

with observed dimensionless fetch in the range of 100 to 5000, while from the consistency relation in Eq. (2.103) a more sensitive dependence on dimensionless fetch is found, $\epsilon_{10} \sim X_{10}^{1.1}$. The reason for this discrepancy may be that the fetch law for the Phillips parameter α_p was originally obtained from a fit to both laboratory

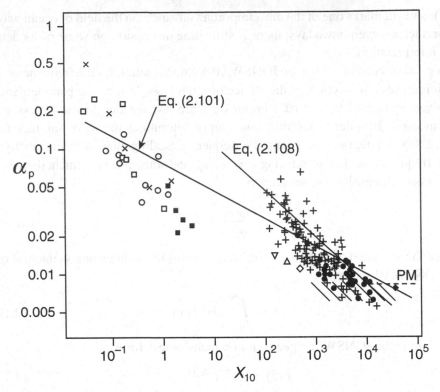

Fig. 2.6. Observed Phillips parameter versus dimensionless fetch from a number of field experiments. The fetch laws of Eqs. (2.101) and (2.108) are also shown. PM denotes the Pierson–Moskowitz limit of $\alpha_p = 0.0083$ (from Hasselmann *et al.*, 1973).

data and field data. However, Donelan *et al.* (1985) have shown that results from the field and the laboratory belong to different families (this is also evident from Fig. 2.6). Therefore, a fit to field data alone is more appropriate. The fetch law for α_p, appropriate for field data, may be obtained from the consistency relation (2.103), the fetch law for ν_{10} and the empirical fetch law for ϵ_{10}. Using the observed value of the shape parameter $\lambda = 0.25$ one finds

$$\alpha_p = 0.2 X_{10}^{-1/3}, \tag{2.108}$$

hence, compared with the original JONSWAP law in Eq. (2.101), the consistency condition gives a more sensitive dependence of α_p on fetch. According to Fig. 2.6, the fetch law in Eq. (2.108) is in fair agreement with the JONSWAP observations for α_p; this fetch law is also almost identical to the one obtained by Günther (1981) who made a fit using the JONSWAP field data only. In terms of the wave-age parameter χ_{10} the modified JONSWAP laws for dimensionless energy and the

Phillips parameter become

$$\epsilon_{10} = 0.002\,24\chi_{10}^3,$$
$$\alpha_p = 0.0093\chi_{10}^{-1}. \tag{2.109}$$

Notice that compared with Eq. (2.96) the above fetch law for the dimensionless wave variance shows a somewhat weaker dependence on wave age, but it is in good agreement with the so-called '3/2 power law' of Toba (1972). Also, combining two data sets, namely the one from JONSWAP and one called 'KNMI' (Sanders, 1976), Janssen *et al.* (1987) found that the dimensionless energy is proportional to the third power of the wave age, but one needs to scale the relevant parameters with the friction velocity and not with the wind speed.

2.6.2 Toba's spectrum and friction-velocity scaling

Another point of discussion regarding the JONSWAP parametrization of fetch-limited windsea spectra is the assumed Phillips power law for the high-frequency part of the spectrum. As already mentioned, Phillips had argued that in the so-called 'saturation range' (typically between $1.5 \times \omega_p$ and $3.5 \times \omega_p$) the spectral density $F(\omega)$ is saturated at a level determined by wave breaking. As a consequence, the saturation level is exclusively determined by the frequency ω and acceleration of gravity g. The resulting spectral shape becomes

$$F(\omega) = \alpha_p g^2 \omega^{-5}, \tag{2.110}$$

but from JONSWAP it was found that α_p varied with fetch instead of being a constant.

An alternative approach to the notion of the saturation range was proposed by Toba (1972, 1973). It is based on his 3/2 power law for the significant wave height H_S and the significant period. In terms of wave variance ϵ and peak frequency ω_p, Toba found from laboratory data that

$$\epsilon \sim gu_*\omega_p^{-3}, \tag{2.111}$$

where u_* is the friction velocity of wind over the water surface. On similarity grounds, Toba (1973) deduced that the saturation spectral density becomes

$$F(\omega) = \alpha_T gu_*\omega^{-4}, \tag{2.112}$$

where α_T is the Toba constant. Initially, empirical support for the Toba spectrum (2.112) was obtained in laboratory experiments (Toba, 1973) and in field experiments (Kawai *et al.*, 1977). However, Toba's work only received recognition in the mid 1980s with the presentation of further empirical evidence supporting

Eq. (2.112) by Mitsuyasu *et al.* (1980), Kahma (1981), Forristall (1981) and Donelan *et al.* (1985), and with theoretical work of Kitaigorodskii (1983) and Phillips (1985).

Toba's proposal for the spectrum in the saturation range has considerable implications for wave prediction. Apart from the obvious implication on the spectral shape, his work also implied that the appropriate velocity scale is not the wind speed at an arbitrarily chosen height of 10 m, but the friction velocity u_* defined as

$$u_* = \sqrt{\tau}. \tag{2.113}$$

Here, $\tau = -\langle u'w' \rangle$ is the kinematic stress in the surface layer above the waves, while u' and w' denote the horizontal and vertical velocity fluctuations around the average plane-parallel flow. These fluctuations are associated with both air turbulence and wave-induced motion.

Intermezzo   The kinematic stress plays a prominent role in the surface layer above the ocean waves. Close to the surface, the mean flow is to a good approximation a plane-parallel flow for which vertical gradients are far more important than horizontal gradients. Scaling considerations applied to the Navier–Stokes equation show then that the friction term is dominant near the surface, so that Eq. (2.1) reduces to

$$\frac{\partial U}{\partial t} = \frac{\partial \tau_x}{\partial z},$$

$$\frac{\partial V}{\partial t} = \frac{\partial \tau_y}{\partial z}, \tag{2.114}$$

where the mean flow is denoted by **U**. In steady-state circumstances the (kinematic) stress is independent of height z, and is therefore a good candidate for the wind-speed scale in growth laws.

In the atmosphere, it is often assumed that the stress is proportional to the vertical gradient of the horizontal velocity, i.e.

$$\tau = \nu_T \frac{\partial \mathbf{U}}{\partial z}, \tag{2.115}$$

in analogy with the expression for the viscous stress. However, there is also an important difference with the case of viscosity where ν is usually taken constant. In the turbulent surface layer, eddies are taking care of the momentum transport and the size of the eddies increases with the distance from the surface. As a consequence, the eddy viscosity ν_T increases with height. General ideas from turbulence theory (Tennekes and Lumley, 1974) suggest that

$$\nu_T = l^2 \left| \frac{\partial \mathbf{U}}{\partial z} \right|, \tag{2.116}$$

with l the so-called 'mixing length'. This approach was first put forward by Prandtl in 1925. In practice, near the surface l is usually taken as

$$l = \kappa z, \tag{2.117}$$

where $\kappa = 0.4$ is the von Kármán constant. This expression for the mixing length vanishes when the surface is approached and causes a singularity in the mean velocity profile, as will be seen in a moment. This singularity may be removed by introducing additional physics. In a very thin layer above the surface, with a height proportional to the molecular viscosity of air ν_a, viscous processes play an important role (for an account of the interaction of turbulence and viscous processes above a flat plate see van Driest (1951)). In addition, for airflow over the ocean surface, short gravity waves are able to extract considerable amounts of momentum from the air because these waves grow the fastest. These waves may become steep so that 'micro-scale' breaking and flow separation may also be important. These processes, the list of which is by no means complete, are no doubt very complicated. If however, one is only interested in the mean flow away from the surface then these processes may be represented by introducing a height scale z_0, called the 'roughness length', at which the mean flow velocity vanishes. The boundary condition therefore becomes

$$\mathbf{U}(z_0) = 0. \tag{2.118}$$

Considering now the case of air flowing in the x-direction, the steady-state solution of the system of equations (2.114)–(2.118) becomes

$$U(z) = \frac{u_*}{\kappa} \ln(z/z_0), \quad z > z_0, \tag{2.119}$$

giving the well-known logarithmic wind profile in the constant-stress layer above the surface. The big unknown in the wind-profile equation is the roughness length z_0, and in the past this quantity has been determined experimentally from profile measurements or from a combination of measuring the surface stress through eddy-correlation techniques and of measuring wind speed at a certain height. Errors in z_0 are large, however, because the roughness length depends in an exponential way on wind-speed and the friction velocity. Nevertheless, as already mentioned in the introduction in Chapter 1, Charnock (1955) was able to propose on dimensional grounds a parametrization of the roughness length which is still in use today. He argued that the short gravity waves are mainly responsible for the momentum transfer from air to ocean, and therefore the scaling parameters are acceleration of gravity g and friction velocity u_*. As a result, the expression for the roughness length over ocean waves becomes

$$z_0 = \alpha_{CH} u_*^2 / g, \tag{2.120}$$

and profile measurements resulted in an estimate of the Charnock constant $\alpha_{CH} = 0.0112$.

Alternatively, the roughness length z_0 may be obtained by means of a detailed study of the processes that extract momentum from the airflow. An example of this is presented in Chapter 3 where the consequences of the growth of surface gravity waves on the mean airflow are studied. The roughness length turns out to be similar to the one proposed by Charnock, except that, in agreement with Stewart (1974), the Charnock parameter α_{CH} is found to depend on the stage of development of windsea, i.e. the wave age χ.

Before closing this intermezzo, it should be mentioned that in practice the drag coefficient C_D is frequently used to relate the surface stress to the wind speed at a given height. It is defined by

$$\tau = C_D U^2(z), \tag{2.121}$$

and for a logarithmic wind profile of the form given in Eq. (2.119) it immediately follows that

$$C_D = \frac{\kappa^2}{\ln^2(z/z_0)}. \tag{2.122}$$

Its value clearly depends on the choice of z_0 and on the observation height z.

Using Charnock's relation, the drag coefficient is found to increase (almost) linearly with wind speed. Since C_D depends on wind speed, it makes a difference whether one uses wind speed or friction velocity as the wind-speed scale in the growth laws for windsea. For example, the saturation wave height scales, using Eqs. (2.88) and (2.93), with the square of the wind-speed scale. Therefore, adopting U_{10} scaling, the significant wave height scales with U_{10}^2 for large duration and fetch, while, adopting friction-velocity scaling, the saturation wave height scales with u_*^2. This would give a more sensitive dependence on wind speed U_{10} as C_D increases with wind speed. Blake (1991) reanalysed a data set collected by Ewing and Laing (1987) and found indeed a more sensitive dependence of the saturation wave height on wind speed.

Finally, another argument in favour of friction-velocity scaling can be inferred from Fig. 2.4b. This plot shows the wave-variance fetch law using u_* scaling for stable and unstable stratification. Evidently, the differences between stable and unstable cases are much smaller than for U_{10} scaling. It is conjectured in Chapters 3 and 5 that the remaining differences may be explained by considering effects of gustiness, which occur in particular for unstable stratification.

Let us now return to Toba's proposal for the high-frequency part of the wave spectrum. In order to investigate whether the Toba spectrum fits the high-frequency part of the spectrum better, Battjes *et al.* (1987) decided to reanalyse the original JONSWAP data set. Instead of the JONSWAP spectral shape of Eq. (2.98) these authors fitted the spectra by means of the expression

$$F(\omega) = F_T(\omega, \omega_p, \alpha_T) \times \gamma^{\Gamma(\omega, \omega_p, \sigma)}, \tag{2.123}$$

where

$$F_T(\omega, \omega_p, \alpha_T) = \alpha_T g u_* \omega^{-4} \exp\left\{-\left(\frac{\omega_p}{\omega}\right)^4\right\}, \tag{2.124}$$

while the expression for Γ was identical to the original JONSWAP proposal of Eq. (2.99). The Toba spectrum F_T depends explicitly on the friction velocity, and since no observed data on the *in-situ* stress were available, Charnock's relation (2.120) with Charnock parameter $\alpha_{CH} = 0.0144$ was used to determine the friction velocity from the surface wind speed at 10 m height. A detailed error analysis showed convincingly that the Toba spectrum fitted the observed spectra better than the Phillips spectrum. Moreover, the Toba parameter α_T showed a much weaker dependence on the wave age than the Phillips parameter α_p. In fact, Battjes *et al.*

(1987) found from their so-called 'full-frequency-range analysis' for α_p that

$$\alpha_p = 0.0057 \chi_{10}^{-1.24}, \tag{2.125}$$

in close agreement with the result we found using invariance of spectral shape (see Eq. (2.109)), while for α_T they found

$$\alpha_T = 0.127, \tag{2.126}$$

i.e. the Toba parameter is a constant. However, it should be noted that when the Toba parameter was obtained from the high-frequency part of the spectrum ($\omega > 1.5\omega_p$) a weak dependence on wave age with power -0.23 was obtained. Finally, the remaining parameters of the proposed spectral shape in Eq. (2.124) are practically constant:

$$\gamma = 3.64, \quad \sigma_a = 0.12, \quad \sigma_b = 0.17. \tag{2.127}$$

The reanalysis of the JONSWAP data set by Battjes *et al.* (1987) therefore indicates that the high-frequency part of the wave spectrum can be better approximated with an ω^{-4} tail than with an ω^{-5} tail. The Toba parameter α_T is found to be virtually independent of the wave age. This finding is consistent with Toba's hypothesis that α_T should be a universal constant. However, it differs from the data of Donelan *et al.* (1985) whose results indicate a noticeable increase of α_T with wave age. The expression for the high-frequency spectrum F_D given by these authors can be written as

$$F_D(\omega, \omega_p, \alpha_D) = \alpha_D g^2 \omega_p \omega^{-4}. \tag{2.128}$$

Comparison with Toba's formulation (2.124) shows that

$$\alpha_T = C_D^{-1/2} \chi_{10} \alpha_D,$$

where C_D is the drag coefficient. When the direction of wave propagation at the spectral peak is in the wind direction, Donelan *et al.* give the following parametrization of α_D:

$$\alpha_D = 0.006 \chi_{10}^{-0.55}, \quad 0.2 < \chi_{10} < 1.2,$$

and, as a consequence, the Toba parameter would increase with wave age according to

$$\alpha_T = 0.006 C_D^{-1/2} \chi_{10}^{0.45}, \tag{2.129}$$

ignoring a possible dependence of the drag coefficient on the wave age.

The reason for this different behaviour of the Toba parameter on wave age (see Eqs. (2.126) and (2.129)) is not clear. It has been suggested, however, that there is a remarkable geographical dependence in these results. Phillips (1985) points out that data from the Pacific give considerably lower values for the Toba parameter than the nonPacific data. A similar remark may apply to the JONSWAP data (North Sea) and the Donelan *et al.* (1985) data which are taken from Lake Ontario. In order to investigate a possible regional dependence of the results, Bidlot (private communication, 2002) determined the Toba parameter from a National Data Buoy Center (NDBC) data set by averaging the quantity

$$\alpha_T = \omega^4 F_{obs}(\omega)/gu_*,$$

between the angular frequencies $1.3\omega_p$ and $2.5\omega_p$ with F_{obs} being the observed spectrum. The mean value of the Toba parameter was close to Eq. (2.126) while, in agreement with Battjes *et al.* (1987), α_T showed hardly any dependence on wave age. The study by Bidlot is quite extensive; it consists of around 150 000 spectra from different regions of the northern hemisphere, i.e. the North Pacific, the Great Lakes and the east coast of the United States. There are certainly regional differences, but these differences can also be explained by the different types of buoys being employed which may, because of their different sizes, have a different high-frequency response. Also air–sea stability effects may play a role.

Therefore, in view of the overwhelming evidence for the constancy of the Toba parameter, it is concluded that a description of fetch-limited spectra by means of the Toba formulation (2.123) and (2.124) is to be preferred. This description of the wind-wave spectrum is only valid for the low-frequency part of the spectrum; as argued below it only holds for angular frequencies below $3\omega_p$.

For a given frequency spectrum and the usual dispersion relation it is straightforward to obtain, by means of Eq. (2.89), the wavenumber spectrum. In wavenumber space the Toba spectrum is found to be proportional to $k^{-2.5}$. However, as pointed out by Banner (1990b), for short waves the linear dispersion relation may not be applied, because these waves, which are riding on top of the long waves, experience a Doppler shift caused by the orbital motion of the long waves. According to Banner (1990b) this affects the high-frequency spectrum in the range $\omega > 3\omega_p$. For short waves there is therefore a preference to determine wavenumber spectra. However, this is a very hard task, and usually requires unconventional measurement techniques, such as from optics. For frequencies above $3\omega_p$ there is therefore little direct evidence of the wavenumber spectrum but it cannot continue with a slope of -2.5 for this would imply short-wave spectral levels far higher than those observed even in very strong forcing in laboratory tanks. Furthermore, the mean-square slope would greatly exceed optical and radar estimates (Cox and Munk, 1954; Jackson *et al.*, 1992). Kitaigorodskii (1983) has postulated a gradual transition in the spectral slope from quasi-saturated ($k^{-2.5}$) to fully saturated (k^{-3}). In other words, for a $k^{-2.5}$ spectrum the local wave slope would increase with wavenumber until the waves become unstable and break. For larger frequencies the spectral shape would be determined by wave breaking giving Phillips' k^{-3} spectrum. Nowadays there is a considerable amount of experimental evidence (Forristall, 1981; Birch and Ewing, 1986; Banner, 1990b; Hwang *et al.*, 1996; Hwang, 1997; Hara and Karachintsev, 2003) that such a transition indeed occurs.

2.6.3 The angular distribution of windsea

The directional properties of ocean waves in open sea have mostly been studied with heave, pitch and roll buoys. These can be used to estimate integral directional parameters (Longuet-Higgins *et al.*, 1963; Kuik *et al.*, 1988) or the buoy data may be used to reconstruct a low-resolution image of the directional energy distribution (see, for example, Longuet-Higgins *et al.*, 1963; Krogstad *et al.*, 1988). Imaging techniques and array techniques provide high-resolution estimates of the two-dimensional spectrum (e.g. synthetic aperture radar (SAR), Brüning *et al.*, 1988; the airborne topographic mapper (ATM), Hwang *et al.*, 2000a,b; or wave gauge arrays, Donelan *et al.*, 1985).

In a pioneering study of the directional properties of ocean waves, Longuet-Higgins *et al.* (1963) found that the width of the angular distribution of wind waves was narrowest near the peak of the wave spectrum and that away from the peak the directional spread increased. Subsequent studies (Hasselmann *et al.*, 1973, 1980; Mitsuyasu *et al.*, 1975; Donelan *et al.*, 1985) confirm this universal characteristic of developing wind waves. Most parametrizations of directional spread are nowadays in terms of the frequency normalized with the peak frequency, i.e. ω/ω_p, suggesting that the directional distribution is predominantly controlled by nonlinear wave–wave interactions (Hasselmann *et al.*, 1980). As an example we quote the parametrization of Donelan *et al.* (1985). In this approach the two-dimensional frequency spectrum $F(\omega, \theta)$ becomes

$$F(\omega, \theta) = \frac{1}{2} F(\omega) \cosh^{-2}[\beta(\theta - \bar{\theta}(\omega))], \tag{2.130}$$

where $\bar{\theta}$ is the mean wave direction and

$$\beta = \begin{cases} 2.61(\omega/\omega_p)^{+1.3}, & 0.56 < \omega/\omega_p < 0.95, \\ 2.28(\omega/\omega_p)^{-1.3}, & 0.95 < \omega/\omega_p < 1.6, \\ 1.24, & \omega/\omega_p > 1.6. \end{cases} \tag{2.131}$$

The Donelan *et al.* (1985) data set only extended to $\omega/\omega_p = 1.6$, thus for frequencies larger than 1.6 a constant value of $\beta = 1.24$ was assumed. Based on high-frequency stereo-photography, Banner (1990b) found that β was not a constant for these higher frequencies as specified by Donelan *et al.* (1985) and he proposed

$$\beta = 10^{(-0.4 + 0.8393 \exp[-0.5670 \log(\omega/\omega_p)^2])}, \quad \omega/\omega_p > 1.6. \tag{2.132}$$

All early parametrizations of the directional distribution shared the property that for high frequencies there is a single maximum in the wind direction, i.e. the distribution is unimodal (see for example Eq. (2.130)). A numerical study of the energy balance equation (Banner and Young, 1994) suggested, however, that for high frequencies there is a bimodal angular distribution which is caused by non-linear four-wave interactions. This result prompted a renewed interest in directional aspects of ocean waves.

The early investigations into the directional distribution of ocean waves were based on the analysis of temporal measurements and did not find any indication of bimodality. Bimodal features have been extracted from these data only recently, using a maximum-entropy method (MEM) or a maximum-likelihood method (MLM) (Young *et al.*, 1995; Ewans, 1998). These results highlight a major difficulty in resolving directional properties from a small number of sensor elements. Depending on the analysis method, significant quantitative differences occur. For example, Ewans (1998) shows a comparison of the bimodal analysis using an MLM and an MEM; the MEM shows a much more pronounced bimodality than the MLM.

Bimodal directional distributions have been observed from spatial measurements using an aerial stereo-photographic technique (Phillips, 1958), an airborne radar system (Jackson *et al.*, 1985) and land-based imaging radar (Wyatt, 1995). In contrast to the analysis of temporal measurements from wave-gauge arrays or buoys, a standard two-dimensional Fourier transformation is sufficient to reveal the bimodal nature of the direction distribution. In addition, present-day technology has advanced considerably resulting in a much more accurate determination of the sea surface topography and hence giving a confident estimate of bimodality. Therefore, Hwang *et al.* (2000b) were able to study in great detail the directional properties of wind-generated ocean waves and they were able to obtain a reliable parametrization of bimodality. A fair agreement with the theoretical results of Banner and Young (1994) was found.

2.6.4 Sea-state dependence of surface stress

As explained in the Intermezzo in Section 2.6.2 the key parameter that determines the surface stress for given wind is the Charnock parameter α_{CH}. Hence, many researchers have made attempts to measure surface stress and wind speed in order to obtain, using Eq. (2.122), the Charnock parameter. But since z_0 depends in an exponential way on these parameters, there are very high demands on the measurement accuracy of U_{10} and u_*. In addition, wind speed and friction velocity are normally measured in the vicinity of large bodies such as measurement platforms and ships. The presence of such large bodies inevitably leads to flow distortion with the consequence that corrections are needed for flow direction, wind speed and friction velocity. These corrections are not easy to obtain and are of a considerable size so that observations from a well-exposed anemometer (such as employed by Donelan (1982) or Smith *et al.* (1992)) are preferred. Ship observations normally suffer from flow distortion (Yelland and Taylor, 1996; Eymard *et al.*, 1999).

Furthermore, the roughness over the oceans is very small ($z_0 \approx 0.1$ mm) compared with that over land ($z_0 \approx 10$ cm). As a consequence, the constant-stress layer over the oceans is thin and Donelan (1990) finds that the assumption that the measured stress (at an observation height of 10 m) is the surface stress always introduces a systematic and wind-speed-dependent underestimation of the surface stress, which in some cases is in excess of 30 per cent.

Finally, the preferred method for measuring the surface stress is the eddy-correlation technique. This method measures all components of the velocity, subtracts the mean flow, and the kinematic stress follows then from the correlations between the components of the (turbulent) fluctuations. Here the kinematic stress tensor is defined as

$$\tau_{\alpha\beta} = -\langle u'_\alpha w'_\beta \rangle, \qquad (2.133)$$

where u_α is the velocity component in the α-direction and a prime refers to the fluctations around the mean. The angle brackets refer to an averaging over a time interval, which is usually about 20 min long. This method works well on stationary platforms but it presents problems on board a ship because the ship's motion needs to be eliminated (nowadays this is indeed possible but it requires some sophisticated hardware). Therefore, some researchers prefer to use a measurement technique that is not sensitive to these low-frequency motions, for example the so-called 'inertial-dissipation technique'. The latter technique is based on the assumption of an inertial subrange in the spectrum of the turbulent velocity fluctuations so that there is a balance between production of turbulent kinetic energy and dissipation. However, in the presence of swells outrunning relatively weak winds the velocity spectra no longer have universal shapes, so the classical Monin–Obukhov similarity theory is no longer valid (Drennan *et al.*, 1999). Also, Janssen (1999) has pointed out that growing wind waves may play a role in the turbulent kinetic-energy budget through the divergence of the pressure–velocity correlation, giving for large winds an increase in stress of the order of 20 per cent. Results obtained with the inertial-dissipation method are therefore uncertain, because this method is based on a number of assumptions which are not always satisfied.

Because of all these complications it should be clear that the sea-state dependence of the surface stress is a controversial issue. Furthermore, there is the additional problem of what parameter gives an appropriate characterization of the sea state. Wind waves presumably cause most of the aerodynamic drag over the oceans. In particular the short waves mainly contribute because they have the largest wind-induced growth. Therefore, it may be argued that the surface stress depends on the steepness of the short waves, which to a good approximation is given by the mean-square slope. However, the latter parameter is not easy to measure. Now, for the case of windsea a proxy for the state of the high-frequency waves is provided by the wave-age parameter. As we have adopted Toba's formulation for the windsea spectrum, which depends on the friction velocity, it makes sense to use from now on the wave age χ_*, defined as

$$\chi_* = c_p/u_* \tag{2.134}$$

where c_p is the phase velocity of the peak of the spectrum. For young wind waves ($\chi_* \simeq 10$) waves are steep, while more gentle waves are found for old windsea ($\chi_* \simeq 30$). Since steeper waves are associated with a rougher surface, it is expected that the Charnock parameter will decrease with increasing wave age.

Despite all the aforementioned complications, this was indeed found by Donelan (1982) and Donelan *et al.* (1993) for the short fetches that occur on Lake Ontario and by the HEXOS group (Smith *et al.*, 1992) in the North Sea. In the case of HEXOS, only windsea cases were selected. Stresses were obtained by means of the eddy-correlation technique. There exists a close agreement between the two data sets, resulting in similar relations between the Charnock parameter and the wave age. This is remarkable because the Lake Ontario data have a typical phase speed of the order of 3 to 4 m/s while the North Sea data have phase speeds of around 10 m/s. For reference, only the HEXOS fit for the Charnock parameter is quoted:

$$\alpha_{CH} = 0.48\chi_*^{-1}, \ 5 < \chi_* < 25, \tag{2.135}$$

supporting the notion that indeed the airflow becomes smoother for older windsea.

The result in Eq. (2.135) marks another important advance in the field of ocean waves, and in particular regarding our understanding of small-scale air–sea interaction for the case of windsea. The early field experiments of Donelan (1982) and the HEXOS group (Smith *et al.*, 1992) consisted of a small number of observations. In view of the large scatter in the Charnock parameter, the significance of the result in Eq. (2.135) was questioned. However, additional field campaigns, such as WAVES87 and SWADE (see, for example, Drennan *et al.*, 1999) and ASGAMAGE (Oost *et al.*, 2002) have confirmed that indeed, the airflow becomes smoother for older windsea. The more recent data sets show, however, a more sensitive dependence of the Charnock parameter on the wave age than found by Eq. (2.135). As an illustration of the sea-state dependence of the roughness, Fig. 2.7 shows the results

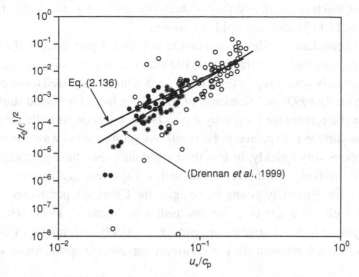

Fig. 2.7. Dimensionless roughness $z_0/\epsilon^{1/2}$ versus inverse wave age $1/\chi_*$ for WAVES87 short fetch (open circles) and long fetch (filled circles), and SWADE data (asterisks); all pure windsea data (from Drennan *et al.*, 1999).

of Drennan *et al.* (1999). These authors have the plausible view that the roughness parameter z_0 is a fraction of significant wave height and therefore they plot the roughness scaled with the standard deviation of the surface elevation, $z_0/\epsilon^{1/2}$, as a function of the inverse of the wave age $1/\chi_*$. Donelan (1990) was able to obtain high correlations between these two parameters for his 1970s Lake Ontario data set. In fact,

$$z_0/\epsilon^{1/2} = 1.84\chi_*^{-2.53}. \qquad (2.136)$$

Note that Eq. (2.136) is directly related to Eq. (2.135). In order to see this, one starts from the left-hand side of Eq. (2.136) and one uses Charnock's relation and the expression for the Charnock parameter (Eq. (2.135)) to eliminate z_0. The wave variance ϵ is then eliminated using Toba's 3/2 power law (Eq. (2.111)). As a consequence, one finds that

$$z_0/\epsilon^{1/2} \sim \chi_*^{-5/2},$$

in very good agreement with Donelan's result (Eq. (2.136)). This last fit is also plotted in Fig. 2.7 and shows a good agreement with the more recent data from the WAVES87 and the SWADE campaigns, except for relatively old windsea ($\chi_* \simeq 50$) when the observed roughness is much smaller than the parametrization (2.136). However, Drennan *et al.* (1999) have shown that in those circumstances the Monin–Obukhov similarity theory does not apply, e.g. the logarithmic wind profile of Eq. (2.119) does not hold any more.

In order to conclude our discussion on the sea-state dependence of the roughness over windsea, we emphasize that Eq. (2.135) has a restricted validity, i.e. it does not hold for extremely young sea states with $\chi_* < 5$. This is discussed more extensively by Donelan *et al.* (1993) and Komen *et al.* (1998). In fact, if the wind starts blowing over a flat surface, there are no gravity waves present. Hence, initially the roughness of the ocean surface is expected to be small. As soon as short waves are generated (which happens very quickly, in less than a minute), they have extracted momentum from the airflow, and therefore the airflow experiences a rougher surface. In other words, for extremely young wave ages, the Charnock parameter is expected to increase with wave age (see the discussion in Nordeng, 1991). However, the extremely young windsea state disappears very quickly, in a matter of 10 minutes, and is therefore not relevant for the relatively large-scale applications we have in mind.

On the other hand, in the case of mixed sea states, consisting of windsea and one or more swells, it is not at all evident that the wave-age parameter is an appropriate measure for the state of the high-frequency waves. Swells propagate over large distances, they are therefore nonlocal in the sense that they do not depend on the local friction velocity; they may interact with the windsea in such a way that the high-frequency waves are damped or, in the case of opposing swell, get larger steepness. On the open ocean, where mixed sea states are frequently occurring, a simple relation between the Charnock parameter and the wave age is therefore not expected, unless wind waves are dominating the sea state. This usually happens for fairly strong winds. Thus, although Yelland and Taylor (1996) were unable to find a sea-state dependence of the Charnock parameter from their open-ocean data in the southern Atlantic, by restricting to wind speeds above 10 m/s Janssen (2001) did find a relationship similar to that found by Donelan *et al.* (1993) and the HEXOS group.

2.7 Summary of results

Starting from the Lagrangian of linear deep-water waves we have obtained the appropriate evolution equations for wave groups in slowly varying circumstances caused by slowly varying currents or water depth. There are, however, other causes for the evolution of wave groups. Waves may grow because of the action of wind and they may lose energy because of dissipation due to, for example, white capping. Furthermore, finite-amplitude ocean waves are subject to four-wave interactions. As long as these perturbations are small they can be added and in the context of a statistical description of ocean waves, the action density N obeys the action balance equation

$$\frac{d}{dt} N = S_{in} + S_{nl} + S_{ds}, \tag{2.137}$$

where in the case of spherical coordinates the operator d/dt is given by Eq. (2.81), while the source terms on the right-hand side represent the physics of wind input S_{in}, non-linear four-wave interactions S_{nl} and dissipation S_{ds}. In the next two chapters these source terms will be discussed in some detail.

3

On the generation of ocean waves by wind

In this chapter the theory of the generation of ocean waves by wind will be developed resulting in an expression for the wind-input source function S_{in} of the action balance equation. As will be seen from the subsequent discussion, this problem has led to many debates and much controversy. There may be several reasons for this. On the one hand, from the theoretical point of view it should be realized that one is dealing with an extremely difficult problem because it involves the modelling of a turbulent airflow over a surface that varies in space and time. Although there has been much progress in understanding turbulence over a flat plate in steady-state conditions, modelling attempts of turbulent flow over (nonlinear) gravity waves are only beginning and, as will be seen, there is still a considerable uncertainty regarding the validity of these models.

On the other hand, from an experimental point of view it should be pointed out that it is not an easy task to measure growth rates of waves by wind. First of all, one cannot simply measure growth rates by studying time series of the surface elevation since the time evolution of ocean waves is determined by a number of processes such as wind input, nonlinear interactions and dissipation. In order to measure the growth of waves by wind one therefore has to make certain assumptions regarding the process that causes wave growth. The commonly adopted cause of wave growth is the work done by the pressure on the surface. This assumption seems plausible in view of the work by Miles (1957). Second, because of the small air–water density ratio the growth rates are small which means that a very accurate determination of amplitude and phase of the wave-induced pressure fluctuations is required.

Nevertheless, considerable progress has been made over the past 40 years or so and therefore a brief review of the main achievements will be given now. The history of the subject of wind-wave generation started in the beginning of the twentieth century when Jeffreys (1924, 1925) assumed that air flowing over the ocean surface was sheltered by the waves on their lee side. This would give a pressure difference, so that work could be done by the wind. Subsequent laboratory measurements on solid waves showed that the pressure difference was much too small to account for the observed growth rates. As a consequence, the sheltering hypothesis was abandoned, and one's everyday experience of the amplification of water waves by wind remained poorly understood. This changed in the mid 1950s, when Phillips (1957) and Miles (1957) published their contributions to the theory of wave generation by wind. Both theories had in common that waves were generated by a resonance phenomenon: Phillips considered the resonant forcing of surface waves by turbulent pressure fluctuations, while Miles considered the resonant interaction between the wave-induced pressure fluctuations and the free surface waves. Phillips' mechanism gives rise to a linear growth of the spectrum in time, but it turned out to be ineffective. This is because the effect is proportional to the variance spectrum of the turbulent pressure fluctuations at the resonant frequency, independent of the wave spectrum, and this is of the order of the square of the air–water density ratio. Miles' mechanism looked more promising. It is proportional to the wave spectrum itself, which implies exponential growth, and it is of the order of the density ratio of air and water.

Although Miles' work aroused renewed interest in the problem, there was also considerable confusion and controversy. One of the main reasons for the controversy was that Miles' theory oversimplified the problem by following the quasi-laminar approach. This approach assumes that the airflow is inviscid and that air turbulence does not play a role except in maintaining the shear flow. Another reason for the controversy is that Miles neglected nonlinear effects, such as wave–mean flow interaction, which are expected to be important at the height where the phase speed of the surface waves matches the wind speed (the so-called 'critical height'). In addition, early field experiments, in particular by Dobson (1971), gave rates of energy transfer from wind to waves that were an order of magnitude larger than those predicted by Miles (1957). As already pointed out, the measurement of the energy transfer from wind to waves is a difficult task as it involves the determination of a small difference between wave-induced pressure fluctuation and the time derivative of the surface elevation signal. More recent field experiments (Snyder, 1974; Snyder *et al.*, 1981; Hasselmann and Bösenberg, 1991) show order of magnitude agreement with Miles' theory, although the theory still predicts energy-transfer rates that are smaller than the measured values, especially for relatively low-frequency waves with a phase speed that is close to the wind speed at 10 m height.

The state of affairs regarding the theory of wind-wave generation remained, however, unsatisfactory. The quasi-laminar approach was criticized because the effect of turbulence on the wave-induced motion was neglected, and nonlinear effects such as wave–mean flow interaction were not considered.

There have been several attempts to overcome these shortcomings by means of numerical modelling of the turbulent boundary-layer flow over a moving water surface. With suitable turbulence closure assumptions the interaction of the wave-induced flow with the mean flow and the boundary-layer turbulence can then be simulated explicitly. One such approach (see, for example, Gent and Taylor, 1976; Makin and Chalikov, 1979; Riley *et al.*, 1982; Al-Zanaidi and Hui, 1984; Jacobs, 1987; Chalikov and Makin, 1991) considers the direct effects of small-scale turbulence on wave growth. Mixing-length modelling or turbulent energy closure is then assumed to calculate the turbulent Reynolds stresses. The resulting diffusion of momentum is then so large that essentially Miles' critical mechanism is thought to be ineffective, but, surprisingly, the results for growing waves are not very different from the results obtained in quasi-laminar theory. Hence, small-scale eddies and finite wave steepness have only a small direct effect on wave growth. On the other hand, in adverse winds or when waves are propagating faster than the wind speed, these theories give a considerable wave damping, which contrasts with Miles' theory.

Observations give a somewhat conflicting picture. For waves that propagate faster than the wind, Snyder *et al.* (1981) and Hasselmann and Bösenberg (1991) report insignificant damping rates. In adverse winds, Dobson (1971) observed damping rates that were an order of magnitude smaller than growth rates under corresponding conditions. Stewart and Teague (1980) observed the decay of swell and found damping rates of about 15 per cent of the corresponding growth rates. In the laboratory, Young and Sobey (1985) found only a weak attenuation for adverse winds. On the other hand, Mizuno (1976) and Donelan (1983) found damping rates that were comparable with corresponding growth rates.

The above turbulence models rely on the analogy with molecular processes. Van Duin and Janssen (1992) pointed out that this approach fails for low-frequency waves. Mixing-length modelling assumes that the momentum transport caused by turbulence is the fastest process in the fluid. This is not justified for low-frequency waves which interact with large eddies whose eddy-turnover time may become larger than the period of the waves. In other words, during a wave period there is not sufficient time for the eddies to transport momentum. For these large eddies (which are identified here with gustiness) another approach is needed. Nikolayeva and Tsimring (1986) considered the effect of gustiness on wave growth, and a considerable enhancement of energy transfer was found, especially for long waves with a phase speed comparable with the wind speed at 10 m height.

Belcher and Hunt (1993) have pointed out that mixing-length modelling is even inadequate for slowly propagating waves. They argue that, far away from the water surface, turbulence is slow with respect to the waves, so that again large eddies do not have sufficient time to transport momentum. This results then in a severe truncation of the mixing length in the so-called 'outer layer' of the flow. In fact, the greater part of the flow may now be regarded as approximately inviscid and the energy transfer from wind to slow waves only occurs in a thin layer above the surface.

However, because of the severe truncation in mixing length the growth rate for the slow waves is considerably smaller than obtained with the usual mixing-length model. Note that the main mechanism for wave growth in the Belcher and Hunt model is the so-called 'non-separated sheltering': the Reynolds stresses close to the surface cause a thickening of the boundary layer on the leeside of the waves which results in flow separation when the slope is large enough. This mechanism is akin to Jeffreys' (1924, 1925) sheltering hypothesis, which was originally developed for separated flows over moving waves of large slope.

In short, the developments over the past 40 years may be summarized as follows. Miles' quasi-laminar theory was the first model to give an explanation of the growth of waves by wind. Because of the neglect of the effects of turbulence on the wave-induced motion the quasi-laminar model has been criticized as being unrealistic, therefore questioning the relevance of the critical-layer mechanism for wind-wave growth. First attempts to describe the effects of turbulence by means of a mixing-length model have been criticized as well, however, mainly because the eddies in the outer layer in the air are too slow to transfer a significant amount of momentum on the time scale of the wave motion. Nevertheless, according to rapid distortion models, such as that of Belcher and Hunt, the critical-layer mechanism is only relevant for very fast-moving ocean waves with a dimensionless phase speed, defined as c/u_*, of the order of 30. But models are based on assumptions and may not necessarily represent reality. Recently, Sullivan *et al.* (2000) studied the growth of waves by wind in the context of an eddy-resolving numerical model. Although the Reynolds number was, compared with nature, too small by an order of magnitude, clear evidence for the existence of a critical layer was found for a wide range of dimensionless phase speeds. As expected from the Miles mechanism, a rapid fall-off of the wave-induced stress was seen at the critical layer. Furthermore, nowadays, there is even direct evidence of the existence and relevance of the critical-layer mechanism from *in-situ* observations (Hristov *et al.*, 2003) obtained from FLIP (a floating instrument platform created by two Scripps scientists some 40 years ago). This is quite a challenge because one has to extract a relatively small wave-coherent signal from a noisy signal. Nevertheless, for the range $16 < c/u_* < 40$, Hristov *et al.* (2003) could see a pronounced cat's-eye pattern around the critical height where the wave-induced stress showed a jump. There was good agreement between observed and modelled wave-induced profiles. Note that it is not clear from observations whether there is a critical layer for dimensionless phase speeds less than 16. These conditions can only be observed when measurements are taken close enough to the ocean surface, in between the ocean waves.

Hence, after all, there is some justification for the validity of Miles' critical-layer mechanism. For this reason, I therefore concentrate on this approach, which allows, at least partly, for an analytical treatment. It has the additional advantage that it can easily be extended to include effects of gustiness; this is known to be important for low-frequency waves.

For a given wind profile, quasi-laminar theory is fairly successful in predicting growth rates and wave-induced profiles. It ignores, however, a possible change of wind profile while the ocean waves are evolving. The momentum transfer from wind to waves may be so large that the associated wave-induced stress becomes a substantial fraction of the turbulent stress (Snyder, 1974; Snyder *et al.*, 1981). The velocity profile over sea waves is controlled by both turbulent and wave-induced momentum flux. Therefore, deviations from the profile of turbulent airflow over a flat plate are to be expected. In addition, the energy transfer from the air to the waves may be affected by the sea state, so that one expects a strong coupling between the turbulent boundary layer and the surface waves.

Observations confirm this expectation. The most direct evidence for the dependence of airflow on the sea state comes from the observed dependence of the drag coefficient on the wave age. Measurements by, for example, Donelan (1982) and Smith *et al.* (1992) have confirmed the dependence of the drag coefficient on wave age. For a fixed wind speed at 10 m height, Donelan found that, depending on the sea state, the drag coefficient may vary by a factor of 2.

The theory of the interaction of wind and waves was elaborated by Fabrikant (1976) and Janssen (1982). The so-called 'quasi-linear theory' of wind-wave generation keeps track of the slow evolution of the sea state and its effects on the wind profile. At each particular time the wave growth follows from Miles' theory. It turns out that quasi-linear theory permits an explanation of the observed dependence of the airflow on the sea state.

The structure of this chapter is as follows. First, we discuss the linear, quasi-laminar theory of wind-wave generation, based on Miles' shear-flow mechanism. Basically, this mechanism of the generation of waves by wind is a resonant interaction of the gravity waves with a plane-parallel flow. Resonance occurs at a critical height z_c which follows from $U(z_c) = c(k)$. Here, U is the air velocity and $c(k)$ the phase velocity of a wave with wavenumber k. Only those waves grow for which the curvature of the velocity profile at the critical height is negative. Results of this theory are compared with available field and laboratory measurements.

After the discussion of the quasi-laminar approach an account of the effects of turbulence is given. First, a brief discussion of the effects of small-scale turbulence is given, followed by the Belcher and Hunt criticism of why mixing-length modelling is not appropriate in rapidly varying circumstances such as occur for airflow over growing wind waves. The approach of Belcher and Hunt allows for a determination of the range of validity of Miles' theory. Depending on the assumption regarding how rapid eddies transfer momentum, Miles' approach may be justified for typical ocean conditions. Second, a treatment of the effects of gusts on wave growth is presented. This treatment assumes that the period of the waves is shorter than a relevant time scale of gustiness. Then, if the probability distribution of the fluctuations in stress due to the gusts is known, an appropriate averaging of Miles' results will immediately give the effect of gusts on wave growth. In particular, the growth rate of low-frequency waves will be affected to a considerable extent.

The final sections of this chapter consist of a discussion of the quasi-linear theory of wind-wave generation, culminating in a derivation of the relation between drag coefficient and the sea state. This relation is then validated against existing field data.

3.1 Linear theory of wind-wave generation

A treatment of Miles' theory is given in which it is emphasized that wind-wave generation is closely related to the instability of a plane-parallel shear flow. The principal difference from shear flow over a flat plate (Drazin and Reid, 1981) is that the lower boundary (the air–sea interface) is allowed to evolve in time and space.

Our starting point is the set of equations for an adiabatic fluid with an infinite sound speed. Hence,

$$\nabla \cdot \mathbf{u} = 0,$$
$$\frac{d}{dt}\mathbf{u} = -\frac{1}{\rho}\nabla p + \mathbf{g}, \tag{3.1}$$
$$\frac{d}{dt}\rho = 0,$$

where all symbols have their usual meanings. We would like to study the stability of the equilibrium solution (see Fig. 3.1) of Eq. (3.1) which corresponds to a flat air–sea interface. Hence, the solution is independent of horizontal position and only depends on height z.

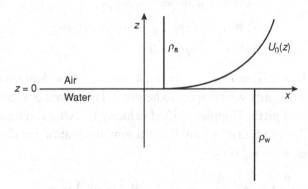

Fig. 3.1. Equilibrium profiles of density and velocity.

The equilibrium of interest is

$$\mathbf{u}_0 = U_0(z)\mathbf{e}_x, \quad \mathbf{g} = -g\mathbf{e}_z,$$

$$\rho_0 = \rho(z), \quad p_0(z) = g \int dz \, \rho_0(z), \tag{3.2}$$

where $\mathbf{e}_x$ and $\mathbf{e}_z$ are unit vectors in the x- and z-directions. Thus, we deal with a plane-parallel flow whose speed and density only depend on height z. The equations for an adiabatic fluid do not explain the height dependence of wind speed and density because effects of small-scale turbulence are not taken into account. In addition, effects of turbulence on the wave-induced motion in the air are not considered. This is a weak point of the present approach, but in Section 3.3 it is argued that this may be justified.

In order to investigate stability, the steady state of Eq. (3.3) is perturbed by writing $\rho = \rho_0 + \rho_1$, etc. with $\rho_1 \ll \rho_0$. Linear equations for the perturbations are then obtained by Taylor expanding Eq. (3.1) around the equilibrium. The steady state is regarded as stable when the perturbations remain small for all times. Otherwise, the steady-state is unstable and as a result gravity waves will be generated at the air–sea interface. The evolution in time of the perturbations is obtained by taking normal-mode solutions of the form $\rho_1 \sim \exp i(kx - \omega t)$, where k is the wavenumber and ω the angular frequency of the wave. The steady state of Eq. (3.3) is unstable when $\Im(\omega) > 0$, because the perturbations then grow exponentially in time ($\Im$ stands for imaginary part).

We shall only consider propagation in one direction. Using Squire's theorem (Drazin and Reid, 1981), results may be generalized immediately to propagation in two dimensions. For waves propagating at an angle with respect to the wind, not the wind itself but the effective wind $\mathbf{k} \cdot \mathbf{U}_0 / k$ is then seen by the waves. Linearization of Eq. (3.1) then gives for normal modes

$$
\begin{aligned}
iku + \partial w / \partial z &= 0, \\
ikWu + wW' &= -ikp_1/\rho_0, \\
ikWw &= \rho_1 p_0'/\rho_0^2 - (\partial p_1/\partial z)/\rho_0, \\
ikW\rho_1 + w\rho_0' &= 0,
\end{aligned}
\tag{3.3}
$$

where $\mathbf{u}_1 = (u, 0, w)$, $W = U_0 - c$, $c = \omega/k$ and the prime denotes differentiation of an equilibrium quantity with respect to height z. The prominent role of the critical layer is evident through the Doppler-shifted velocity W. After some algebra one then arrives at the following Sturm–Liouville differential equation for the displacement of the streamlines $\psi = w/W$:

$$\frac{d}{dz}\left(\rho_0 W^2 \frac{d}{dz}\psi\right) - \left(k^2\rho_0 W^2 + g\rho_0'\right)\psi = 0, \tag{3.4}$$

which is subject to the boundary condition of vanishing displacement at infinite height or depth,

$$\psi \to 0, \quad |z| \to \infty. \tag{3.5}$$

The boundary-value problem (3.4) and (3.5) determines, in principle, the real and imaginary parts of the complex phase speed $c = \omega/k$, giving the growth rate $\gamma = \Im(\omega)$ of the waves.

The (unstable) waves resulting from the above boundary-value problem are called internal gravity waves. By application of the Miles theorem (Miles, 1961), one immediately infers the possibility of instability of the equilibrium given in Fig. 3.1. This theorem states that a sufficient condition for stability is that $dU_0/dz \neq 0$ and that the Richardson number $Ri = g\rho'/(\rho W'^2)$ is smaller than $-1/4$ everywhere. The latter condition is clearly violated in air so there is a possibility of instability, that is there is possibly an energy transfer from the shear flow $U_0(z)$ to the gravity waves.

Whether there is instability or not can only be decided by solving the boundary-value problem. This will only be done for the special case of surface gravity waves and not for the more general case of internal waves. Surface gravity waves are obtained by choosing an appropriate density profile. To that end (see Fig. 3.1), the density profile is chosen to be constant in air and water with a jump at the interface at $z = 0$. Because the air density ρ_a is much smaller than the water density ρ_w, there is a small parameter $\epsilon = \rho_a/\rho_w \approx 10^{-3}$ in the problem, and therefore it will be possible to construct an approximate solution of the eigenvalue problem (3.4) and (3.5).

In water (no current, a constant density) the eigenvalue problem simplifies considerably. We have for $z < 0$

$$\frac{d^2}{dz^2}\psi_w = k^2\psi_w, \tag{3.6}$$

which gives, using the boundary condition at infinity, the solution

$$\psi_w = Ae^{kz}. \tag{3.7}$$

The boundary condition at the interface between air and water is derived from an integration of Eq. (3.4) from just below (-0) to just above ($+0$) the water surface. Note that at $z = 0$ the density profile shows a jump so that $\rho_0' = (\rho_a - \rho_w)\delta(z)$ where $\delta(z)$ is the Dirac delta function. Requiring now that the displacement ψ of the streamlines be continuous across the interface at $z = 0$, we obtain from Eq. (3.4)

$$\rho_0 W^2 \frac{d}{dz}\psi \Big|_{-0}^{+0} = \int_{-0}^{+0} dz \left[\rho_0 k^2 W^2 + g\rho_0'\right]\psi. \tag{3.8}$$

Since in the limit only the integral involving ρ_0' gives a contribution, we obtain, using Eq. (3.7), the following dispersion relation for the phase speed of the waves:

$$c^2 = \frac{g(1-\epsilon)}{k - \epsilon\psi_a'(0)}, \quad \epsilon = \frac{\rho_a}{\rho_w}, \tag{3.9}$$

where, without loss of generality, we have taken the amplitude $A = 1$ as we are dealing with a linear problem. In air we take a constant density so that the eigenvalue problem of Eqs. (3.4) and (3.5) simplifies to

$$\frac{d}{dz}\left(W^2 \frac{d}{dz}\psi_a\right) = k^2 W^2 \psi_a,$$

$$\psi_a(0) = 1; \quad c^2 = \frac{g(1-\epsilon)}{k - \epsilon\psi_a'(0)}, \tag{3.10}$$

$$\psi_a \to 0, \quad z \to \infty.$$

Note that the air–water density ratio only occurs in the dispersion relation. In the absence of air we obtain the usual dispersion relation for deep-water gravity waves, that is $c^2 = g/k$. The effect of air on the surface waves is small since $\epsilon \ll 1$. We therefore solve the dispersion relation in an approximate manner with the result

$$c = c_0 + \epsilon c_1 + \cdots, \tag{3.11}$$

where $c_0 = \sqrt{g/k}$ and $c_1 = \frac{1}{2}c_0(\psi_a'/k - 1)$. As a result, the problem in Eq. (3.10) now reduces to

$$\frac{d}{dz}\left(W_0^2 \frac{d}{dz}\psi_a\right) = k^2 W_0^2 \psi_a,$$

$$\psi_a(0) = 1, \tag{3.12}$$

$$\psi_a \to 0, \quad z \to \infty.$$

where $W_0 = U_0 - c_0$ is now known. As c_0 is known already, the solution of the differential equation is simplified considerably. In addition, we now have an explicit expression for the growth rate γ_a of the amplitude of the waves

$$\frac{\gamma_a}{\omega_0} = \epsilon\Im\left(\frac{c_1}{c_0}\right) = \frac{\epsilon}{2k}\Im(\psi_a') = \frac{\epsilon}{4k}\mathcal{W}(\psi_a, \psi_a^*)\Big|_{z=0} \tag{3.13}$$

where the Wronskian $\mathcal{W}$ is given as

$$\mathcal{W}(\psi_a, \psi_a^*) = -i\left(\psi_a'\psi_a^* - \psi_a\psi_a'^*\right). \tag{3.14}$$

Finally, before we give some interpretation of the result in Eq. (3.13), we remark that it is rather common to use the vertical component of the wave-induced velocity instead of the displacement of the streamlines $\psi \sim w/W$. In terms of the normalized vertical velocity $\chi = w/w(0)$, the eigenvalue problem in Eq. (3.12) becomes

$$W_0 \left(\frac{d^2}{dz^2} - k^2 \right) \chi = W_0'' \chi,$$

$$\chi(0) = 1, \tag{3.15}$$

$$\chi \to 0, \; z \to \infty.$$

and the growth rate of the wave is given by

$$\frac{\gamma_a}{\epsilon \omega_0} = \frac{1}{4k} \mathcal{W}(\chi, \chi^*) \Big|_{z=0}, \tag{3.16}$$

where the Wronskian is now given by $\mathcal{W} = -i(\chi'\chi^* - \chi\chi'^*)$.

Regarding Eq. (3.15) we remark that the differential equation, known as the Rayleigh equation, has a singularity at $W_0 = U_0 - c_0 = 0$. Since $W_0 = 0$ defines the critical height z_c (i.e. the height where the phase speed of the wave matches the wind speed) it is now clear that the resonance at the critical height plays a special role in the problem of wind-wave generation.

Furthermore, we remark that the Wronskian $\mathcal{W}$ is related to a physical quantity, known as the wave-induced stress $\tau_w = -\langle u_1 w_1 \rangle$. In order to see this, remember that we are dealing with normal modes of the type $u_1 = u \exp(i\theta) +$ c.c., so that

$$\tau_w = -\langle (u \exp(i\theta) + \text{c.c.})(w \exp(i\theta) + \text{c.c.}) \rangle = -uw^* + \text{c.c.}$$

Using $\nabla \cdot \mathbf{u} = 0$ to eliminate u we thus have

$$\tau_w = -\frac{i}{k} \left[w^* w' - w w'^* \right], \tag{3.17}$$

and this indeed corresponds to the Wronskian of Eq. (3.16). Therefore, the result in Eq. (3.16) is an elegant one as it relates the growth of the waves to the wave-induced stress.

Another reason why the result in Eq. (3.16) is an elegant one is that the Wronskian $\mathcal{W}$ plays a special role in the theory of second-order differential equations. Namely, by means of the Rayleigh equation (3.15) it may be shown that the Wronskian is independent of height except at the critical height, where it may show a jump (see Fig. 3.2).

We can check this by calculating the derivative of $\mathcal{W}$ with respect to z,

$$\frac{d}{dz} \mathcal{W} = -i(\chi^*\chi'' - \chi\chi''^*).$$

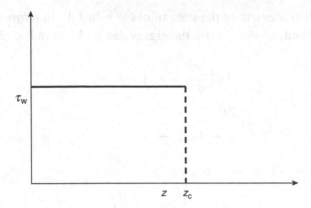

Fig. 3.2. Wave-induced stress shows a jump at critical height.

Using the Rayleigh equation to eliminate the second derivatives this becomes

$$\frac{d}{dz}\mathcal{W} = -i\left(k^2 |\chi|^2 + \frac{W_0''}{W_0} |\chi|^2 - \text{c.c.}\right). \tag{3.18}$$

Now, for $z \neq z_c$ ($W_0 \neq 0$) we immediately find that the right-hand side of Eq. (3.18) vanishes. Hence, $\mathcal{W}$ and the wave-induced stress are independent of height. Since the wave-induced velocity vanishes for large height, we conclude that $\mathcal{W}$ vanishes for $z > z_c$, but it still may have a finite value below the critical height because at the surface the vertical velocity is finite. In fact, the jump at $z = z_c$ may be obtained from Eq. (3.18) by a proper treatment of the singularity at the critical height.

To that end we consider the singular function

$$\frac{1}{W_0} \rightarrow \frac{1}{W_0 - i\Delta},$$

and we take the limit for positive vanishing Δ. Then,

$$\frac{1}{W_0} \rightarrow \frac{\mathcal{P}}{W_0} + \pi i \delta(W_0),$$

where the symbol $\mathcal{P}$ denotes the principal value of the singular function. Equation (3.18) now becomes

$$\frac{d}{dz}\mathcal{W} = 2\pi W_{0c}'' |\chi_c|^2 \delta(W_0) \tag{3.19}$$

where the subscript c refers to evaluation at the critical height z_c ($W_0 = 0$).

Integration of Eq. (3.19), using the boundary condition that $\mathcal{W}$ vanishes for $z > z_c$, gives

$$\mathcal{W} = -2\pi \frac{W_{0c}''}{|W_{0c}'|} |\chi_c|^2, \quad \text{for } z < z_c, \tag{3.20}$$

and therefore the growth rate of the waves becomes

$$\frac{\gamma_a}{\epsilon \omega_0} = -\frac{\pi}{2k} \frac{W_{0c}''}{|W_{0c}'|} |\chi_c|^2. \tag{3.21}$$

This is Miles' classical result for the growth of surface gravity waves due to shear flow. From Eq. (3.21) we obtain the well-known result that only those waves are unstable for which the curvature $W_0'' = U_0''$ of the wind profile at the critical height is negative. This is the case, for example, for a logarithmic profile.

Before we proceed, a few remarks need to be made. The first one is of a somewhat technical nature and is related to the choice of sign of the parameter Δ in the singular function $1/W_0$. It should be pointed out that it is important to make a proper choice for the sign of Δ since the final results regarding the stability or instability of the flow will depend on it. Hence, special care regarding the treatment of the poles is needed. The right choice of the sign of Δ follows from the solution of the initial-value problem by means of Laplace transformation and by applying asymptotic techniques to find the large-time behaviour of the solution. In case of instability, the large-time solution is determined by the most unstable normal mode, and a prescription of how to treat the singularity may be inferred: one simply treats the singularity in the limit of vanishing positive imaginary part of the phase speed c (as we have done). More details may be found in Drazin and Reid (1981) while in the plasma physics context a detailed account is given by Davidson (1972).

A second remark concerns the relation between Miles' instability and the possible instability of a shear flow in a pipe or over a flat plate. In the inviscid context (Drazin and Reid, 1981) a necessary condition for instability to occur is that there is a point of inflexion (i.e. a point where the mean flow curvature vanishes). Such a condition is not immediately apparent in the instability of shear flow over water waves (see Eq. (3.21)). However, if air and water are regarded as one flow then it is clear that there is at least one inflexion point located at the air–sea interface (because there is no water current) hence there is no conflict with the above necessary condition for instability.

Furthermore, note that there is an important consequence of the instability of surface waves. While the waves are growing and therefore receive momentum and energy from the airflow, a slowing down of the airflow by the gravity waves will result. The waves will therefore give rise to a force which is the gradient of the wave-induced stress τ_w. Since the wave-induced stress is proportional to the Wronskian $\mathcal{W}$, which has a step-function discontinuity at the critical height, the force is a δ-function. This suggests an important limitation of linear theory because a considerable wave–mean flow interaction would result, giving rise to a modified mean flow. More details of this wave–mean flow interaction, which is a key issue of this book, will be discussed in Section 3.4.

In order to solve the boundary-value problem of Eq. (3.15), we finally have to specify the shape of the wind profile. In the case of neutrally stable conditions (no density stratification by heat and moisture) the wind profile has a logarithmic height dependence. This is covered in the Intermezzo of Section 2.6.2 in the discussion on momentum transport by turbulence (see Eq. (2.119)). Hence,

$$U_0(z) = \frac{u_*}{\kappa} \log(1 + z/z_0), \tag{3.22}$$

which follows from the condition that the momentum flux in the surface layer is a constant for steady conditions. Recall that $\kappa \simeq 0.40$ is the von Kármán constant which is supposed to be a universal constant, the friction velocity u_* is a measure for the momentum flux in the surface layer and the roughness length z_0 is a parameter that reflects the momentum loss at the sea surface. It is given by the Charnock relation (2.120). We repeat it here for convenience,

$$z_0 = \alpha_{CH} u_*^2 / g, \tag{3.23}$$

with α_{CH} being the Charnock parameter. Although in the present linear treatment it is assumed that the Charnock parameter α_{CH} is a constant (we take $\alpha_{CH} = 0.0144$), it can be argued that α_{CH} is not a constant but is dependent on the sea state. This issue will be discussed in great detail when wave–mean flow interaction is discussed.

For a given wind profile (Eq. (3.22)) the boundary-value problem of Eq. (3.15) may now be solved. Originally, Miles (1957) applied a variational approach to obtain an approximate solution. However, when compared with the numerical results of Conte and Miles (1959) this approximation gave growth rates that were too large by a factor of 3. Prompted by asymptotic matching results of van Duin and Janssen (1992), Miles (1993) revisited this problem and realized that the overestimation was caused by the neglect of some higher-order terms. The improved approximation gave good agreement with Conte and Miles (1959). Since, however, the approximation of Miles (1993) is formally valid for 'slow' waves, only the numerical results will be compared with observations.

3.2 Numerical solution and comparison with observations

The numerical solution of Eq. (3.15) seems to be complicated because of the presence of the singularity at the critical height. However, Conte and Miles (1959) and Miles (1959) have shown how this singularity may be treated and they found a very efficient method to solve the boundary-value problem of Eq. (3.15) avoiding the relatively expensive shooting methods that are required for regular boundary-value problems. In this approach one writes

$$\chi = \rho e^{i\theta}, \tag{3.24}$$

and one finds that ρ and the gradient of θ, θ', are related to the Wronskian $\mathcal{W}$ in the following manner:

$$W = 2\rho^2\theta'. \tag{3.25}$$

Using Eq. (3.25) to eliminate θ' one obtains the amplitude ρ from the real part of the Rayleigh equation of Eq. (3.15),

$$\rho'' - \rho\theta'^2 = \rho\left[k^2 + W_0''\frac{\mathcal{P}}{W_0}\right], \tag{3.26}$$

where again $\mathcal{P}$ denotes the principal value. The integration of Eq. (3.26) is performed by means of a fourth-order Runge–Kutta method, except for the region in the neighbourhood of the critical height where the solution is continued analytically using the so-called 'Frobenius solutions'. Defining the distance from the critical layer by $y = z - z_c$ one obtains from the Rayleigh equation two independent solutions

$$\chi_1 = C_0 y\left(1 + \frac{1}{2}C_0 y + \cdots\right),$$

$$\chi_2 = 1 + C_0 y \log y + \cdots, \tag{3.27}$$

where $C_0 = W_{0c}''/W_{0c}'$. Clearly, χ_1 is an entire function of y whereas χ_2 is a multi-valued function of y. Thus, a branch cut in the complex y plane has to be introduced in order to make χ_2 single valued.

The choice of the proper branch cut is determined by the same rule as is applied to the treatment of the singular function $1/W_0$, namely the limit of vanishing positive imaginary part of the phase speed c is taken. Hence, we study the root of

$$W_0 = U_0 - c_0 - ic_1(c_1 \downarrow 0). \tag{3.28}$$

Expanding the wind speed around the critical height z_c, $U_0 = U_{0c} + yU_{0c}'$, where of course $U_{0c} = c$, we obtain $y = ic_1/U_{0c}'$. Hence, in the limit $c_1 \downarrow 0$, y approaches the real axis from above (below) for positive (negative) U_{0c}'. With this convention χ_2 is uniquely determined.

In the neighbourhood of the critical height the general solution of Eq. (3.15) is given by a linear combination of the two Frobenius solutions in Eq. (3.27),

$$\chi = A(\chi_2 + B\chi_1), \tag{3.29}$$

where A and B are still arbitrary constants, and with this solution the Wronskian $\mathcal{W}(z = z_c)$ may be evaluated giving the same result as found in Eq. (3.20). Note that for $y > 0$, χ_1 and χ_2 are real, hence the gradient in the phase θ' vanishes. This is consistent with the vanishing of the Wronskian, and hence the wave-induced stress, for $z > z_c$.

The general solution in Eq. (3.29) may be used as a starting point for the numerical method. Above the critical height two independent solutions are generated by integrating the Rayleigh equation (3.26) from $z = z_c + \delta$, with $\delta \ll 1$, to large heights using the value and the derivative of χ_1 and χ_2 at $z_c + \delta$. Imposing the boundary conditions at 'infinity' then gives the unknown coefficient B. Similarly, A may be determined by integrating Eq. (3.26) from $z_c - \delta$ to the surface, using the value and the first derivative of Eq. (3.29) at $z_c - \delta$ (respecting the branch cut). Imposing the boundary condition at $z = 0$ then yields A and hence the growth rate of the ocean waves by wind.

Before we compare the numerical results with observations of the growth rate, dimensionless quantities are introduced in order to see which dimensionless parameters determine the problem. Since we are dealing with gravity waves it is natural to use the acceleration of gravity g as a basic scaling quantity. On the other hand, since we are dealing with wind-generated waves another basic scaling quantity must be some measure of the strength of the airflow. Considering airflows that are steady with respect to the typical period of the waves, we choose as a scaling quantity the friction velocity because this is the only relevant airflow quantity that is independent of height z. As we have already discussed in Chapter 2 the growth of waves by wind has frequently been analysed in terms of the wind speed U_{10} at the standard height of 10 m. The reason for this usage was a practical one since it is not easy to measure the surface stress. In general, however, the use of U_{10} does not seem appropriate because implicitly a rather arbitrary height scale, namely 10 m, is introduced that bears no relation to any relevant length scale in the physical problem (see also Kitaigorodskii, 1962). From now on we therefore adopt friction-velocity scaling.

Thus, we scale velocities with u_* and, in agreement with the Charnock relation (3.23) lengths are scaled with u_*^2/g, hence

$$z_* = gz/u_*^2, z_{0*} = gz_0/u_*^2,$$
$$c_* = c/u_*, U_{0*} = U_0/u_*, \tag{3.30a}$$

while the dimensionless wavenumber becomes

$$k_* = ku_*^2/g. \tag{3.30b}$$

As a consequence, the boundary-value problem in Eq. (3.15) has in terms of these dimensionless quantities the same form, while from Eqs. (3.22) and (3.23) the dimensionless wind profile becomes

$$U_{0*} = \frac{1}{\kappa} \log \left(1 + \frac{z_*}{z_{0*}} \right). \tag{3.31}$$

For a given wave characterized by its dimensionless phase speed c_* we can solve

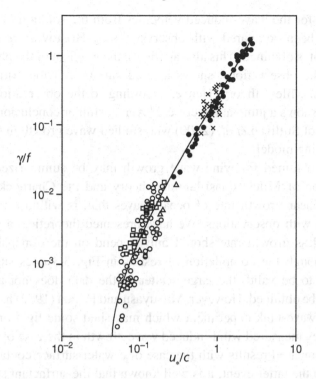

Fig. 3.3. Comparison of growth rates according to Miles' theory with observations compiled by Plant (1982). Continuous line, Miles' theory; open symbols, field data; filled symbols and ×, laboratory data (from Komen *et al.*, 1994).

for the dimensionless growth rate γ/ω_0 of the energy of the waves, which is twice the growth rate of the amplitude of the waves,

$$\gamma/\omega_0 = 2\gamma_a/\omega_0. \tag{3.32}$$

The dimensionless growth rate depends in general on the dimensionless roughness length z_{0*}. However, remarkably, with Charnock's relation we have $z_{0*} = \alpha_{CH}$ which for the moment is regarded as a constant, independent of u_*. Therefore, for a neutrally stable airflow with a logarithmic wind profile, the growth rate γ/ω_0 only depends on c_*.

The numerical results for the dimensionless growth rate γ/f are plotted as a function of the inverse of the dimensionless phase speed u_*/c in Fig. 3.3. For comparison, we have also shown measurements, compiled by Plant (1982), of wave growth from the field (Snyder *et al.*, 1981) and the laboratory (Plant and Wright, 1977). These observations show a considerable scatter but in the mean there is a fair agreement between Miles' quasi-laminar theory and observations. This remark applies to both the laboratory data ($u_*/c > 0.2$) and the field data ($u_*/c < 0.2$).

The profiles for the wave-induced velocities from the solution of the Rayleigh equation have been compared with observations by Hristov *et al.* (2003) and a good agreement is obtained. This also applies to the gradient of the phase θ' which, according to the observations, shows a rapid variation at the critical height in agreement with Miles' theory. Hence, according to the observations the wave-induced stress shows a jump at the critical layer. A similar conclusion also follows from the work of Sullivan *et al.* (2000) who studied wave growth in the context of an eddy-resolving model.

The results obtained for wind-wave growth may be summarized as follows: the combination of Miles' quasi-laminar theory and the Charnock relation has resulted in a linear growth rate of ocean waves that, is within a factor of 2, in fair agreement with observations. We have presented theoretical arguments why the dimensionless growth rate should only depend on the dimensionless phase speed c_*. Although the compilation of results in Fig. 3.3 does suggest that u_* scaling seems to be valid, the large scatter in the data does not allow definite conclusions to be obtained. However, Mitsuyasu and Honda (1982) have performed an interesting wave-tank experiment which may shed some light on the issue of u_* scaling. They measured wind-induced wave growth in the case of a clean water surface and compared results with the case of a water surface contaminated with a surfactant. In the latter event, it is well known that the surfactant suppresses the short waves, thus one deals with a smoother airflow. For the same wind speed at a certain reference height it was found that in the case of a clean surface the growth rate was higher than in the case of a contaminated surface. Mitsuyasu and Honda explained this difference in terms of the difference in stress in the two cases. In other words, when the results of the relative growth rate were plotted as a function of the inverse of the dimensionless phase speed c_*, the observed growth rates collapsed to one curve, while in terms of wind-speed scaling a considerable scatter was found. These results (see also the discussion in Section 2.6) support scaling in terms of the friction velocity, but more evidence on this issue is clearly desirable.

A valid question to ask is what happens in the case where the ocean waves are travelling faster than the 'wind' or when ocean waves are propagating against the wind. Miles' quasi-laminar theory gives in the first case a vanishingly small growth rate while in the case of an adverse wind there is no wave growth because of the absence of a critical layer. However, on intuitive grounds one would expect a damping of the ocean waves. For example, when waves are faster than the wind one could imagine that there is a transfer of momentum from waves to wind, accelerating the airflow. It should be noted that, for an adverse wind, the momentum transfer from waves to airflow would result in a slowing down of the airflow. Clearly, the damping of ocean waves is not modelled by Miles' mechanism but the question is whether the damping is important. Hence the need to discuss possible observational evidence of wave damping by wind.

Is there observational evidence of the damping of waves that propagate faster than the wind? The Bight of Abaco results (Snyder *et al.*, 1981) and a similar experiment performed in the North Sea (Hasselmann and Bösenberg, 1991) report insignificant damping rates in these circumstances. What about evidence for wave damping in an adverse wind? Observational evidence on this subject is scant. Dobson (1971) observed the pressure on the surface of a group of waves advancing against a light wind, and found damping rates that were an order of magnitude smaller than growth rates under corresponding conditions. Stewart and Teague (1980) observed the decay of swell with decametre radar and found rates of about 15 per cent of the corresponding growth rates. Their measurements do not yield the wind input directly, but rather the overall balance including nonlinear interactions and dissipation.

The laboratory measurements of Young and Sobey (1985) showed no significant surface pressure–wave slope correlation in an adverse wind and the weak attenuation they found was attributed to the wave-coherent tangential stress. On the other hand, Mizuno (1976) and Donelan (1983) from similar laboratory measurements found damping rates that were comparable with the corresponding growth rates. Several researchers have studied the propagation of ocean waves into an area of calm or light wind. The direct loss of momentum from the waves to the air above them must lead to acceleration of the airflow in the direction of the waves. Such a 'wave-driven' wind has been reported by Harris (1966), Davidson and Frank (1973) and Holland (1981). More details of this interesting phenomena are discussed in Donelan (1990). It is not clear, however, whether this momentum transfer from water to air is caused by a damping of the water waves. Currents caused by wave-induced motions and other causes may give rise to a tangential stress on the air resulting in wind. In addition, as follows from the observations of Holland (1981), the wave-driven wind only occurs in a thin layer above the water surface, so that this phenomenon has presumably no impact on the dynamics of the atmosphere.

Thus, in conclusion, one could summarize by saying that observational evidence for wave damping when waves propagate faster than the wind is inconclusive. Evidence of damping of waves travelling against the wind is also conflicting. Nevertheless, the evidence of a wave-driven wind seems intriguing indeed. These examples serve to illustrate that Miles' theory is, of course an idealization of reality. It is a linear theory that disregards effects of turbulence on the wave-induced motion. These effects will be discussed in the following subsections. However, Miles' theory seems to give reasonable results for the major problem of the generation of ocean waves by wind. The possible damping in the case of waves travelling faster than the wind appears to be small, while it may be doubted whether wave-driven winds are relevant for ocean-wave prediction and air–sea interaction.

3.3 Effects of turbulence

In our discussion on the effects of turbulence we shall make a fairly loose distinction using the scale of turbulence, namely we shall discuss effects of small-scale, high-frequency turbulence separately from the effects of large-scale turbulence which has a typical time scale that is much longer than the period of the ocean waves. This distinction in turbulence with respect to the period of the ocean waves seems to be useful, because it has implications for the modelling of turbulence. For example, high-frequency, small-scale turbulence is so fast with respect to the waves that it is always in equilibrium with the shear flow. In that event, some form of mixing-length modelling of the effect of turbulent eddies on the wave-induced motion seems appropriate, because, in analogy with molecular viscosity, the small-scale eddies are the 'fastest' process in the system. In that event, the turbulent transfer of momentum depends on the size of the eddies. On the other hand, this is certainly not the case for large-scale turbulence, because these eddies simply do not have the time to transport a significant amount of momentum. The large-scale turbulence will, from now on, be called 'gustiness'. Gusts in the wind require a different treatment.

3.3.1 *Effects of small-scale turbulence*

Many authors have devoted their attention to the study of the effect of small-scale turbulence on wave growth. This line of research was started in the mid 1970s by Chalikov (1976) and Gent and Taylor (1976), when it was felt that the approach of Miles (1957) was inadequate because the effect of eddies on the wave-induced motion was disregarded. Chalikov and Makin produced a series of papers in which they addressed the growth of a single wave and effects of nonlinearity, the structure of airflow over a spectrum of waves and effects of atmospheric stability. Their work culminated in a determination of the drag coefficient over sea waves (Chalikov and Makin, 1991). Although more sophisticated turbulence closures were considered, they concluded that a simple mixing-length model gave sufficiently reliable results.

More-complicated turbulence models were advocated by Gent and Taylor (1976) and Al-Zanaidi and Hui (1984). The former used a one-equation model whereas the latter used a two-equation model of Saffman and Wilcox (1974). Results for wave growth do not seem to differ much from those of Chalikov and Makin so that their conclusion on the choice of turbulence model may be justified.

Thus far we have discussed results of numerical calculations of wave growth by wind. However, there have also been a few attempts to obtain an analytical expression for the growth rate. Using a simple eddy-viscosity model Jacobs (1987) obtained a very elegant expression for the growth rate by means of matched asymptotic expansions, where the small parameter is the drag coefficient. Van Duin and Janssen (1992) realized that Jacob's approach was not quite consistent. They noted that at least three layers (a viscous layer, an intermediate layer and an outer layer) were needed to obtain a valid asymptotic expansion. The end result was, however, identical to the one obtained by Jacobs. We shall discuss some of the details of the analytical approach because the result for the growth rate is similar in flavour to the numerical approaches. In addition, the result is almost identical to an expression for the growth rate suggested by Stewart (1974) on semi-intuitive grounds.

The governing equations for air were taken as

$$\left(\frac{\partial}{\partial t} + \mathbf{u} \cdot \boldsymbol{\nabla}\right)\mathbf{u} = -\frac{1}{\rho_a}\boldsymbol{\nabla}p + \boldsymbol{\nabla} \cdot v\left\{\boldsymbol{\nabla}\mathbf{u} + (\boldsymbol{\nabla}\mathbf{u})^{\mathrm{T}}\right\}, \qquad (3.33)$$

where all symbols have been defined before, except for the superscript T which denotes the transpose and v which is the kinematic viscosity. It is emphasized that the last term in the Reynolds equation (3.33) attempts to model the effects of turbulent eddies in terms of a stress tensor that depends on the local gradient of the flow. This model of turbulence is, however, only valid when eddies are in equilibrium with the shear flow which consists of a mean flow and a wave-induced motion. In other words, this model is only valid for small-scale eddies with a much shorter eddy-turnover time than the typical period of the waves.

The combined effects of molecular viscosity and turbulence are thus taken into account by assuming that the kinematic viscosity is given by

$$v = v_a + v_e, \qquad (3.34)$$

where v_a is the constant molecular viscosity and v_e is the eddy viscosity which may be time and space dependent. The turbulence was modelled using the eddy viscosity

$$v_e = lu_*, \qquad (3.35)$$

where the mixing length l is given as

$$l = \kappa(z - \eta), \qquad (3.36)$$

with η the displacement of the air–water interface.

After a lengthy analysis, details of which can be found in van Duin and Janssen (1992), the growth rate of the waves is found to depend on the wind speed at height $1/k$ (see also Al-Zanaidi and Hui, 1984) and the friction velocity. The relative growth rate becomes

$$\frac{\gamma}{\omega} = 2\kappa\epsilon\frac{u_*}{c}\left[\frac{V}{c}\cos(\theta) - 1\right], \qquad (3.37)$$

where θ is the angle between wind and wave propagation direction, and $V = U_0(1/k)$. Although Eq. (3.37) has been derived under a number of restrictive assumptions from an asymptotic analysis, the result itself, in particular the functional form, is very appealing. Recalling that wave growth is related to the wave-induced stress $-\langle u'w' \rangle$, it is intuitively clear that u' scales with the wind speed at height $1/k$ $(U_0(1/k))$, while w' scales with the friction velocity u_* (Stewart, 1974), suggesting the above expression for wave growth.

Formally, the results only apply for slowly moving ocean waves, $c/V < 1$, while also the drag coefficient $C_D(k) = (u_*/V)^2$ at height $1/k$ should be much smaller than 1. Jenkins (1992) has shown that, in particular, the second condition is restrictive. By decreasing the roughness in his numerical model of airflow over gravity waves which includes small-scale turbulence he found agreement between Eq. (3.37) and his results for the growth rate. However, a good agreement was only obtained for roughness lengths that are much smaller than those observed over the oceans. For realistic roughness lengths, Eq. (3.37) is found to underesti-mate wave growth by at least a factor of 2. Nevertheless, Eq. (3.37) captures the essentials of the growth rate of waves by wind from models that include effects of small-scale turbulence. For high-frequency waves this expression shows, apart from a logarithmic dependence, a similar scaling with friction velocity as found in Miles' theory. For low-frequency waves that are propagating faster than the wind ($c/V > 1$ for $\theta = 0$), the present results show a considerable damping of the waves.

In particular, a problem with the result of Eq. (3.37) is that around the transition point of growth to damping, γ/ω depends linearly on u_*/c. Hence, damping rates are of a similar magnitude to corresponding growth rates. As mentioned earlier, observations from major field campaigns do not seem to support this as insignificant damping rates were found. It is, however, important to realize that mixing-length models may only have a restricted validity. Van Duin and Janssen (1992) pointed out that such models do not adequately describe the growth of low-frequency waves because the eddy-turnover time becomes larger than the wave period.

The following discussion is of a rather qualitative nature, but the essential mes-sage is in good agreement with detailed numerical simulations of turbulent airflow over a single gravity wave using a second-order turbulence scheme (Mastenbroek, 1996). The first systematic criticism on the application of a mixing-length model to the problem of airflow over water waves was given by Belcher and Hunt (1993). This followed earlier work on flows over hills by, for example, Jackson and Hunt (1975) since there are certain similarities between the 'hill' problem and the water-wave problem. In these circumstances mixing-length modelling has a restricted validity because the closure scheme in Eqs. (3.34)–(3.36) assumes that the turbulence is so fast that the eddies have sufficient time to transfer an appreciable amount of momentum on the relevant macroscopic time and length scales as imposed by the gravity waves. In that event, the turbulence is in equilibrium with the flow (which also includes the gravity waves) and the turbulent stress then depends on the local velocity gradient. In the presence of surface gravity waves this equilibrium condition is only satisfied in a relatively thin layer above the water surface. In order to see this, one introduces the advection time scale T_A, which is basically the travel time of an eddy over a gravity wave. Hence,

$$T_A = \frac{1}{k \, |U_0(z) - c|}, \tag{3.38}$$

which involves the difference between the speed $U_0(z)$ of an eddy and the phase speed c of the gravity wave. Secondly, an eddy-turnover time scale T_L is defined which measures, according to Belcher and Hunt (1993), the time it takes for eddies to decorrelate and interact with each other. Thus T_L is the time scale for the turbulence to come into equilibrium with the surrounding flow. In a constant-stress surface layer, the length scale of the largest eddies is κz while the motion in the eddies has a speed that scales with the friction velocity. The eddy-turnover time scale therefore becomes

$$T_L = \frac{\kappa z}{u_*}. \tag{3.39}$$

Belcher and Hunt then argue that the eddies are in equilibrium with the waves when the eddy-turnover time scale T_L is smaller than the advection time scale T_A, or $T_L < T_A$. This implies that the eddies are in local equilibrium with the flow in a layer above the waves with thickness z_t given by

$$k z_t = \frac{2\kappa u_*}{|U_0(z_t) - c|}. \tag{3.40}$$

Here, the $\mathcal{O}(1)$ proportionality constant has been chosen in an appropriate fashion. Except for very slow waves ($c/u_* \simeq 1$) it is seen that mixing-length modelling is only valid in a relatively thin layer above the water surface, with a thickness of $\mathcal{O}(C_D^{1/2}(z_t))$ where $C_D(z_t)$ is the drag coefficient at height z_t. On the other hand, for heights larger than z_t the eddies do not have sufficient time to transport a significant amount of momentum on the wave time scale. Therefore, following work on flows over hills, Belcher and Hunt suggest the use of a truncated mixing-length model, which in practice means that above $z = z_t$, effects of turbulence on the wave-induced motion are disregarded, hence the airflow is treated as being inviscid. The layer with $z < z_t$ will henceforth be called the 'turbulent surface layer' or the 'inner layer'.

Note that, in the case of ocean waves, the equation for the inner-layer depth reveals some interesting structure (Belcher and Hunt, 1998). The solutions to Eq. (3.40) are plotted in Fig. 3.4 which shows the dependence of the dimensionless thickness $k z_t$ on the dimensionless phase speed c/u_*. When $k z_c < 2e\kappa^2$ there is just one solution.

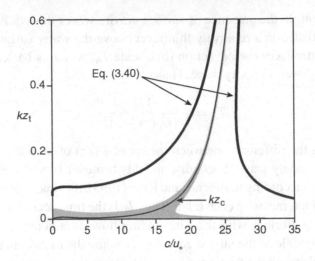

Fig. 3.4. Dimensionless thickness kz_t of inner region as a function of dimensionless phase speed c/u_*, according to Belcher and Hunt (1993), see Eq. (3.40), and according to modified distortion theory, Eq. (3.48) (indicated by shading). For comparison, the dimensionless critical height kz_c is also shown. The Charnock parameter is 0.0144.

For small critical height, corresponding to $c/u_* < 15$, one then finds approximately

$$kz_t = \frac{2\kappa^2}{\log(2\kappa^2/(kz_c))}. \tag{3.41}$$

In this case the structure of the airflow is then similar to flow over hills, that is a local-equilibrium inner region near the surface, $z < z_t$, which contains the critical height, and a rapid distortion outer region with $z > z_t$. The difference from flow over hills is that the roughness length z_0 is replaced by the critical height z_c. As the dimensionless phase speed increases, the critical height moves away from the surface resulting in a thicker inner region. This continues to hold until kz_c reaches the critical point. For $kz_c > 2e\,\kappa^2$ there are three solutions, but the flow can again be considered to have a two-layer structure with an inner region, $z < z_t$, whose depth is given by the smallest solution to Eq. (3.40), and an outer region, $z > z_t$, which now contains the critical height surrounded by the other two solutions of Eq. (3.40). Turbulence modelling follows as for slow waves, with the additional observation that fluid elements do not spend long enough in the critical layer to come in local equilibrium so that rapid distortion effects need to be accounted for as well.

Belcher and Hunt (1993) determined the growth rate for slowly moving waves and from an asymptotic analysis it follows that their result is formally smaller than the mixing-length result (Eq. (3.37)). The mixing-length result of Eq. (3.37) is formally of the order Δ, where Δ is a small parameter defined as $\Delta(z = 1/k) =$

$C_D^{1/2}(k) = u_*/V$. This is immediately seen by scaling all velocities in Eq. (3.37) with V. Furthermore, from the analysis of van Duin and Janssen (1992) it follows that the growth rate of the waves is proportional to the gradient of the mixing length l (Eq. (3.36)). Now, in the truncated mixing-length model of Belcher and Hunt the eddy viscosity is only finite in a thin layer of the order Δ, and therefore in such a model the growth rate is expected to be of the order Δ^2. Indeed, the growth rate obtained by Belcher and Hunt scales as $(u_*/c)^2$, rather than $u_* V/c^2$ as found from Eq. (3.37). In other words, the growth rate of the wave energy becomes

$$\gamma/\omega = \epsilon\beta \left(\frac{u_*}{c}\right)^2, \tag{3.42}$$

and from the numerical work of Mastenbroek (1996) it follows that for slow waves the growth parameter β is $\mathcal{O}(15)$. For comparison, the observations compiled by Plant (1982) give $\beta \simeq 32$, while Miles' theory (see Fig. 3.3) gives good agreement with observations.

Mastenbroek (1996) also determined from his numerical model the growth parameter for waves with a phase speed of the order of the wind speed. For these fast-moving waves the truncation of the mixing length results in a large reduction of the damping rate, compared with the standard mixing-length result of Eq. (3.37). Numerically, it is found that $\beta = \mathcal{O}(-4)$, while a mixing-length model gives $\beta = \mathcal{O}(-15)$.

The truncation of the mixing length has important consequences for the height profile of the (wave-induced) Reynolds stress. Thus, according to the truncated mixing-length model of Belcher and Hunt the wave-induced stress is only finite in the inner region of $\mathcal{O}(\Delta)$, while it becomes vanishingly small in the outer region. Observations of the Reynolds stress of flow over a hill indeed show that this quantity is only finite in the inner region. The standard mixing-length model disagrees with these observations because this model shows considerable Reynolds stresses far away from the surface (see Belcher and Hunt, 1993). The profile of the wave-induced stress was also measured for airflow over moving water waves in an experiment performed in the Marseilles wave tank by Mastenbroek *et al.* (1996). In this case the wave-induced stress was also found to be vanishingly small over a large part of the airflow giving some support, but not in a convincing way, to the suggestion that ordinary mixing-length models are not adequate over water waves.

An important question to ask now is what is the role of the critical layer in the context of the Belcher and Hunt model for turbulent flow over gravity waves? Returning to Fig. 3.4, in addition to the curve that gives the dependence on the dimensionless depth of the inner layer kz_t as a function of the dimensionless phase speed c/u_*, the dimensionless critical height kz_c is also shown. For a wide range of phase speeds, $1 < c/u_* < 25$, the critical layer is inside the inner region $z < z_t$. This implies that

effects of turbulence are expected to be important near the critical layer, except for
the fast moving waves with dimensionless phase speeds larger than 25. Supported
by the numerical work of Mastenbroek (1996), Belcher and Hunt (1998) claim that
the turbulence gives rise to such a large momentum diffusion across the critical
layer that the critical height plays no significant dynamical role. Therefore, the
wave-induced stress should vanish at the inner-layer depth z_t, and not at the critical
height z_c as predicted by Miles' theory. However, this prediction of the Belcher
and Hunt model seems to be at variance with the direct numerical simulations of
Sullivan *et al.* (2000) and with the *in-situ* observations of Hristov *et al.* (2003).

The simulations by Sullivan *et al.* (2000) give convincing evidence that around
the critical height a region of closed streamlines (or cat's-eye pattern) exists. This
region was found to be dynamically important, even at low to moderate values of
dimensionless phase speed, $4 < c/u_* < 12$. More importantly, the wave-induced
stress showed a dramatic reduction in magnitude near the critical height. Similarly,
for the range $16 < c/u_* < 40$, Hristov *et al.* (2003) observed a pronounced cat's-
eye pattern around the critical height where the wave-induced stress showed a
jump. These studies seem to suggest that the critical layer plays an essential role in
understanding wind-induced wave growth.

It is of some interest to try to understand why results from the Belcher and Hunt
approach are in contrast with the above recent findings on wave growth. There may
be two reasons for this. One point is related to the effect of turbulence on the critical
layer, and the second point concerns the relevant time scale of momentum transport
by eddies.

Effects of viscosity (as well as nonlinearity and unsteadiness) on the critical-layer
dynamics have been studied in great detail. We concentrate here on linear theory
only. A good overview of the effects of molecular viscosity is found in Drazin and
Reid (1981). The relevant evolution equation in the case of turbulent viscosity is
given in van Duin and Janssen (1992). It reads in terms of the stream function Ψ,
implicitly defined by $u = \partial \Psi / \partial z$, and $w = -\partial \Psi / \partial x$,

$$\left(\frac{\partial}{\partial t} + U_0 \frac{\partial}{\partial x} \right) \nabla^2 \Psi - \frac{d^2 U_0}{dz^2} \frac{\partial \Psi}{\partial x} = \nu \nabla^4 \Psi + \cdots \qquad (3.43)$$

The left-hand side of this equation is, for normal modes, identical to the Rayleigh
equation (3.15), while in the right-hand side only the most dominant 'viscous'
term near the critical height has been retained. Hence, when momentum transport
by eddies is important the Rayleigh equation (3.15) is replaced by a fourth-order
Orr–Sommerfeld equation and for finite viscosity the singularity at the critical
height is removed. Viscosity then gives rise to a broadening of the critical layer and
the wave-induced stress profile becomes smooth across the critical height, rather
than showing a sudden jump as in inviscid theory. However, if the width of the
critical layer is small, still considerable variations in wave-induced stress occur

near the critical height. Now, the extent of the broadening of the critical layer can be estimated by balancing the advective term with the dominant term due to eddy viscosity, assuming that the relevant length scale for estimating gradients is the distance to the critical height, $\delta = z - z_c$. Expanding the mean velocity U_0 around the critical height z_c, one finds

$$\delta^3 = \frac{\nu}{kU'_c}, \tag{3.44}$$

which is in accord with the usual scaling that the thickness of the critical layer is proportional to $R^{-1/3}$, with $R = LU/\nu$ the Reynolds number and L a typical length scale (Drazin and Reid, 1981). The difference is, however, that the molecular viscosity is replaced by the eddy viscosity. For the most important case, when the critical layer is inside the inner layer, the eddy viscosity is given by Eq. (3.35). For a logarithmic wind profile one therefore finds

$$k\delta = (\kappa k z_c)^{2/3}, \tag{3.45}$$

which for small critical heights (i.e., slow waves) results in much smaller broadening of the critical layer than would result from an estimate using the inner-layer depth. In other words, although the wave-induced stress does not show a jump at the critical height, the above discussion suggests that even in the presence of eddies a considerable variation in wave-induced stress is expected near a height that is tied in with the critical height. For slow waves the wave-induced stress is expected to vanish at heights much smaller than the inner-layer depth.

The second point of concern regarding the Belcher and Hunt approach is related to the estimation of the time scale of momentum transport by eddies. In the truncated mixing-length modelling discussed so far the turbulent time scale is taken to be equal to the eddy-turnover time scale given in Eq. (3.39). However, in the context of flow over hills the truncated mixing-length models of Jackson and Hunt (1975) (and as a consequence the one suggested by Belcher and Hunt (1993)) have been criticized because observations at Askervein top do suggest a much thinner inner-region layer than would follow from Eq. (3.40) (see Beljaars and Taylor, 1989; Walmsley and Taylor, 1996). The point is that the use of the eddy-turnover time as the time scale for momentum transport by eddies may give a momentum-transfer time scale that is too short by an order of magnitude. In order to understand this point, let us consider the solution of the initial-value problem

$$\frac{\partial}{\partial t} U_0 = \frac{\partial}{\partial z} \left(\nu \frac{\partial}{\partial z} \right) U_0,$$

$$\nu \frac{\partial}{\partial z} U_0 = u_*^2, \; z = 1/k, \tag{3.46}$$

$$U_0(z_0) = 0,$$

with z_0 the roughness length and $\nu = \kappa u_* z$, while the initial value follows from

$$U_0(z, t = 0) = g(z),$$

where $g(z)$ is arbitrary except that it satisfies the no-slip condition at $z = z_0$.

The eddy-turnover time scale of Eq. (3.39) may be obtained by estimating the gradient of the wind profile as $\partial U_0/\partial z \simeq U_0/z$. It should be noted, however, that for a logarithmic wind profile the gradient is overestimated by an order of magnitude since for $U_0 = (u_*/\kappa) \log(z/z_0)$ we have $\partial U_0/\partial z = u_*/\kappa z$. As a consequence, the time scale for momentum transport would become

$$T_M = \frac{\kappa z}{\Delta(z) u_*},\tag{3.47}$$

where now the small parameter $\Delta(z) = u_*/U_0(z)$. The time scale T_M would only be of comparable magnitude to T_L for heights very close to the water surface, but away from the surface $T_M \gg T_L$.

The above estimation of the time scale T_M is based on a fairly loose argument. In order to get more confidence in Eq. (3.47) we discuss the actual solution of the initial-value problem in Eq. (3.46). The initial-value problem may be solved by means of the method of separation of variables, and a solution is found in terms of Bessel functions. The resulting eigenvalue problem gives rise to two types of modes. The first kind is damped with a time scale that corresponds to the usual eddy-turnover time (Eq. (3.39)). For large times, these modes are heavily damped and therefore do not contribute to the solution. Note that for small roughness the damping rate of these modes does not depend on the roughness length and, there-fore, these modes do also exist in homogeneous flows. The second kind of mode only arises in flows with a boundary layer and therefore depends on the surface roughness. For small roughness (i.e. small compared with the wavelength) the damping rate of these modes corresponds to the time scale T_M of Eq. (3.47) and they are therefore weakly damped. Hence, for large times the modes of the second kind will dominate the solution of the initial-value problem in Eq. (3.46). In other words, the time scale for momentum transfer is indeed given by Eq. (3.47).

Now, using T_M as a measure for the momentum-transfer time scale by eddies, it is suggested that mixing-length modelling is valid if $T_M < T_A$. This condition is satisfied in a thin layer above the water surface with thickness z_t given by

$$k z_t = \frac{2\kappa u_* \Delta(z_t)}{|U_0(z_t) - c|}.\tag{3.48}$$

Hence, with the use of T_M, the inner region becomes much thinner. This is illustrated by Fig. 3.4 where we have shown the thickness of the inner layer according to the estimate in Eq. (3.48); this can be compared with the proposal of Belcher and Hunt,

Eq. (3.40). From Fig 3.4 it is clear that according to Eq. (3.48) the critical layer plays a more prominent role in the problem of the wind-induced growth of waves. For $c/u_* > 5$ to 10, the wind-induced stress will show a jump near the critical layer as the inner-layer depth follows the critical layer closely.

We conclude from the above discussion that, in particular, the issue of the momentum-transport time scale may bridge the gap between the results found by Belcher and Hunt (1993, 1998) and the recent findings by Sullivan *et al.* (2000) and Hristov *et al.* (2003). The implication is that for dimensionless phase speeds c/u_* larger than 10, the critical-layer mechanism plays a prominent role in understanding the growth of waves by wind. Note that this conclusion can also be reached by applying the scaling relation in Eq. (3.48) to the Reynolds equation (3.33) and to lowest significant order the boundary-value problem of Eq. (3.15), involving the singular Rayleigh equation, will be recovered. Finally, it is of interest to study what happens for really slow waves, $1 < c/u_* < 5$. In those circumstances the critical layer is well inside the inner region so one would expect that effects of turbulence are relevant. However, for slow waves there are considerable complications regarding the physics of wind-wave generation. For example, effects of molecular viscosity in the air may be relevant, as follows for instance from the work of Benjamin (1959), Valenzuela (1976), Kawai (1979) and van Gastel *et al.* (1985). In addition, short gravity waves extract a considerable amount of momentum from the air just above the surface, and hence the presence of these short waves may have a considerable impact on the production of air turbulence and hence on the size of the eddies. In analogy with the damping effect of viscosity on turbulent eddies (van Driest, 1951), this may result in a reduced mixing-length. Close to the surface the mixing length could therefore have been overestimated considerably, resulting in even larger momentum-transfer time scales and an even thinner inner region.

Although the quasi-laminar approach of Miles has been severely criticized in the past, the present results seem to indicate that in a large domain above the water surface the wave-induced air motion may indeed be regarded as approximately inviscid. Hence, for $c/u_* > 10$, Miles' mechanism seems to provide an adequate model. For small, dimensionless phase speed the situation is less clear. The inviscid Miles model, although perhaps formally not valid for slow waves, gives good agreement, regarding the growth rate of waves by wind, with the observations shown in Fig. 3.3. From a pragmatic point of view it is therefore tempting to apply the critical-layer model even for slow waves. Although these slow waves are not relevant for wave prediction, they are important for estimating the slowing down of the wind by growing waves because these waves have the largest growth rate. An accurate estimation of wave growth of the slow waves is important and Miles' critical-layer model seems to do an adequate job.

3.3.2 Effects of gustiness on wave growth

In Section 3.2 we have seen that within a factor of 2 there is a fair agreement between Miles' quasi-laminar theory and field observations of Snyder *et al.* (1981) and Hasselmann and Bösenberg (1991). However, in the low-frequency range with waves having a phase speed that is about the same as the wind speed at 10 m height, theory predicts energy-transfer rates that are smaller than measured values.

As we have seen in Section 3.3.1, attempts to include small-scale turbulence have not resulted in increased energy-transfer rates. On the other hand, as will be seen in a moment, large-scale turbulence (or gustiness) may have a considerable impact on wave growth. Gustiness requires, however, a different treatment from small-scale turbulence since the gusts have time scales that are much longer than the typical wave time scale or the momentum-transfer time scale T_M of Eq. (3.47).

Let us first describe what is meant by gustiness in the context of wind-wave growth. The atmosphere shows variability basically at all scales, ranging from micro-scale turbulence to synoptic scales. Numerical models of the atmosphere model synoptic variability well, but because of finite spatial resolution there is a lack of variability at the small scales. Likewise, when carrying out observations of winds and waves we tend to characterize the atmospheric state by average quantities such as a 20-min average wind speed; however, from experimental practice we know that there may be considerable fluctuations around the mean wind speed. The size of these fluctuations depends on the air–sea temperature difference, and we will refer to the unresolved part of the atmospheric variability as gustiness.

In order to investigate the impact of gustiness on wave growth it is assumed that the gustiness time scales are much longer than the wave time scale, $1/\omega$, or the momentum-transfer time scale T_M. As a consequence, the surface wind is in quasi-equilibrium and the fluctuations are so slow that to a good approximation the wind profile is logarithmic. The gustiness is then reflected by means of the variability in the surface stress or friction velocity.

Note that the present approach differs from the attempt made by Nikolayeva and Tsimring (1986) to study the effects of gustiness on wave growth. These authors applied a so-called 'kinetic model for fluid turbulence', proposed by Lundgren (1967), to the problem of wind-generated water waves and a substantial enhancement of energy transfer due to gustiness was found, in particular for the low-frequency waves. Gustiness was modelled by means of a height-independent perturbation of the wind profile. Therefore, particularly near the surface this resulted in large deviations from the equilibrium, logarithmic profile. This seems unlikely because the small-scale turbulence has ample time to restore the equilibrium profile.

In order to appreciate the effects of gusts on wave growth, Miles' theory of wind-wave generation is extended by allowing the mean air-velocity profile to be a slowly varying function of time. By means of an elaborate analysis it can be shown that Miles' expression for growth of waves by wind still holds. Details of this analysis, which relies on the use of the multiple-time-scale method, are given in Janssen (1986). The rate of change of the wave spectrum F due to wind is then

$$\frac{\partial}{\partial t} F(\omega) = \gamma F(\omega), \tag{3.49}$$

where we recall that γ, the growth rate of the wave energy, is twice the growth rate of the amplitude:

$$\gamma = -\pi \epsilon c \frac{W_{0c}''}{|W_{0c}'|} |\chi_c|^2. \tag{3.50}$$

The dimensional considerations of Section 3.2 (see also Miles, 1957) have shown that the growth rate of gravity waves due to wind only depends on two parameters, namely the dimensionless phase speed c/u_* and the Charnock parameter α_{CH}, or alternatively, the profile parameter $\Omega_M = \kappa^2 g z_0 / u_*^2$ as introduced by Miles (1957). Therefore,

$$\gamma = \gamma(u_*/c, \Omega_M).$$

In the context of our model, gusts correspond to variability in u_*. Hence, one may regard the friction velocity as a stochastic variable with a steady part $\bar{u}_*$ and a fluctuating part δu_*. Thus,

$$u_* = \bar{u}_* + \delta u_* \tag{3.51}$$

so that

$$\gamma = \bar{\gamma} + \delta \gamma. \tag{3.52}$$

Note that the assumed vanishing of the ensemble average of the fluctuations in the friction velocity does not imply that the ensemble average $\langle \delta \gamma \rangle$ is zero. The reason for this is that the growth rate γ is always positive. This is, in particular, relevant for those low-frequency waves whose phase speed is close to the wind speed. For these long waves a positive fluctuation in u_* will result in enhanced wave growth but a negative fluctuation will not give rise to diminished growth. The growing waves act as a rectifier and therefore gustiness may have a considerable impact on wave growth.

Consider now the solution of the stochastic equation

$$\frac{\partial}{\partial t} F = (\bar{\gamma} + \delta \gamma) F. \tag{3.53}$$

When the correlation time τ_c of the random process $\delta \gamma$ is small compared with the relevant time scale of spectral change, the evolution equation of the ensemble average of the wave spectrum $\langle F \rangle$ may be readily obtained. Normally, i.e. when the random process depends in a smooth way on the random variable, the effect of the random fluctuations on the mean spectrum is only quadratic in the amplitude of the fluctuations. However, growing waves act as a rectifier, hence already an effect linear in $\delta\gamma$ is found. Thus, in lowest order one finds

$$\frac{\partial}{\partial t}\langle F \rangle = \langle \bar{\gamma} + \delta\gamma \rangle \langle F \rangle, \tag{3.54}$$

Therefore, to a good approximation the effect of gusty winds on wave evolution can be taken into account by determining the average of the growth rate where the weight is given by the probability distribution function (pdf) of the gusts. There is ample evidence (see, for example, Smith *et al.*, 1990) that fluctuations of wind speed U_{10} and direction around an average value are well represented by a Gaussian distribution. Not much is known about the distribution function for the friction velocity, but it can be shown that, if the fluctuation levels are small, the pdf for u_* is close to a Gaussian as well. Hence, introducing the gustiness level σ_u, the pdf for the friction velocity becomes

$$p_{u_*}(x) = \frac{1}{\sigma_u\sqrt{2\pi}} \exp\left\{-\frac{(x - \bar{u}_*)^2}{2\sigma_u^2}\right\}, \tag{3.55}$$

so that the average of the growth rate becomes

$$\langle \gamma \rangle = \langle \bar{\gamma} + \delta\gamma \rangle = \int_{-\infty}^{+\infty} \mathrm{d}x \, \gamma(x/c, \Omega_{\mathrm{M}}) \, p_{u_*}(x). \tag{3.56}$$

To quantify the effects of gusts we shall use a relatively simple fit of the growth rate, valid for ocean waves. To that end we take the empirical fit by Snyder *et al.* (1981) which was adapted by Komen *et al.* (1981) to accommodate friction-velocity scaling. It reads

$$\frac{\gamma}{\omega} = \max\left\{0.2\epsilon\left(28\frac{u_*}{c} - 1\right), 0\right\}.$$

It is important to note that the empirical fit has a positive growth rate for $28u_*/c > 1$, while the growth rate vanishes in the opposite case. As seen in Fig. 3.5 the growth rate shows a kink at $28u_*/c = 1$.

Substitution of the above expression for the growth rate into Eq. (3.56) results in the average growth rate

$$\frac{\langle \gamma \rangle}{\epsilon\omega} = 0.2\int_{x_0}^{\infty} \mathrm{d}x \, p_{u_*}(x)\left[28\frac{x}{c} - 1\right],$$

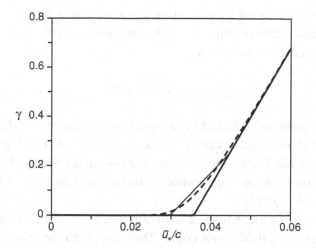

Fig. 3.5. Growth rate for a gustiness level $\sigma_u/\bar{u}_* = 0.2$ (dashed line). The thick continuous line describes wave growth in the absence of gustiness while the thin line corresponds to the approximation in Eq. (3.58).

where $x_0 = c/28$. Evaluation of the integral gives the final result

$$\frac{\langle \gamma \rangle}{\epsilon \omega} = 0.2 \left[\frac{28\sigma_u}{c\sqrt{2\pi}} \exp \left\{ -\frac{1}{2\sigma_u^2} (\bar{u}_* - c/28)^2 \right\} \right.$$
$$\left. + \frac{1}{2} \left(\frac{28\bar{u}_*}{c} - 1 \right) \left(1 - \mathrm{erf} \left(\frac{c/28 - \bar{u}_*}{\sigma_u\sqrt{2}} \right) \right) \right], \qquad (3.57)$$

where

$$\mathrm{erf}(z) = \frac{2}{\sqrt{\pi}} \int_0^z dt \, e^{-t^2}$$

is the error function. Equation (3.57) shows that, as expected, the effect of gustiness on wave growth is proportional to the standard deviation σ_u of the friction velocity. It is therefore rather large, especially for the low-frequency waves with u_*/c around $1/28$. This is illustrated in Fig. 3.5 where we have plotted the growth rate as function of $\bar{u}_*/c$ for a fluctuation level $\sigma_u/\bar{u}_*$ of 0.2 while, as a reference, the growth rate in the absence of gustiness is also shown. It is clear from the figure that, in particular, the low-frequency waves are affected by the variability in the wind. Hence, in the later stages of wave growth, which is determined by the growth of the low-frequency waves, wave evolution is expected to be affected by gustiness.

In general, it is not straightforward to obtain a simple expression for the effect of gusts on wave growth. A possible line of attack is to start from Eqs. (3.55) and (3.56) and to derive an expansion for small gustiness levels σ_u. This requires differentiation of the growth rate $\gamma(u_*/c)$ with respect to u_*. As shown by Miles

(1997) this can be avoided by evaluating the exponential integral using Gauss–Hermite quadrature. One obtains ($n = 2$ in Abramowitz and Stegun, 1965, Section 25.4.46) the extremely simple form

$$\langle \gamma \rangle = \frac{1}{2}\left[\gamma(\bar{u}_* + \sigma_u) + \gamma(\bar{u}_* - \sigma_u)\right], \tag{3.58}$$

in other words, gustiness effects follow by taking the average of the growth rates at $\bar{u}_* + \sigma_u$ and at $\bar{u}_* - \sigma_u$. It can be checked, as illustrated in Fig. 3.5, that the approximation of Eq. (3.58) already gives the essential effect. If higher accuracy is needed one may use an approximation involving three zeros of the Hermite polynomials (Miles, 1997).

Note that with this approach it is also in principle possible to take the meso-scale variability of the wind field into account. Therefore, both the temporal and spatial variability of the wind field can and should be treated, since wave-height results may be affected significantly by taking effects of gusts into account. However, this assumes a good knowledge of the variability of the wind field, which to a large extent depends on synoptic conditions. An easy way out of this is to increase the spatial resolution of a wave prediction system and its forcing wind field, because in this way wind variability is automatically included. However, this is an expensive solution which presently is not feasible in the context of global modelling, so it is expected that for the near future one has to rely on a parametrization of gustiness.

A recent account of the impact of wind variability on wave-prediction results was given by Abdalla and Cavaleri (2002). They followed a somewhat different approach to studying effects of gustiness. Wind variability was modelled in a real-istic manner by taking the mean wind speed at every 15-min time step and adding a perturbation which was randomly drawn from the Gaussian distribution of the wind. Correlations in time were introduced to ensure the right temporal coherence so that the frequency spectrum of the perturbations agreed with observations. No correlation in space was introduced because the spatial resolution of the numerical wave prediction model was in the range of 25 to 50 km. The gustiness level was obtained from an observed correlation between the standard deviation of the wind and the air–sea temperature difference. Simulations with a single-grid-point model showed a large, realistic variability in the growth curves for significant wave height (see also Section 5.3). Simulations for the North-Atlantic area showed considerable impact on wave height, in particular regarding the statistics for extreme wave height.

Finally, the effects of gusts on wave growth are particularly relevant in the later stages of wave growth. Therefore, it seems to be important to analyse observations of, for example, fetch-limited wave growth in terms of the gustiness level. We recall Fig. 2.4b which shows fetch-limited wave variance data stratified against stability. In particular for large fetches, there is a discrepancy between stable and unstable groups. It is expected that the introduction of a gustiness-level parameter into the analysis of wave observations may explain to a large extent this discrepancy.

3.4 Quasi-linear theory of wind-wave generation

In the previous sections we have explored some of the linear theories of wind-wave generation. Normally, linearization is a good approximation in ocean wave dynamics. In the case of Miles' theory however, we encounter a complication because the dynamics inside the critical layer is not described by linear theory. This follows from the δ-function behaviour of the vortex force $d\tau_w/dz$ which implies that in linear theory the critical layer is regarded as being infinitesimal, although in practice the width of this layer will be finite. Inside the critical-layer a pattern of closed streamlines will form, the so-called 'Kelvin cat's-eyes pattern' (see Lighthill, 1962; Phillips, 1977). Neglecting turbulence for the moment, the closed streamline pattern will give rise to a smoothing of the vorticity distribution inside the critical layer in such a way that for large times the curvature of the wind profile vanishes. Hence, in an inviscid fluid, stabilization of a single wave results and the momentum transfer from air to the gravity wave would be quenched. In the context of the generation of a single wave by wind, this important result was obtained by Reutov (1980) who exploited an analogy with the problem of the resonant interaction of plasma waves and electrons. The nonlinear evolution equation for the amplitude of the water wave was solved numerically and showed the usual exponential growth for small times followed by a damped oscillation around the saturation level. This example of the evolution of a single wave in an inviscid fluid illustrates that it is important to investigate the possible consequences of wave–mean flow interaction on the evolution of surface waves. In this section we will present some of the results for a continuous spectrum of surface gravity waves. It is emphasized, however, that for a sufficiently broad spectrum, wave–mean flow interaction is of a different nature because we are dealing with a continuum of critical layers with random phase. Because of the random-phase approximation, neighbouring critical layers will counteract each other so there will be a considerable reduction in the impact of critical-layer dynamics on the evolution of the mean wind and surface waves. The resulting effect will be determined in Section 3.4.2. First we will discuss the critical-layer dynamics for the case of a single wave in an inviscid fluid, at the same time providing a physical picture of Miles' instability mechanism.

3.4.1 Critical-layer dynamics

Let us discuss in some detail what is happening near a critical layer in an inviscid fluid, confining ourselves to the case of the propagation (in one dimension only) and growth of a single gravity wave. For a more detailed mathematical discussion we refer to the original paper by Reutov (1980).

The equations of motion of a two-dimensional inviscid incompressible fluid may be written in the following concise manner:

$$\frac{d}{dt}\zeta = 0, \quad \zeta = \left(\frac{\partial^2}{\partial x^2} + \frac{\partial^2}{\partial z^2}\right)\Psi, \tag{3.59}$$

where Ψ is the stream function, defined in such a way that $u = \partial\Psi/\partial z$, and $w = -\partial\Psi/\partial x$, and ζ is the vorticity. Furthermore, d/dt denotes the rate of change moving with the fluid, $d/dt = \partial/\partial t + \mathbf{u} \cdot \nabla$. Therefore, in two dimensions vorticity is conserved when following a fluid element. Hence, vorticity may be used to label such a fluid element.

As remarked already, the flow inside the critical layer forms a pattern of closed streamlines, the Kelvin 'cat's-eyes' pattern. This pattern may be obtained as follows. In a frame moving with the speed c of the gravity wave, the stream function Ψ may be written as

$$\Psi = \int_{z_c}^{z} dz\,(U_0 - c) + A(z)\cos(kx), \tag{3.60}$$

where the first term corresponds to the equilibrium flow and the second term denotes the perturbation by the (unstable) gravity wave. Close to the critical height z_c one may perform a Taylor expansion with the result

$$\Psi = \frac{1}{2}y^2 U_{0c}' + A_c\,\cos(kx), \tag{3.61}$$

where $y = z - z_c$ and A_c is the value of the amplitude at the critical height. The streamlines now follow from the condition that $\Psi = \Psi_0$ ($=$ constant), or,

$$y = \pm\sqrt{\frac{2}{U_{0c}'}\left[\Psi_0 - A_c\cos(kx)\right]}, \tag{3.62}$$

and the resulting streamline pattern, the cat's-eyes, is given in Fig. 3.6. In an inviscid fluid, the fluid elements follow the streamlines and, since we have restricted ourselves to the two-dimensional case, vorticity is conserved for every fluid element. It is thus of interest to determine their trajectories. These follow from Hamilton's equations

$$\frac{dx}{dt} = u = \frac{\partial}{\partial z}\Psi, \quad \frac{dz}{dt} = w = -\frac{\partial}{\partial x}\Psi. \tag{3.63}$$

Using the approximate expression for the stream function in Eq. (3.61) and

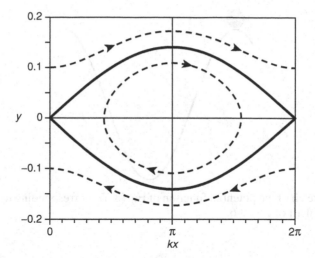

Fig. 3.6. Sketch of Kelvin's cat's-eye pattern, giving lines of constant stream function around the critical height. The thick line denotes the separatrix between critical-layer (trapped) fluid elements and outer-layer (untrapped) fluid elements.

elimination of z gives an evolution equation for x,

$$\frac{d^2 x}{dt^2} = k U'_{0c} A_c \, \sin(kx), \tag{3.64}$$

which may be solved exactly in terms of elliptic functions, hence the time evolution of the fluid elements for a steady gravity wave may be determined. There are two cases of special interest, namely the case of fluid elements at the centre of the cat's-eye, $kx = \pi$, and the case of fluid elements close to the separatrix. Close to the centre of the cat's-eye the sine function in Eq. (3.64) may be linearized and one finds

$$\frac{d^2 x}{dt^2} = -k U'_{0c} A_c (kx - \pi), \tag{3.65}$$

hence the fluid elements oscillate with angular frequency

$$\omega_B = k \sqrt{U'_{0c} A_c}, \tag{3.66}$$

where it was tacitly assumed that the shear in the mean velocity at the critical height is positive. Note that in plasma physics ω_B is usually called the 'bounce frequency' or the 'trapping time', that is the time it takes to capture a particle in a potential well. Hence, near the centre of the cat's-eye, fluid elements execute an oscillation with a frequency that depends on the amplitude of the surface waves. In contrast, close to the separatrix it takes an infinitely long time for the fluid elements to cross the cat's-eye, as one would expect from a resonance phenomenon. In order to see

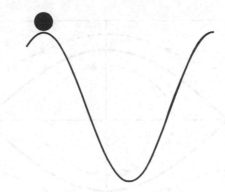

Fig. 3.7. Sketch of the potential function of Eq. (3.67) corresponding to the Kelvin cat's-eye pattern of Fig. 3.6.

this, Eq. (3.64) is multiplied by dx/dt and integrated with respect to time to obtain the conservation law

$$\frac{1}{2}\left(\frac{dx}{dt}\right)^2 + \mathcal{V}(x) = \text{constant}, \qquad (3.67)$$

where the 'potential' $\mathcal{V}(x) = A_c U'_{0c} \cos kx$. We remark that with $dx/dt \simeq y U'_{0c}$, Eq. (3.67) is identical to Eq. (3.61). The shape of the potential is given in Fig. 3.7 and we note that fluid elements at the centre of the cat's-eye correspond to particles at the bottom of the potential well, while the fluid elements near the separatrix of Fig. 3.6 correspond to the particles at the top of the potential well; it is well known that it takes a very long time for these particles to travel from one top to the next one. Therefore, the oscillation frequency of the fluid elements inside the cat's-eye depends on the position in the cat's-eye.

This property furnishes an efficient mechanism for the stabilization of a water wave that grows due to shear flow instability. We have seen in Section 3.1 that the growth rate of Miles' instability is proportional to the curvature in the equilibrium wind profile, which is just the gradient of the equilibrium vorticity with respect to height. At the initial stage of the instability, the width of the cat's-eye, being proportional to the square root of the amplitude, will be small and, therefore, the evolution in time of the amplitude of the gravity wave will be determined by the fluid elements just outside the cat's-eye region. The reason for this is that these are the fluid elements that are in resonance with the wave because their travel time across the cat's-eye is very long, and for this reason these fluid elements take care of the momentum transfer from mean flow to the gravity wave. In order to understand this better we follow Lighthill (1962) by calculating the momentum transfer in the following manner. The momentum equations for an inviscid fluid may be written as

$$\frac{\partial}{\partial t}\mathbf{u} = -\zeta \times \mathbf{u} - \nabla\left(\frac{p}{\rho} + \frac{\mathbf{u}\cdot\mathbf{u}}{2}\right), \tag{3.68}$$

where $\zeta = \nabla \times \mathbf{u}$ is the vorticity and $\zeta \times \mathbf{u}$ is the vortex force. In order to determine how much momentum the mean airflow might possibly lose, Eq. (3.68) is averaged over a wavelength to obtain for the rate of change of the mean air speed U_0

$$\frac{\partial}{\partial t}U_0 = -\overline{\zeta w},$$

hence the mean momentum loss is determined by the average vortex force. Since ζ consists of a mean part ζ_0 and a fluctuation ζ_1, and since for a parallel flow w only has a fluctuating part, we find

$$\frac{\partial}{\partial t}U_0 = -\overline{\zeta_1 w};$$

ζ_1 may be obtained from conservation of vorticity in a two-dimensional flow, see Eq. (3.59). Separating into mean and fluctuating parts we find from Eq. (3.59) in the linear approximation

$$\zeta_1 = -\frac{w}{ikW}\frac{\partial}{\partial z}\zeta_0, \tag{3.69}$$

where, as before, $W = U_0 - c$ is the Doppler-shifted air velocity. The mean vortex force now becomes

$$\overline{\zeta_1 w} = \frac{i}{k}\,|w|^2\,\frac{\partial}{\partial z}\zeta_0\left(\frac{1}{W} - \frac{1}{W^*}\right) \tag{3.70}$$

and with the calculus for the singular function $1/W$, developed in Section 3.1, we have

$$\overline{\zeta_1 w} = -\frac{2\pi}{k}\,|w|^2\,\delta(W)\frac{\partial}{\partial z}\zeta_0, \tag{3.71}$$

and, because of the appearance of the δ-function, only the resonant fluid elements take part in the momentum transfer from mean flow to gravity waves.

In passing, it is worth noting that the mean vortex force is proportional to the gradient of the mean vorticity, that is to the curvature of the wind profile. As a consequence, the rate of change of the mean air flow is also proportional to the curvature of the wind. In the context of a spectrum of gravity waves, this result will be discussed in somewhat more detail in Section 3.4.2.

Thus far we have made plausible that in the initial stages of wave growth the evolution in time of the wave energy is determined by the resonant fluid elements just outside the cat's-eye. However, in the course of time the width of the cat's-eye will grow, trapping more and more resonant fluid elements. This trapping mechanism will eventually lead to stabilization of the instability because fluid elements, having their own vorticity, will oscillate more rapidly near the centre of the cat's-eye than at the border. Hence, after a sufficiently long time, a vigorous mixing of the vorticity will occur in such a way that the vorticity becomes an erratic function of position. In a coarse-grain sense vorticity may then be regarded as constant in the cat's-eye region, and since a constant vorticity means zero curvature of the wind profile, the momentum transfer from mean flow to the wave is quenched. The mixing of vorticity inside the cat's-eye occurs on the time scale of the inverse of the bounce frequency ω_B (Eq. (3.66)) and when the waves are infinitesimal the mixing is not effective. However, in the course of time the waves grow exponentially resulting in an increasing bounce frequency. Stabilization now occurs when the bounce frequency matches the linear growth rate γ, which is proportional to the density ratio ϵ, or,

$$\omega_B \sim \gamma = \mathcal{O}(\epsilon). \tag{3.72}$$

As the bounce frequency depends on the square root of the amplitude of the waves, the consequence is that the saturation amplitude of the water waves scales with the square of the density ratio, and is therefore quite small.

Using matched asymptotic expansion techniques, Reutov (1980) was able to give a detailed mathematical description of the mixing of vorticity inside the critical layer and he found the saturation level given in Eq. (3.72). The work of Reutov has resulted in a considerable step forwards in the understanding of the critical-layer dynamics in growing water waves and it is a pity that this work is hardly known in the wave community. Unfortunately, the resulting saturation level of the water waves is very small, and therefore the question is whether the mixing of vorticity is relevant for ocean-wave growth. As already pointed out, in practice we deal with a spectrum of waves with random phase and therefore we have a continuum of critical layers with random phase which will tend to counteract each other. In that event, a considerable reduction of the effect of critical-layer dynamics on the evolution of surface gravity waves is expected. The resulting effect will be determined in the following subsection.

3.4.2 Quasi-linear theory; many waves

Gravity waves receive energy and momentum from the airflow and one should expect that this results in a slowing down of the airflow. In other words, surface gravity waves and their associated momentum flux may contribute in controlling the shape of the wind profile over the oceans. The common belief in the field was (Phillips, 1977) that air turbulence was dominant in shaping the wind profile and the effect of surface gravity waves was considered to be small. However, Snyder *et al.* (1981) found that the momentum transfer from wind to waves might be considerable, in particular for young windseas, because the related wave-induced stress is a substantial fraction of the total stress in the surface layer. There may, therefore, be considerable deviations in the velocity profile over sea waves from the usual profile of turbulent airflow over a flat plate. One of the consequences is that the drag coefficient at 10 m height should depend on the sea state. Experimental evidence for this was found by Donelan (1982) and later confirmed by Maat *et al.* (1991) and Smith *et al.* (1992) during the HEXOS campaign.

The implication is that a theory had to be developed that takes into account the consequences of growing waves on the mean flow. This theory was independently obtained by Fabrikant (1976) and Janssen (1982), while Miles (1965) also studied wave–mean flow interaction in this context. The first two authors utilized an analogy that exists between resonant wave–mean flow interaction in a fluid and the interaction of plasma waves and particles.

The linear interaction of plasma waves with particles was first succesfully investigated by Landau (1946). The rate of change of the energy of the particles was found to be proportional to the derivative of the particle distribution function at the point of resonance, that is where the particle velocity matches the phase velocity of the plasma wave. For a plane-parallel flow, the energy increment is, as we have seen, proportional to the derivative of the vorticity at the critical height. Also, in a plane-parallel flow, a pattern of closed streamlines is found near the point of resonance, Kelvin's cat's-eye pattern, and the same feature is found in the phase space orbits of trapped particles in a given monochromatic wave.

The linear theories of resonant interaction of gravity waves with a flow and plasma waves with particles are only valid on a short time scale. Because of the exponential growth of the waves, nonlinear effects may become important in the course of time. The study of nonlinear effects on the interaction of plasma waves and particles was started by Vedenov *et al.* (1961) and Drummond and Pines (1962). One of the main results of these investigations was that the plasma waves modify the particle distribution function in such a way that the complete system, that is both waves and modified particle distribution, stabilizes for large times. Later these results were extended to include three-wave interactions and nonlinear wave–particle interactions (see, e.g. Davidson, 1972). The plasma waves were assumed to have a sufficiently broad spectrum such that the random-phase approximation applies. This formalized the picture of a spectrum determined by an energy balance equation, where the shape of the wave spectrum is determined by several processes such as a generation term, a nonlinear interaction term and a dissipation term. Incidentally, in the field of surface gravity waves, development of the nonlinear theory took place in reverse order. Phillips (1960) and Hasselmann (1962) initiated the

theory of resonant four-wave interactions, while much later the possible relevance
of wave–mean flow interactions was recognized by Fabrikant (1976) and Janssen
(1982).

Let us now return to the problem of the effect of water waves on the mean flow. A
multiple-time-scale technique is used to obtain dynamical equations for the slowly
varying energy density of the water waves and the air velocity. The growth of the
water waves due to the atmospheric input occurs on a long time scale since this
energy transfer is proportional to the ratio of air density to water density, which is
small. Hence, there are at least two time scales, namely one related to the relatively
rapid water-wave oscillations and one of the order of the energy-transfer time to the
water waves. Another reason for the use of the multiple-time-scale method is that an
iterative solution of a set of nonlinear equations (in this case the Euler equations plus
boundary conditions) usually gives rise to secular terms in time. The introduction
of different time scales then provides freedom to prevent secularity. In fact, the
condition resulting from the elimination of secularity gives rise to the evolution
equations for the slow time dependence of the wave energy and the air velocity. For
a lucid account of the multiple-time-scale method please consult Whitham (1974)
or Davidson (1972). In addition, we are concerned with a statistical description of
the interaction of air and water waves, that is we consider the evolution in time of
ensemble averages of quantities such as the energy density. To this end, the nonlinear
set of equations (3.1) is solved iteratively by means of a systematic expansion of
the relevant quantities in powers of a small parameter. Finally, averaging is applied
to obtain equations for the averaged quantities. Thus a weakly nonlinear system
is considered for which the random-phase approximation is assumed to be valid
(Hasselmann, 1967; Davidson, 1972). Note that the random-phase approximation
implies that the wave spectrum should be sufficiently broad, a condition that can be
understood on intuitive grounds. Another reason for a sufficiently broad spectrum
is to avoid the trapping phenomenon discussed in Section 3.4.1. Trapping of fluid
elements requires a more complex treatment.

It is expected that there are at least two time scales in the problem determined
by the density ratio of air to water ϵ. In order to proceed we also need to choose the
order of magnitude of the amplitude of the oscillations. A convenient choice would
be $\epsilon^{1/2}$. This choice follows from Stewart (1967), where it was observed that a
substantial amount of energy is contained in the water waves, or

$$\rho_{\mathrm{w}} \langle w^2 \rangle = \mathcal{O}(\rho_{\mathrm{a}} U_{10}^2).$$

Here, the angle brackets denote as usual an ensemble average. Consequently, $\langle w^2 \rangle = \mathcal{O}(\epsilon U_{10}^2)$, suggesting that the amplitude of the waves is $\mathcal{O}(\epsilon^{1/2})$. This choice of small
amplitude parameter is, however, not entirely appropriate, because the consequence
is that formally the effects of nonlinear wave–wave interactions can be ignored.

Anticipating the main result of the next chapter, namely that the nonlinear four-wave interactions are $\mathcal{O}(N^3)$, one finds that if the wave variance (and hence the action density N) is $\mathcal{O}(\epsilon)$, then the nonlinear transfer is found to be $\mathcal{O}(\epsilon^3)$ while the wind input is $\mathcal{O}(\epsilon N) = \mathcal{O}(\epsilon^2)$. Instead, insisting on a balance between wind input and nonlinear interactions a more appropriate order of magnitude for the amplitude of the waves is $\epsilon^{1/4}$. This is formally substantially larger than suggested by Stewart (1967), but results regarding wave–mean flow interaction are not really different from the results obtained by Janssen (1982) who used $\epsilon^{1/2}$ as a small amplitude parameter. The scaling relation is only relevant when one is interested in the dependence of the wave spectrum on the air–sea density ratio. Hence, we anticipate that

$$\langle w^2 \rangle \sim \epsilon^{1/2}, \tag{3.73}$$

and it is, therefore, tempting to expand the surface elevation, the velocity and the pressure in powers of $\Delta = \epsilon^{1/4}$. Thus,

$$\eta = \sum_{l=1} \Delta^l \eta_l, \quad \mathbf{u}_\mathrm{w} = \sum_{l=1} \Delta^l \mathbf{u}_{\mathrm{w},l}, \quad p_\mathrm{w} = \sum_{l=1} \Delta^l p_{\mathrm{w},l},$$

$$\mathbf{u}_\mathrm{a} = \mathbf{U}_0 + \sum_{l=1} \Delta^l \mathbf{u}_{\mathrm{a},l}, \quad p_\mathrm{a} = \epsilon \sum_{l=1} \Delta^l p_{\mathrm{a},l}. \tag{3.74}$$

We remark that the series for the air pressure starts with a term $\mathcal{O}(\epsilon)$ since

$$p_\mathrm{a} = \mathcal{O}(\rho_\mathrm{a} U_\mathrm{a}^2). \tag{3.75}$$

Furthermore, we note that we anticipate much larger amplitude waves in the case of a continuous wave spectrum compared with the single-wave case (see Section 3.4.1).

A straightforward iterative solution of a set of equations may, as already pointed out, give rise to secular terms in time in the series solution (Davidson, 1972) so that small terms grow indefinitely and may become as large as the lowest-order solution. This is clearly not desirable, and for this reason we introduce different time scales such that there is sufficient freedom to prevent secularity. To that end, it is sufficient to assume that average quantities such as $\langle w^2 \rangle$ are allowed to depend on the time scales $\tau_0 = t$, $\tau_2 = \Delta^2 t$, ... Hence,

$$\frac{\partial}{\partial t} \langle w^2 \rangle = \sum_{l=0} \Delta^{2l} \frac{\partial}{\partial \tau_{2l}} \langle w^2 \rangle. \tag{3.76}$$

The τ_0 time scale takes account of the relatively rapid wave oscillations, while growth of the waves due to the atmospheric input occurs on the τ_4 scale, since the energy input is proportional to $\epsilon = \Delta^4$. By construction, four-wave nonlinear transfer occurs on the τ_4 scale as well. The intermediate time scale is needed to accommodate for the fairly rapid nonresonant three-wave interactions and to accommodate the feedback of the growing ocean waves on the wind profile. We emphasize that it is the condition resulting from the elimination of secular behaviour on the short time scale τ_0 that gives us the slow time dependence of the wave spectrum and the consequent feedback on the mean air velocity U_0. This then assures momentum and energy conservation at the air–sea interface.

The discussion of the effect of gravity waves on the mean flow is limited to the case of constant density in air and water. Extension to the case of stratification in the atmosphere is, in principle, feasible, but has not been done yet. Likewise, only one-dimensional propagation will be studied, and therefore, possible interesting couplings between wave growth and vortex stretching will not be explored here.

Neglecting currents in the water, the water motion is assumed to be irrotational. For simplicity, we only consider the deep-water case so that the water velocity vanishes as $z \to -\infty$. The basic equations for air and water become

$$\text{Air:} \; \nabla \cdot \mathbf{u} = 0, \; \frac{\mathrm{d}}{\mathrm{d}t}\mathbf{u} = -\frac{1}{\rho_a}\nabla p + \mathbf{g} \, (z > \eta(x, t)),$$
$$\text{Water:} \; \nabla^2 \phi = 0 \, (z < \eta(x, t)), \tag{3.77}$$

where we note that the water velocity may be derived from a potential ϕ, $\mathbf{u}_w = \nabla \phi$, since the water is irrotational. For a divergence-free flow this then results in Laplace's equation for the potential.

The motion of the interface is given by the kinematic condition

$$\left(\frac{\partial}{\partial t} + \mathbf{u} \cdot \frac{\partial}{\partial \mathbf{x}}\right)\eta = w, \; z = \eta(x, t), \tag{3.78}$$

while at the interface we have continuity of the pressure

$$p_a = p_w, \; z = \eta(x, t). \tag{3.79}$$

For an irrotational flow the pressure may be obtained from Bernoulli's law, which follows from an integration of the Euler equations over depth. Thus the water pressure p_w at the surface is given by

$$p_w = -\rho_w\left[\frac{\partial}{\partial t}\phi + \frac{1}{2}(\nabla\phi)^2 + g\eta\right], \; z = \eta(x, t). \tag{3.80}$$

The potential equation for water may be solved relatively easily by means of Fourier transformation. With the boundary condition that $\phi \to 0$, $z \to -\infty$, we have

$$\phi = \int \mathrm{d}k \, \hat{\phi}(k, t)\exp(ikx + |k|\, z). \tag{3.81}$$

In addition writing for η and p_a

$$\eta = \int dk\ \hat{\eta}(k,t)\exp(ikx),$$

$$\tag{3.82}$$

$$p_a = \int dk\ \hat{p}_a(k,t)\exp(ikx),$$

we obtain from Eqs. (3.78)–(3.80) the following equation for the Fourier transform of η:

$$\frac{\partial^2}{\partial t^2}\hat{\eta} + \sigma^2\hat{\eta} = -|k|\,\hat{p}_a/\rho_w + NL, \quad z = 0, \tag{3.83}$$

where $\sigma^2 = g\,|k|$ and NL represents all nonlinear terms, including three- and four-wave interactions, which will be discussed in Chapter 4. Since only the lowest significant order of the physical processes, such as wind input, nonlinear transfers and wave dissipation, needs to be considered we can study them in isolation. In the present context the term NL will be dropped. Using the series given in Eq. (3.74) and the multiple-time-scale expansion (3.76) we obtain from Eq. (3.83) to lowest significant order in Δ

$$\frac{\partial^2}{\partial \tau_0^2}\hat{\eta}_1 + \sigma^2\hat{\eta}_1 = 0, \tag{3.84}$$

i.e. on the fast time scale we deal with free gravity waves because the air pressure term is, as already noted in Eq. (3.75), of $\mathcal{O}(\epsilon)$. One may proceed in this fashion to obtain the effect of nonlinearity and the atmospheric input on the evolution in time of $\hat{\eta}$. However, we are not interested in the detailed evolution of the amplitudes and phases of the surface waves. In practice, a statistical description of the sea surface suffices since, at best, the wave spectrum is needed. Noting that the energy of a harmonic oscillator is given by

$$F = \frac{1}{2}\frac{\rho_w}{k}\left(\left|\frac{\partial\hat{\eta}}{\partial t}\right|^2 + \sigma^2\,|\hat{\eta}|^2\right), \tag{3.85}$$

we obtain from Eq. (3.83) the energy balance equation

$$\frac{\partial}{\partial t}F = -\frac{1}{2}\left(\hat{p}_a^*\partial\hat{\eta}/\partial t + \text{c.c.}\right), \tag{3.86}$$

where, as usual, the asterisk and c.c. denote complex conjugation. In lowest order we deal with free gravity waves and the energy density of the waves may be simplified considerably. Thus,

$$F = \Delta^2 F_2 + \Delta^4 F_4 + \cdots, \tag{3.87}$$

where

$$F_2 = \rho_w g \, |\hat{\eta}_1|^2. \tag{3.88}$$

Note that in this chapter the wave spectrum F includes the factor $\rho \hat{w} g$. Of course, on the fast time scale τ_0 the energy density of the waves is conserved, i.e. $\partial F_2 / \partial \tau_0 = 0$, and the slow time dependence of F_2 follows from the requirement that there be no secularity in F_4 on the τ_0 scale, hence

$$\frac{\partial}{\partial \tau_4} F_2 = -\frac{1}{2} \left(\hat{p}_{a,1}^* \partial \hat{\eta}_1 / \partial \tau_0 + \text{c.c.} \right). \tag{3.89}$$

According to Eq. (3.89) the energy density F_2 changes in time owing to linear effects only, because, as it will turn out, its right-hand side is proportional to the wave spectrum multiplied by twice the growth rate (for the amplitude) of linear theory (see Eqs. (3.13), (3.21) and (3.32)). However, owing to the resulting energy transfer from wind to waves, the wind profile (and hence the growth rate) may change in the course of time and it is therefore common to use the term quasi-linear approximation (Drummond and Pines, 1962; Bernstein and Engelmann, 1966; Davidson, 1972). Here, we shall only discuss the change of the wave spectrum owing to the resonant interaction of a single wave with the mean flow. Multiple wave and mean-flow interactions (e.g. between wave groups and the wind) will be disregarded since formally their effect is of higher order in Δ (Tsimring, 1983).

The next step we have to deal with is to relate the pressure term to the surface elevation. Since we disregard effects of turbulence for the moment, the dynamical equations for air read

$$\nabla \cdot \mathbf{u} = 0, \quad \frac{d}{dt} \mathbf{u} = -\frac{1}{\rho_a} \nabla p + \mathbf{g} \ (z > \eta(x, t)), \tag{3.90}$$

and for simplicity we write

$$\mathbf{u}_a = \mathbf{U} + \delta \mathbf{u}, \quad p_a = P_a + \delta p_a, \tag{3.91}$$

where $\delta \mathbf{u}$ and δp_a represent the fluctuating parts of the series given in Eq. (3.74) (hence $\langle \delta \mathbf{u} \rangle = \langle \delta p_a \rangle = 0$), while $\mathbf{U}$ and P_a denote the 'steady-state' parts. By means of the decomposition (3.91) and ensemble averaging we then obtain from Eq. (3.90) an equation for $\mathbf{U}$,

$$\frac{\partial}{\partial t} U_\beta + \frac{\partial}{\partial x_\alpha} \langle \delta u_\alpha \delta u_\beta \rangle = -\frac{1}{\rho_a} \frac{\partial}{\partial x_\beta} \langle P_a \rangle + g_\beta, \tag{3.92}$$

while for the fluctuation we have

$$\frac{\partial}{\partial t} \delta u_\beta + \frac{\partial}{\partial x_\alpha} \left(U_\alpha \delta u_\beta + \delta u_\alpha U_\beta \right) = -\frac{1}{\rho_a} \frac{\partial}{\partial x_\beta} \delta p_a + \frac{\partial}{\partial x_\alpha} T_{\alpha\beta}, \tag{3.93}$$

where $T_{\alpha\beta} = \delta u_\alpha \delta u_\beta - \langle \delta u_\alpha \delta u_\beta \rangle$. Here the subscripts α and β denote the various components of the vector quantities $\mathbf{U}$, $\delta\mathbf{u}$ and $\mathbf{g}$, and the summation convention is assumed. Finally,

$$\frac{\partial}{\partial x_\alpha}\delta u_\alpha = 0. \tag{3.94}$$

Elimination of the pressure fluctuation δp_a from Eqs. (3.92), (3.93) and (3.94) gives an equation for the vertical component of the perturbed velocity, δw:

$$\left[\left(\frac{\partial}{\partial t} + U\frac{\partial}{\partial x}\right)\Delta - U''\frac{\partial}{\partial x}\right]\delta w = \frac{\partial}{\partial x}(\nabla \times \nabla \cdot \mathbf{T}), \tag{3.95}$$

where the prime again denotes differentiation with respect to z. In obtaining Eq. (3.95) we have assumed that $\mathbf{U}$ points in the x-direction and is a function of z and t only. Note that the left-hand side operator of Eq. (3.95) is nothing but the Fourier transform of the Rayleigh equation (3.15).

Let us concentrate now on the set of equations (3.92), (3.94) and (3.95). Equation (3.92) describes the rate of change of the steady-state velocity $\mathbf{U}$ due to the wave-induced stresses $\langle\delta u_\alpha\delta u_\beta\rangle$ as the pressure $\langle P\rangle$ follows the hydrostatic law, given by the last line of Eq. (3:2). In order to determine these stresses we need to solve Eq. (3.95). This will be done in an iterative fashion because the amplitude of the fluctuations is assumed to be small.

In agreement with Eq. (3.74) we expand $\langle\delta u_\alpha\delta u_\beta\rangle$ in powers of Δ,

$$\langle\delta u_\alpha\delta u_\beta\rangle = \Delta^2\langle\delta u_\alpha\delta u_\beta\rangle_2 + \cdots, \tag{3.96}$$

while we also expand the mean velocity U according to

$$U = U_0 + \Delta^2 U_2 + \cdots \tag{3.97}$$

Substitution of Eqs. (3.76), (3.96) and (3.97) in the x-component of Eq. (3.92) then gives the hierarchy of equations

$$\frac{\partial}{\partial\tau_0}U_0 = 0, \quad \frac{\partial}{\partial\tau_0}U_2 = -\frac{\partial}{\partial\tau_2}U_0 - \frac{\partial}{\partial x_\alpha}\langle\delta u_\alpha\delta u\rangle_2. \tag{3.98}$$

The first equation tells us that U_0 is independent of the fast time scale τ_0. Integration of the second equation of Eq. (3.98) with respect to the fast time scale τ_0 gives

$$U_2(\tau_0) - U_2(0) = -\tau_0\left[\frac{\partial}{\partial\tau_2}U_0 + \frac{\partial}{\partial x_\alpha}\langle\delta u_\alpha\delta u\rangle_2\right], \tag{3.99}$$

where, as will be shown, $\langle\delta u_\alpha\delta u\rangle_2$ is independent of τ_0. In order to avoid secularity of U_2 on the τ_0 time scale, the right-hand side of Eq. (3.99) should vanish, hence

$$\frac{\partial}{\partial\tau_2}U_0 = -\frac{\partial}{\partial x_\alpha}\langle\delta u_\alpha\delta u\rangle_2, \tag{3.100}$$

resulting in an equation for the slow time evolution of the mean flow U_0. We emphasize that in order to obtain the τ_2 dependence of U_0 we only need the wave-induced stress to lowest significant order. Now, the crucial point is that in contrast to the usual fluid-turbulence problem an explicit expression of the wave-induced stress may be given. In fact, for a single wave we have given the appropriate expression for the wave stress already in Eq. (3.17). Here, we shall derive the wave stress for a continuous wave spectrum. In contrast to the case of a single wave, where the wave stress shows a jump at the critical height, we shall find that for a spectrum of waves the wave stress is a continuous function of height z.

Note that because of our choice of amplitude scaling, see Eq. (3.74), the wind profile adjusts itself, compared with the time scale of wave evolution (which is τ_4), very rapidly to the given sea state, namely on the τ_2 time scale. Strictly speaking, to be consistent with the wave-evolution time scale, one should therefore continue the analysis of the mean-flow equation up to the next order, i.e. on the τ_4 time scale. Here, we shall not be concerned with this because our main interest is in lowest-order significant results.

In order to calculate $\langle \delta u_\alpha \delta u \rangle_2$ we solve Eq. (3.95). To lowest order we find

$$
\left(\frac{\partial}{\partial t} + U \frac{\partial}{\partial x} \right) \Delta \delta w_1 = U'' \frac{\partial}{\partial x} \delta w_1,
\tag{3.101}
$$

i.e. δw_1 satisfies the well-known Rayleigh equation.

The boundary conditions for δw_1 follow from the usual requirements that the interface shall remain a streamline and that the fluctuation δw_1 vanishes at infinity. To lowest order we therefore obtain

$$
\delta w_1(z = 0) = \frac{\partial}{\partial z} \phi(0); \quad \delta w_1 \to 0, \; z \to \infty.
\tag{3.102}
$$

By means of the lowest-order solution of the surface gravity waves (Eq. (3.81)), the first boundary condition can be written as a linear combination of waves propagating to the right ($\theta_+ = kx - \sigma\tau_0$) and to the left ($\theta_- = kx + \sigma\tau_0$), or

$$
\delta w_1(z = 0) = -i \int_0^\infty dk \, \sigma(k) \left\{ \hat{\eta}_+ e^{i\theta_+} + \hat{\eta}_- e^{i\theta_-} \right\} + \text{c.c.},
\tag{3.103}
$$

hence, the air at $z = 0$ is forced to oscillate in the manner prescribed by Eq. (3.103). As suggested by this boundary condition we therefore try the solution

$$
\delta w_1 = -i \int_0^\infty dk \, \sigma(k) \left\{ \hat{\eta}_+ \chi_+ e^{i\theta_+} + \hat{\eta}_- \chi_- e^{i\theta_-} \right\} + \text{c.c.},
\tag{3.104}
$$

to obtain the following problem for χ_+:

$$
\left(W\Delta - W'' \right) \chi_+ = 0, \quad \chi_+(0) = 1, \quad \chi_+(\infty) = 0,
\tag{3.105}
$$

where $W = U_0 - c_0$ and $\Delta = \partial^2/\partial z^2 - k^2$. Since $c_0 = \sigma/k$ as well as U_0 is positive, resonance of the wave with the airflow is only possible for the χ_+ component of δw_1. From now on, we therefore omit the contribution of the waves propagating to the left, and we drop the subscripts $+$ and $-$.

From incompressibility we have

$$\delta u_1 = \int_0^\infty dk \left\{ c_0 \hat{\eta} \frac{\partial}{\partial z} \chi e^{i\theta} + \text{c.c.} \right\} \tag{3.106}$$

to obtain, using homogeneity of the ensemble of waves, for the wave stress

$$\langle \delta u \delta w \rangle_2 = -i \int_0^\infty dk \frac{F_2}{\rho_w} \left(\chi \frac{\partial}{\partial z} \chi^* - \chi^* \frac{\partial}{\partial z} \chi \right). \tag{3.107}$$

Now, because of homogeneity in the x-direction, we have

$$\frac{\partial}{\partial x_\alpha} \langle \delta u_\alpha \delta u \rangle_2 = \frac{\partial}{\partial z} \langle \delta u \delta w \rangle_2;$$

hence Eq. (3.100) may be written as

$$\frac{\partial}{\partial \tau_2} U_0 = i \int_0^\infty dk \frac{F_2(k)}{\rho_w} \left(\chi \frac{\partial^2}{\partial z^2} \chi^* - \chi^* \frac{\partial^2}{\partial z^2} \chi \right). \tag{3.108}$$

Finally, by means of the Rayleigh equation (3.95) the term between the brackets may be simplified to obtain the main result of this subsection:

$$\frac{\partial}{\partial \tau_2} U_0 = 2\pi \int_0^\infty dk \frac{F_2(k)}{\rho_w} |\chi|^2 W'' \delta(W), \tag{3.109}$$

where $\delta(W)$ is the usual δ-function. Note that Eq. (3.109), which gives the rate of change of the mean flow due to the growth of a continuous spectrum, generalizes the single-wave result from the average vortex force calculation (see Eq. (3.71)). Performing the integration over k, we obtain an equation of the diffusion type:

$$\frac{\partial}{\partial \tau_2} U_0 = D_W \frac{\partial^2}{\partial z^2} U_0, \tag{3.110}$$

where the wave diffusion coefficient D_W is proportional to the surface elevation spectrum $F(k)$,

$$D_W = \frac{\pi c^2 k^2 |\chi|^2}{|c - v_g|} \frac{F_2(k)}{\rho_w g}. \tag{3.111}$$

Here, the wavenumber k has to be expressed as a function of height through the resonance condition $W = 0$ and v_g is the group velocity $\partial\sigma/\partial k$.

Equation (3.111) tells us that the airflow at a certain height z changes with time owing to resonant interaction of a water wave with frequency $\sigma = g/U_0(z)$. Hence, in this fashion there is possibly an energy transfer from the airflow U_0 to the water waves, thus giving a rate of change of the spectrum as in Eq. (3.89). Using now the z-component of Eq. (3.93), we can write the air-pressure fluctuation $\hat{p}_{a,1}$ in terms of the wave-induced velocity,

$$\hat{p}_{a,1} = i\rho_w k \int_0^\infty dz \, W\hat{w}; \qquad (3.112)$$

hence, with $\hat{w} = -i\sigma\hat{\eta}_1\chi$ we obtain from Eq. (3.89)

$$\frac{\partial}{\partial\tau_4} F_2(k) = -\frac{1}{2}k^2 F_2(k)\left[i\int_0^\infty dz \, W\chi + \text{c.c.}\right]. \qquad (3.113)$$

Then, by means of the Rayleigh equation we obtain the well-known result (Miles, 1957)

$$\frac{\partial}{\partial\tau_4} F_2(k) = -\frac{\pi\sigma}{k} |\chi|^2 \frac{W_c''}{|W_c'|} F_2(k), \qquad (3.114)$$

where again the subscript c refers to evaluation at the critical height. To summarize our results, we obtain the following quasi-linear equations for the generation of water waves by wind:

$$\frac{\partial}{\partial t} F(k) = -\epsilon\pi c \, |\chi|^2 \frac{W_c''}{|W_c'|} F(k),$$

$$\frac{\partial}{\partial t} U_0 = D_W \frac{\partial^2}{\partial z^2} U_0, \quad D_W = \frac{\pi c^2 k^2 \, |\chi|^2}{|c - v_g|} \frac{F(k)}{\rho_w g}, \qquad (3.115)$$

$$W\nabla^2\chi = W''\chi, \quad \chi(0) = 1, \quad \chi(\infty) = 0,$$

where we returned to the original variables ($\tau_2 = \Delta^2 t$, $\tau_4 = \Delta^4 t$, $F_2 = F/\Delta^2$).

From the first line of Eq. (3.115) we obtain the well-known result that only those waves are unstable for which the curvature U_0'' of the wind profile, at the critical height is negative. The growth rate of the waves is, however, a function of time, as the wind profile depends on time according to the diffusion equation for U_0, possibly quenching the instability for large time t. Quasi-linear theory therefore emphasizes that there is a mutual interaction between wind and waves. Waves that are generated by the wind will affect the wind profile, in particular when the waves are steep enough (since D_W depends on the wave spectrum F). In fact, with the present choice of the order of magnitude of the wave spectrum, $F = \mathcal{O}(\epsilon^{1/2})$, it is found that the waves are so steep that the wind profile adjusts itself to the given sea state more rapidly than the slow time scale induced by wave growth. In other words, if one is interested in evolution on the long time scale of spectral evolution, the wind profile is always in equilibrium with the given sea state.

Before we proceed, some historical remarks are in order. First, it should be emphasized that the derivation of the wave-induced stress and hence the effect of waves on the wind is only possible because gravity waves have a strong dispersion, that is the phase speed c is not equal to the group speed v_g. This is also clear from the expression for D_W which becomes unbounded for $c \to v_g$. Thus, for turbulent airflow, where the fluctuations are thought to have hardly any dispersion, such a quasi-linear approach is probably not possible. Second, it is noted that previous authors have obtained similar results. Lighthill (1962), who discussed the physical interpretation of Miles' theory of wave generation by wind, obtained a similar result regarding the effect of a single wave on the wind profile (the essence of this derivation, using the vortex force, is given in Section 3.4.1). He did not realize, however, that the wind profile U_0 may be a slowly varying function of time. In addition, Fabrikant (1976) obtained a similar set of equations, although along different lines, while Miles (1965) also discussed the possible interaction between wind profile and waves.

Furthermore, we should address to some extent the objection that the effect of turbulent Reynolds stresses on the wind profile and the growth of the waves has not been included. Keeping in the spirit of Miles' quasi-laminar approach, we only consider the effect of turbulence on the mean flow, disregarding possible interactions between the wave-induced oscillations and turbulence (in agreement with our discussion in Section 3.3.1). Thus, the mean-flow equation becomes

$$\frac{\partial}{\partial t} U_0 = D_W \frac{\partial^2}{\partial z^2} U_0 + \frac{\partial}{\partial z} K(z) \frac{\partial}{\partial z} U_0, \qquad (3.116)$$

where D_W is the wave diffusion coefficient (Eq. (3.111)), while $K(z)$ is the eddy viscosity of air, which may be modelled as $K(z) = \kappa u_* z$. It is emphasized that we have not given a real justification for Eq. (3.116). However, we will see in a while that results compare remarkably well with quasi-linear models including the effects of turbulence on the mean flow and on wave-induced oscillations.

The effect of momentum transport by turbulent eddies is clear. If no water waves are present the well-known logarithmic wind profile is obtained in the steady state as $K \sim z$. In the presence of water waves the eddies will maintain a logarithmic profile if $D_W \ll K$, but if the waves are steep enough the effect of the waves on the wind profile may overcome the effects of turbulence; this may especially occur in the layer just above the water waves because in that layer the size of the eddies is small.

In order to be able to appreciate the effect of waves on the wind let us first disregard the effects of turbulence altogether, and investigate some of the properties of the quasi-linear theory of wind-wave generation. First, we question whether or not the set of equations (3.115) admits a steady state. This question is of interest in the context of stability theory. Assuming that initially we have a wind profile with negative curvature, then according to Miles' theory gravity waves will be generated at an exponential rate. The question one may ask then is whether the growth of the gravity waves will be arrested by nonlinear effects (because the waves extract momentum from the airflow which may lead to a reduction of the curvature of the wind profile, hence a reduced wave growth, etc.), or whether a nonlinear instability will develop which leads to explosive growth. It will be argued here that the latter possibility of explosive growth seems unlikely. In order to see this we derive a balance equation for the enstrophy of the mean flow. We differentiate the diffusion equation for U_0 with respect to height z to obtain a diffusion equation for the mean vorticity $\zeta_0 = \partial U_0 / \partial z$,

$$\frac{\partial}{\partial t} \zeta_0 = \frac{\partial}{\partial z} D_W \frac{\partial}{\partial z} \zeta_0, \tag{3.117}$$

where we note that D_W is always positive. We next multiply Eq. (3.117) by ζ_0 and integrate over height with the result

$$\frac{d}{dt} \int_0^\infty dz \; \zeta_0^2 = 2 \int_0^\infty dz \; \left[\frac{\partial}{\partial z} \left(D_W \zeta_0 \frac{\partial}{\partial z} \zeta_0 \right) - D_W \left(\frac{\partial}{\partial z} \zeta_0 \right)^2 \right]. \tag{3.118}$$

For appropriate boundary conditions (e.g. by realizing that the wave diffusion co-efficient should disappear at zero and ∞) the perfect derivative on the right above integrates to zero, hence

$$\frac{d}{dt} \frac{1}{2} \int_0^\infty dz \; \zeta_0^2 = - \int_0^\infty dz \; D_W \left(\frac{\partial}{\partial z} \zeta_0 \right)^2. \tag{3.119}$$

This equation states that the time derivative of the mean flow enstrophy, which is a positive quantity, is nonpositive. Hence, the mean airflow tends toward a condition where the right-hand side disappears, which requires in the region where $D_W \neq 0$ that

$$\frac{\partial}{\partial z} \zeta_0 = \frac{\partial^2}{\partial z^2} U_0 \to 0, \quad \text{as } t \to \infty. \tag{3.120}$$

Thus, for large times the wind profile becomes linear, implying that according to the first equation of Eq. (3.115) the growth rate of the waves vanishes. Apparently, quasi-linear theory predicts, for large times, a limitation of the amplitude of the initially unstable water waves, i.e. the energy transfer from the airflow to the water waves is quenched. In passing it should be pointed out that Eq. (3.120) holds for a layer above the waves. As a consequence, the critical layers move up in the course of time. This is discussed in more detail in Janssen (1982). In practice, which means when air turbulence is taken into account as well, the reduction of the curvature of the wind profile is only expected to occur just above the water surface where the wave effect may overcome the effects of turbulent eddies. In other words, the reduction of momentum-transfer from air to gravity waves is expected to occur for the high-frequency waves which have their critical-layer just above the water surface.

Second, we note that the set of quasi-linear equations (3.115) admits an infinite set of balance equations, notably

$$\frac{d}{dt}\left[\rho_a \int_0^\infty f(U_0)\, dz + \int_0^\infty f'(U_0)\frac{F}{c}\, dk\right] = 0, \tag{3.121}$$

where $f(U_0)$ is an arbitrary function of U_0, the prime denotes differentiation with respect to U_0 and in the second term of the left-hand side we apply the resonance condition $U_0 = c$ to map the height coordinate z to wavenumber space. Equation (3.121) may be obtained by multiplication of the diffusion equation for U_0 by $f'(U_0)$; then integration with respect to z gives

$$\frac{d}{dt}\int_0^\infty f(U_0)\, dz = \int_0^\infty f'(U_0)D_W\frac{\partial^2 U_0}{\partial z^2}\, dz. \tag{3.122}$$

In the integral on the right-hand side we next convert to an integration over k via the resonance condition $U_0 = c(k)$, then using the expression for D_W and the evolution equation for the wave spectrum, we finally arrive at the conservation law (3.121).

By an appropriate choice of $f'(U_0)$ we are able to express all moments of the spectrum in terms of an integral of $f(U_0)$ in k-space. In particular, we obtain for $f'(U_0) = 1$, conservation of momentum,

$$\frac{d}{dt}\left[\rho_a \int_0^\infty U_0\, dz + \int_0^\infty \frac{F}{c}\, dk\right] = 0, \tag{3.123}$$

and for $f'(U_0) = U_0$, conservation of mechanical energy,

$$\frac{d}{dt}\left[\frac{1}{2}\rho_a \int_0^\infty U_0^2\, dz + \int_0^\infty F\, dk\right] = 0. \tag{3.124}$$

From the conservation laws we see once more that for growing waves the wind profile changes in time.

The property of having an infinite set of conservation laws suggests that the quasi-linear set of equations admits exact solutions. This turns out to be not quite the case, but the time-asymptotic form of the wave spectrum may be obtained (Janssen, 1982). As a result the high-frequency part of the spectrum has an f^{-4} power law, and closely resembles Toba's law, Eq. (2.112). We shall not give the details of this calculation as the quasi-linear theory in Eq. (3.115) only represents one aspect of the physics of wave evolution. Processes such as wave breaking and nonlinear interactions are relevant too, and it is expected that the subtle balance between these processes determines spectral shape. However, the process of quenching of the Miles instability is most certainly relevant, as will be seen in Section 3.4.3. In general, the growth rate of the waves and the related momentum transfer from air to waves will be sea-state dependent. As a consequence, the drag of airflow over sea waves depends on the state of the waves.

3.4.3 Wave-induced stress and the drag of airflow over sea waves

We have seen that the growth of the waves depends on the sea state as the wind profile is affected by the presence of the surface gravity waves. In this subsection we shall ask ourselves the question of to what extent the wind profile depends on the sea state.

To be sure, careful observations from the field (Donelan, 1982; Smith *et al.*, 1992; Oost *et al.*, 2002) do indicate that the roughness length and the drag coefficient depend on the sea state. These experimentalists observed wind and waves from fixed platforms in Lake Ontario and in the North Sea. Here, the surface stress was determined by means of the eddy-correlation technique, while the state of the ocean waves was inferred from observed wavenumber or frequency spectra. The sea state was in essence characterized by the wave-age parameter (see the discussion in Section 2.6.4), although theoretically it seems more natural to choose as measure of sea state the ratio of wave-induced stress to the total surface stress. Experimentally, this is, however, a nearly impossible task. In a short while we will see that this involves the determination of the growth rate of the waves by wind and the determination of the two-dimensional wavenumber spectrum up to wavenumbers in the gravity–capillary range. Therefore, we shall only study the wave-age dependence of, for example, the drag of airflow over ocean waves. However, it is emphasized that the characterization of the sea state by the wave-age parameter has a limited validity; strictly speaking it is only valid for windseas.

Before we solve the quasi-linear equations it seems a good idea to get a feeling for the order of magnitude of the effect of waves on the wind. Hence, the wave-induced stress is determined using a simple model for the growth of waves by wind and of the high-frequency part of the wave spectrum. The wave stress is found to be a substantial fraction of the total stress in the surface layer. In particular, we are interested in the dependence of the wave stress on the energy level of the high-frequency waves and on the directional distribution of the waves.

From the momentum balance (Eq. (3.123)) we infer that the wave-momentum spectrum P is, as expected, given by

$$P = \frac{F}{c}, \tag{3.125}$$

and since the wave stress is given by the rate of change in time of the wave momentum due to wind, we have

$$\tau_w = \int dk \, d\theta \frac{\partial}{\partial t} P \bigg|_{wind}. \tag{3.126}$$

The main contribution to the wave stress is determined by the medium- to high-frequency gravity waves with dimensionless phase speed c/u_* in the range of 1 to 10, as these are the waves with the highest growth rate. The rate of change due to wind of this part of the wave spectrum is fairly well known and can be described by the empirical expression proposed by Plant (1982) which was based on the observations displayed in Fig. 3.3. Hence, writing $\partial P/\partial t = \gamma P$ we take for the growth rate γ,

$$\gamma = \epsilon\beta\omega \left(\frac{u_*}{c}\right)^2 \cos^2\theta, \quad c/u_* < 25, \quad |\theta| < \pi/2, \tag{3.127}$$

while γ vanishes for $c/u_* > 25$ or $|\theta| > \pi/2$. Here, the constant $\beta \simeq 30$ and all other symbols have their usual meaning. Note that at $c/u_* = 25$ the growth rate shows a jump, but, as this occurs for low-frequency waves which have a small growth rate, this is not important for estimation purposes. Since we are only interested in orders of magnitude anyway, a very simple two-dimensional spectral shape will be adopted:

$$F(k, \theta) = \begin{cases} \frac{1}{2}\rho_w g\alpha_p k^{-3}\Phi(k, \theta), & k > k_p \\ 0, & k < k_p \end{cases} \tag{3.128}$$

where k_p is the wavenumber of the peak of the spectrum, while we adopt the very simple directional distribution (see the discussion in Section 2.6.3)

$$\Phi(k, \theta) = \begin{cases} \frac{1}{2\Delta(k)}, & |\theta| \leq \Delta(k) \\ 0, & |\theta| \geq \Delta(k) \end{cases} \tag{3.129}$$

Here, $\Delta(k)$ is the width of the directional distribution which depends according to Donelan *et al.* (1985) on the ratio k/k_p in such a way that near the peak the spectrum is narrow while the spectrum broadens for increasing wavenumber. The following simple form for $\Delta(k)$ is chosen:

$$\Delta(k) = \min \left\{ \pi/2, \alpha\left(\frac{k}{k_p}\right)^{m/2} \right\}, \quad \alpha \text{ and } m \text{ are constants}, \tag{3.130}$$

with a maximum value of $\pi/2$. The ratio of the wave-induced stress τ_w to the total stress $\tau = \rho_a u_*^2$ is now obtained by substituting Eqs. (3.125), (3.127), (3.128) and (3.129) in Eq. (3.126). Performing the integration over the angle θ and introducing the integration variable $y = k/k_p$ we find for windsea with $k_p > g/(25u_*^2)$

$$\frac{\tau_w}{\tau} = \frac{1}{4}\beta\alpha_p \int_1^{\chi^2} \frac{dy}{y}\left[1 + \frac{\sin 2\Delta(k)}{2\Delta(k)}\right],\tag{3.131}$$

where, again, $\chi = c_p/u_*$ is the wave age. It is clear that the integration over wavenumber may give rise to a logarithmic singularity and thus an appropriate high-wavenumber cut-off k_c should be chosen. For present purposes we take as upper limit $c/u_* = 1$ which in terms of the integration variable y translates into an upper bound $y_c = \chi^2$. When discussing the feedback of the waves on the wind profile and the consequences on the growth rate of the high-frequency waves this choice of cut-off value will become clearer.

In order to study the dependence of the wave stress on the sea state consider first the case of a wavenumber-independent width $\Delta(k) = \Delta_0$. In that event Eq. (3.131) may be integrated immediately with the result

$$\frac{\tau_w}{\tau} = \frac{1}{2}\beta\alpha_p \log\chi\left[1 + \frac{\sin 2\Delta_0}{2\Delta_0}\right],\tag{3.132}$$

and it is immediately noted that a narrow angular distribution ($\Delta_0 \to 0$) results in a higher waves tress than a broad distribution ($\Delta_0 \to \pi/2$) by a factor of 2. Furthermore, for older windsea, i.e. increasing χ, more waves will contribute to the wave stress which is reflected by the $\log\chi$ factor. On the other hand, since the JONSWAP study (see Section 2.6.1) it is known that the Phillips parameter α_p decreases with increasing wave age. Therefore, whether the wave stress will decrease or increase with wave age depends to some extent on how rapidly the Phillips parameter varies with wave age.

In Section 2.6 an extensive discussion of the dependence of the Phillips parameter on wave age was given. The reanalysis by Battjes *et al.* (1987) and Günther (1981) of the JONSWAP data set, and the so-called 'KNMI data set' (Janssen *et al.*, 1984) all suggest a fairly sensitive dependence on wave age. We therefore choose the law as found by Battjes *et al.* (1987) and given in Eq. (2.125), which shows the following dependence on wave age,

$$\alpha_p \sim \chi^{-1.24}.\tag{3.133}$$

Combined with the wavenumber-dependent width of the directional spectrum Eq. (3.131) then results, as can be seen in Fig. 3.8, in a very sensitive sea-state dependence of the normalized wave stress. The importance of the wavenumber-dependent width is shown by plotting in Fig. 3.8 the case of a constant width $\Delta = \pi/2$. Comparing the two cases, it is seen that, particularly for young windseas with a narrower angular distribution in the range $c/u_* > 1$, a considerable enhancement of the wave stress is found. Finally, the importance of the sea-state dependence

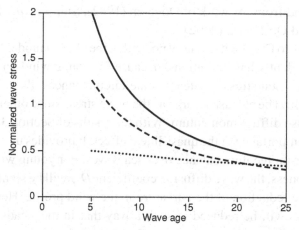

Fig. 3.8. Normalized wave stress versus wave age for different choices of directional distribution and sea-state dependence of the Phillips parameter α_p. Continuous line refers to Eq. (2.125) and wavenumber-dependent width; dashed line refers to Eq. (2.125) and constant width; dotted line refers to Eq. (3.134) and a constant width of $\pi/2$.

of the Phillips parameter is illustrated by plotting the case of the original JONSWAP parametrization of the Phillips parameter,

$$\alpha_p \sim \chi^{-2/3}. \tag{3.134}$$

The result is a wave-induced stress that is virtually independent of the wave age.

It is clear from the present discussion that there are two aspects of the wave spectrum that determine the wave-age dependence of the wave-induced stress. The first factor is the sea-state dependence of the high-wavenumber part of the spectrum, while the second factor is the wavenumber dependence of the angular width of the spectrum. However, uncertainties remain as to what extent the wave stress depends on the sea state. The reason for this is that relations for the spectral width $\Delta(k)$ and for α_p have only been validated for relatively low-frequency waves in the range of 1 to 2.5 times the peak frequency. However, a considerable part of the wave stress, typically 50 per cent, is carried by the high-frequency waves. It is only very recently that Donelan (private communication, 2003) has found that the high-wavenumber part of the spectrum is indeed steeper for young windseas. Nevertheless, we favour the power law (Eq. (3.133)) for the Phillips parameter because, on the one hand, it is at least in agreement with the field data for relatively low-frequency waves and, on the other hand, the wave stress is large for young windsea while it is small for old windsea. This behaviour of the wave stress is in agreement with one's intuition that airflow over young windsea is rougher than over old windsea. Evidence for this may be found from the work of Donelan (1982), Smith *et al.* (1992), Drennan *et al.* (1999), and Oost *et al.* (2002).

Returning now to Fig. 3.8 it is finally noted that the chosen model for wave growth (Eq. (3.127)) evidently has limitations, because for young windsea the wave stress is larger than the total stress. In steady-state circumstances, the total stress in the surface layer should be at least as large as the wave stress since other processes, e.g. viscosity, will also diffuse momentum. With the results of Section 3.4.2 in mind we now realize the importance of the quasi-linear effect. It provides us with a consistent picture of the air-momentum balance over sea waves. For young windsea, having a large wave steepness, the wave diffusion coefficient D_W will be significant resulting in a considerable reduction of the curvature in the wind profile. Hence, the growth rate of the waves will be reduced in such a way that in the steady state the wave stress is less than the total stress in the surface layer.

Let us therefore discuss again the quasi-linear set of equations (3.115), but now including effects of turbulence on the mean flow (Eq. (3.116)). For ease of reference, we reproduce these equations here:

$$\frac{\partial}{\partial t} U_0 = \nu \frac{\partial^2}{\partial z^2} U_0 + \frac{1}{\rho_a} \frac{\partial}{\partial z} \tau_{\text{turb}}, \quad \nu = \nu_a + D_W, \tag{3.135}$$

where

$$D_W = \frac{\pi c^2 k^2 |\chi|^2}{|c - v_g|} \frac{F(k)}{\rho_w g}. \tag{3.136}$$

The rate of change of the wave spectrum is given by

$$\frac{\partial}{\partial t} F(k) \bigg|_{\text{wind}} = \gamma F(k), \tag{3.137}$$

where

$$\gamma = -\epsilon \pi c |\chi|^2 \frac{W_c''}{|W_c'|}. \tag{3.138}$$

The wave-induced vertical velocity χ satisfies the Rayleigh equation

$$W \nabla^2 \chi = W'' \chi, \quad \chi(0) = 1, \quad \chi(\infty) = 0. \tag{3.139}$$

Finally, the turbulent stress τ_{turb} is modelled by means of a mixing-length model

$$\tau_{\text{turb}} = \rho_a l^2 \left| \frac{\partial}{\partial z} U_0 \right| \frac{\partial}{\partial z} U_0, \tag{3.140}$$

with mixing length given by $l = \kappa z$ (where κ is the von Kármán constant).

In order to be able to solve our problem we still have to specify two boundary conditions for the mean flow. For large heights the condition of constant stress is imposed and it is assumed that the waves have no direct impact on the wind profile at those heights, hence,

$$\tau_{turb} \to \rho_a u_*^2, \quad z \to \infty. \tag{3.141}$$

At the lower boundary we choose

$$U_0(z) = 0, \quad z = z_0, \tag{3.142}$$

where following Charnock (1955), we take as roughness length

$$z_0 = \alpha_{CH} u_*^2 / g, \tag{3.143}$$

with $\alpha_{CH} = 0.0144$. This choice of boundary condition requires justification. From observations at sea (Garrat, 1977; Smith, 1980; Large and Pond, 1982; Wu, 1982) it is known that even for old windsea the aerodynamic drag increases with wind speed. This increase in drag must for the greater part be caused by the momentum loss to the short waves such as gravity–capillary waves as the longer waves have such a small steepness (because the Phillips parameter α_p is small for old windsea) that their wave stress is small. It is, therefore, important to take the momentum loss to the short waves into account. However, even nowadays we have no firm knowledge about the spectral shape of the gravity–capillary waves or about the energy and momentum transfer to these waves, at least in field conditions. It is not even firmly established whether gravity–capillary waves contribute at all to the stress in the atmospheric surface layer. For example, Makin *et al.* (1995) found that gravity–capillary waves only support a small fraction of the surface stress. These authors used a spectral form for these waves proposed by Donelan and Pierson (1987). This model of the short-wave spectrum has a k^{-3} power law, while effects of viscosity result in a viscous cut-off, beyond which there is no wave energy. Such a spectral shape would indeed result in a relatively small contribution to the surface stress. However, the laboratory observations of Jähne and Riemer (1990) do not show evidence for a viscous cut-off, while the gravity–capillary waves are steeper because the observed spectra decay over a large range as $k^{-5/2}$. The work of Janssen *et al.* (1998) suggests that the $k^{-5/2}$ behaviour of the spectrum is caused by three-wave interactions. Using the empirical expression for the growth rate suggested by Plant (1982) (Eq. (3.42) with $\beta = 32$), the Jähne and Riemer (1990) spectra suggest that the gravity–capillary waves support about 30 per cent of the total stress.

Although at present we have no firm information on wind input to the gravity–capillary waves and their spectral shape, it is believed that they contribute to some extent to the surface stress. In addition, there are other processes that transfer momentum from air to the water surface which are not described by Miles' shear flow mechanism. For example, flow separation over small-scale breaking waves (Banner and Melville, 1976) may be important while wave–current interactions may also play a role in extracting momentum from the airflow. Because of all these complications, we rely on a simple parametrization. Since the phase speed of these short waves is much smaller than, say, the wind speed at 10 m height, the airflow encounters a water surface with more or less stationary perturbations, that is the airflow 'feels' a water surface with a certain roughness. The choice of roughness length (Eq. (3.143)) is in agreement with the observed increase of drag coefficient with wind speed.

The parametrization of unresolved scales and processes has been further discussed by Chalikov and Makin (1991), Chalikov and Belevich (1993) and Makin *et al.* (1995). For example, the use of the 'background' roughness (Eq. (3.143)), in particular its dependence on u_* and g, suggests that here gravity waves also play an important role; this appears to be inconsistent with the assumption that the gravity waves are already accounted for in the wave stress τ_w. However, it is common practice to assume that the background roughness is proportional to the height of the short gravity waves, which scales with u_*^2/g. An attempt to obtain a consistent theory to calculate the drag over sea waves, avoiding the use of the Charnock-type relation (3.143) for the background roughness, was introduced by Makin *et al.* (1995).

The set of equations (3.135)–(3.143) describes the effect of gravity waves on the airflow. To close this set of equations we need to specify the gravity wave spectrum F since the wave diffusion coefficient D_W depends on spectral shape. Clearly, the evolution of the wave spectrum and the wind speed are coupled and, in principle, one should solve the energy balance equation for the waves together with the momentum equations for wind. This approach will not be pursued here, however, because according to the multiple-time-scale analysis the wind profile changes so rapidly that it is always in equilibrium with the sea state. Hence, we assume that the wave spectrum is given by an empirical relation (the JONSWAP shape, Eq. (2.97)) and we concentrate on the effect of waves on the wind profile. For simplicity, we give the overshoot parameter γ and the width parameter σ the constant values 3.3 and 0.1 respectively. Furthermore, for the Phillips parameter α_p we use a relation proposed by Snyder (1974), namely,

$$\alpha_p = 0.57 \chi^{-3/2}, \tag{3.144}$$

which shows a more sensitive dependence on wave age χ than the relation (3.133) suggested by Battjes *et al.* (1987) or Günther (1981). Note that the present quasi-linear set of equations only considers one-dimensional propagation but, as we have seen, directional effects may give rise to a sea-state dependence of the wave stress as well. Therefore, in order to mimic the effects of the directional distribution, Eq. (3.144) for α_p is chosen rather than Eq. (3.133).

Given the spectral shape we shall search for steady-state solutions of the airflow over sea waves by means of an iteration method that initially takes $D_W = 0$, then

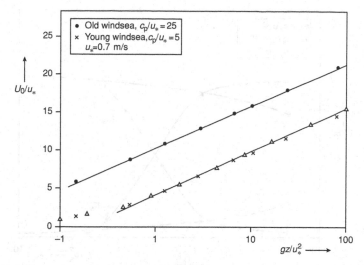

Fig. 3.9. Effect of waves on wind profile for old and young windsea, shown by plotting dimensionless wind speed U_0/u_* as a function of dimensionless height gz/u_*^2. The parametrization (3.147) for young windsea is denoted by $\triangle$ (from Janssen, 1989).

calculates the wind profile to obtain the growth rate γ and the diffusion coefficient D_W and so on. The rate of convergence of this procedure was judged by calculating the total stress

$$\tau_{\text{tot}} = \rho_a \left[\nu_a \frac{\partial}{\partial z} U_0 + l^2 \left| \frac{\partial}{\partial z} U_0 \right| \frac{\partial}{\partial z} U_0 \right] + \tau_w, \tag{3.145}$$

as in the steady state this is given by its asymptotic value for large z, $\tau_{\text{tot}} = \rho_a u_*^2$. Here, τ_w is given by Eq. (3.126).

Details of the numerical procedure may be found in Janssen (1989). Here, only some of the results are discussed. The effect of gravity waves on the wind profile is illustrated in Fig. 3.9, where we have plotted the dimensionless wind speed U_0/u_* as a function of dimensionless height gz/u_*^2 for young and old windsea ($\chi = 5$ and $\chi = 25$ respectively). Since the dimensionless height is plotted on a logarithmic scale, a straight line corresponds to a logarithmic wind profile. Thus, the wind profile is found to be approximately logarithmic except in a region close to the water surface ($gz/u_*^2 < 1$), where considerable deviations from the logarithmic profile are found. Even for the large friction velocity used ($u_* = 0.7$ m/s) this region corresponds to only a few centimetres above the sea surface. Now, by extrapolating the straight lines for young and old windsea towards the surface, the intersection with the x-axis gives the Charnock parameter. It is therefore clear from Fig. 3.9 that the gravity waves extract a significant amount of momentum from the airflow, but young windsea appears to be rougher than old windsea. The reason for this

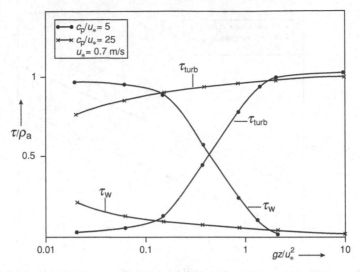

Fig. 3.10. Turbulent stress τ_{turb} and wave stress τ_w as a function of dimensionless height gz/u_*^2 for young and old windsea (from Janssen, 1989).

is that the steepness of the young wind waves is so much larger than that for old waves.

Furthermore, it is of interest to study the distribution of the stress in the surface layer over turbulence, the wave effect and the viscosity. This is shown in Fig. 3.10, where we have plotted the turbulent stress τ_{turb} and the wave stress τ_w as a function of dimensionless height for young and old windsea. The viscous stress is not plotted because it is usually quite small; this is a result of using the Charnock relation (3.143) where the water surface is already rough for $u_* > 0.1$ m/s. For young windsea it is observed that around $gz/u_*^2 \simeq 1$ the wave-induced stress becomes a considerable fraction of the total stress, corresponding to the deviations from the logarithmic wind profile shown in Fig. 3.9. On the other hand, for old windsea, the stress going into the long waves is only 35 per cent of the total stress so that most of the stress is supplied to the very short gravity and capillary waves.

Referring again to Fig. 3.8 we have seen that the wave-induced stress obtained from parametric relations may be a considerable fraction of the total stress or even larger. Figure 3.11 shows what happens according to quasi-linear theory of wind-wave generation. Here, filled circles denote the wave-induced stress calculated according to linear theory and open squares show the results according to quasi-linear theory. We infer from Fig. 3.11 that according to quasi-linear theory the ratio τ_w/τ is reduced considerably for young windsea whereas for old windsea this ratio hardly changes. Apparently, in equilibrium the curvature of the wind profile is reduced for young windsea in such a way that the ratio remains less than unity. This must then be accompanied by a reduction of the growth rate of the waves.

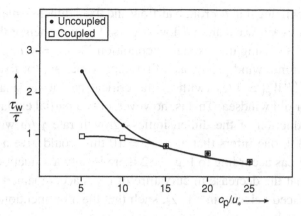

Fig. 3.11. Reduction of the wave-induced stress due to the quasi-linear effect (from Janssen, 1989).

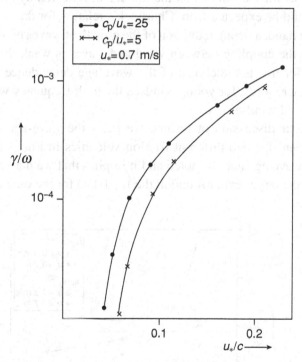

Fig. 3.12. Normalized growth rate γ/ω of the waves versus the inverse of the dimensionless phase speed c/u_* for young and old windsea (from Janssen, 1989).

This is illustrated in Fig. 3.12 where the normalized growth rate γ/ω of the waves versus the inverse of the dimensionless phase speed c/u_* is plotted. Clearly, for fixed phase speed, the growth rate of the waves is larger for old windsea than for young windsea. This reduction of the growth rate for young windsea may be understood by realizing that for rough airflow and fixed dimensionless phase speed, the resonance between wave and airflow occurs at a larger height than for smooth airflow (see Fig. 3.9 using the resonance condition $U_0/u_* = c/u_*$). Consequently, using the logarithmic wind profile as a first approximation, for fixed phase speed the quantity $W_c''/|W_c'| \simeq 1/z_c$ (with z_c the critical height) is smaller for young windsea than for old windsea. This is, however, only a partial explanation because an excessive reduction of the dimensionless growth rate γ/ω would result. By means of Fig. 3.9, one infers that at $c/u_* = 10$ this would give a reduction of a factor of 6, whereas according to Fig. 3.12 there is only a reduction of a factor of 2. It turns out that the decrease in curvature of U_0 is accompanied by an increase of the wave-induced velocity in air, χ_c, such that the aforementioned reduction is partly compensated.

It is concluded that for young windsea there is a strong two-way interaction between wind and waves. This is, on the one hand, reflected by an airflow that is rougher that would be expected from Charnock's relation for the roughness alone, and, on the other hand, a strong reduction of the growth of waves by wind. However, for old windsea the coupling between wind and waves is weak. It is emphasized that the reason for this is the choice of the wave-age dependence of the Phillips parameter. In other words, for young windsea the high-frequency waves are much steeper than for old windsea.

To conclude our discussion of results, we show the wave-age dependence of the drag coefficient for two different friction velocities in Fig. 3.13. A sensitive dependence on wave age may be noted and it implies that we may now suggest an explanation for the large scatter found in the field data for the drag coefficient as a

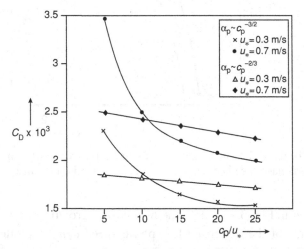

Fig. 3.13. The wave-age dependence of the drag coefficient for two different friction velocities (fro Janssen, 1989).

function of the wind speed at 10 m height. The combination of Snyder's suggestion for wave-age dependence of the Phillips parameter with quasi-linear theory implies a variability of the drag coefficient by a factor of 2, which agrees with results found by Donelan (1982) for the field data from Lake Ontario, and results given by Drennan *et al.* (1999). In other words, the scatter in the field data may be related to a sea-state dependence of the drag.

We have seen that quasi-linear theory of wind-wave generation gives a realistic description of air–sea momentum transfer since it gives a sea-state-dependent drag coefficient. Linear theory cannot account for this. It is reassuring that we can give a description of the momentum transfer that is consistent with observations. Waves play an important role in this process. It seems, therefore, to be a good idea to include this effect in applications such as storm-surge modelling, weather prediction and climate modelling. A more detailed discussion of this will follow in Chapter 5. However, it should be emphasized that theory is still weak on a number of points. We have to parametrize the effects of gravity–capillary waves, flow separation and current instabilities on the mean flow by means of a roughness length and we disregard the effects of turbulence on the wave-induced flow. Concerning the last point, Jenkins (1992) included the effects of small-scale turbulence and found very similar results for the drag coefficient to those found here by means of quasi-linear theory. Also, when allowing for a sensitive wave-age dependence of the Phillips parameter, Makin *et al.* (1995) reach similar conclusions. As already mentioned, by resolving even smaller scales, Makin *et al.* (1995) were able to study the role of gravity–capillary waves. Even more recently, Makin and Kudryatsev (2002) proposed a model of wind–wave interaction that takes the effects of flow separation into account. According to these authors flow separation plays an important role (of the order of 50 per cent) in the air–sea momentum transfer and could provide an alternative explanation of the sea-state dependence of the drag, because this process is most important for waves of around 2 to 3 times the peak frequency. Although flow separation no doubt will play a role, others (Banner, private communication, 2002) question whether this process plays such an important role. Usually it is thought that flow separation only accounts for at most 20 per cent of the drag.

Finally, it is stressed that a rather strong wave-age dependence of the Phillips parameter α_p is needed to explain within the framework of one-dimensional quasi-linear theory the sea-state dependence of the surface stress over the oceans. However, as already discussed, the highly directional character of the wavenumber spectrum combined with the directional dependence of the growth rate of waves by wind results in an additional sea-state dependence of the wave stress. Additional support for the present theoretical framework is presented in Section 3.5. It is shown that a parametrization of quasi-linear theory (using as input the measured wind speed and observed *in-situ* frequency spectra) gives surface stresses that are in good agreement with observed stresses during the HEXOS experiment.

As a last remark, it should be pointed out that according to our results a smoother airflow is found for increasing wave age. This inverse dependence of the drag coefficient on wave age has, however, a restricted range of validity. For extremely young windsea ($\chi < 5$, a situation that seldom occurs in nature, but is frequently encountered in the laboratory) the opposite trend is to be expected, as suggested by Nordeng (1991). The reason for this is that under these extreme conditions the scaling laws for the Phillips parameter (e.g. Eq. (3.144)) are not valid. In fact, for zero wave age the Phillips parameter is seen to become infinite. In its stead, a limitation of α_p is to be expected. As a result, for extremely young windseas the wave stress and therefore the drag coefficient are expected to increase with increasing wave age until the waves reach a wave age of the order of 5 to 10, after which the drag coefficient starts to decrease again. This picture is confirmed by means of results from a simulation of extremely young windsea with the energy balance equation (Komen *et al.*, 1998).

3.5 Parametrization of quasi-linear theory

For applications such as wave modelling, details of which will be discussed in Chapter 5, there is an evident need for parametrization of the source terms in the energy balance equation (2.86). This is so because, for example, the wind input term requires per grid point a considerable amount of central processing unit (CPU) time. The same remark applies to the nonlinear source term to be discussed in Chapter 4. In practice, a typical 1-day, global wave forecast should be completed in a time span of the order of a few minutes, so it should be clear that compromises have to be made regarding the functional form of the source terms in the action balance equation. We therefore discuss now a parametrization of the wind-input source term and we discuss the adequacy of this approximation.

In Section 3.4 we presented the results of the numerical solution of the momentum balance of airflow over growing surface gravity waves, summarizing a series of studies by Janssen (1982, 1989). The main conclusion was that the growth rate of the wind-generated waves depends on the friction velocity divided by the phase speed of the waves and on a number of additional factors such as wind gustiness and wave age. Although this was not discussed in detail, for light winds the growth rate may even depend on atmospheric stability (Janssen and Komen, 1985). In this section we focus on the dependence of wave growth on friction velocity and wave age, and the related dependence of the aerodynamic drag on the sea state. It is also briefly indicated how to extend the present parametrization to include the effects of gustiness.

A realistic parametrization of the interaction between wind and waves was given by Janssen (1991), and we will follow this approach closely. The starting point was the assumption (corroborated by the numerical results shown in Fig. 3.9) that even for young windsea the wind profile has a logarithmic shape, though with a roughness length that depends on the wave-induced stress. Just as in Section 3.4, we introduce a background roughness length

$$z_0 = \hat{\alpha} u_*^2/g, \quad \hat{\alpha} = \text{constant}, \tag{3.146}$$

which reflects the momentum loss by air due to processes not considered (e.g. flow separation). The constant $\hat{\alpha}$ will be fixed at a later stage. Furthermore, the effect of the (short) gravity waves on the wind profile will be modelled by a roughness length z_1. Then, the wind profile that satisfies the boundary condition $U_0(z_0) = 0$ is assumed to be given by

$$U_0(z) = \frac{u_*}{\kappa} \log\left(\frac{z + z_1}{z_0 + z_1}\right). \tag{3.147}$$

Determining the roughness length z_1 from the drag coefficient shown in Fig. 3.13 gives good agreement between Eq. (3.147) and the numerical results for the wind profile, as may be inferred from Fig. 3.9. It is remarked that Jacobs (1987) also found that the effect of waves on a turbulent airflow may be represented by an effective roughness length.

The observation that the wind profile may be described by the logarithmic profile (3.147) considerably simplifies the problem of the parametrization of the growth rate of ocean waves by wind. Following scaling arguments by Miles (1957), the growth rate then only depends on two parameters, namely,

$$x = (u_*/c)\cos(\theta - \phi) \quad \text{and} \quad \Omega_M = g\kappa^2(z_0 + z_1)/u_*^2. \tag{3.148}$$

As usual u_* denotes the friction velocity, c is the phase speed of the waves, ϕ the wind direction and θ the direction in which the waves propagate. Note that we have introduced a slight generalization by allowing ocean waves to propagate at an angle with the wind. In that event the waves only interact with the so-called 'effective component' $U_0 \cos(\theta - \phi)$. In other words, the growth of a wave propagating at an angle $\theta - \phi$ with respect to the wind can be obtained by replacing U_0 by $U_0 \cos(\theta - \phi)$ (see the discussion near Eq. (3.4), Squire's theorem). Furthermore, the profile parameter Ω_M, which apart from a constant is nothing but Charnock's parameter α_{CH}, characterizes the state of the mean flow through its dependence on the roughness length $z_0 + z_1$. Thus, through Ω_M the growth rate depends on the roughness of the airflow, which, in its turn, depends on the sea state. Following scaling arguments by Miles (1957), the growth rate of the waves by wind can then be written as

$$\gamma = \epsilon \omega \beta x^2, \tag{3.149}$$

where γ is the growth rate of the energy of the waves, ω the angular frequency, ϵ the air–water density ratio and β the so-called 'Miles parameter'. In terms of the dimensionless critical height $\mu = kz_c$ (where k is the wavenumber and z_c the critical height defined by $U_0(z = z_c) = c$) Miles obtained for a logarithmic profile in the limit of small μ

$$\beta = \frac{\beta_m}{\kappa^2} \mu \log^4(\mu), \ \mu < 1, \tag{3.150}$$

where β_m is a constant. In terms of wind and wave quantities μ is given by

$$\mu = \left(\frac{u_*}{\kappa c}\right)^2 \Omega_M \exp(\kappa/x), \tag{3.151}$$

and the wind-input source function S_{in} is given by

$$S_{in} = \gamma N \tag{3.152}$$

where γ follows from Eq. (3.149) and N is the action density spectrum.

Clearly, in order to obtain a definite answer for the growth of the waves we need to determine the roughness of the airflow. This will follow from a consideration of the stress balance of airflow over ocean waves. In the steady state, the momentum balance (Eq. (3.135)) may be integrated once with respect to height to obtain

$$\tau_w + \tau_{turb} + \tau_{visc} = \tau. \tag{3.153}$$

Here, $\tau_{visc} = \nu_a \partial U_0 / \partial z$ is the stress caused by molecular viscosity, τ_{turb} is the turbulent stress (as given in Eq. (3.140)) while the wave-induced stress equals

$$\tau_w(z) = -\int_z^\infty dz \, D_W \frac{\partial^2}{\partial z^2} U_0. \tag{3.154}$$

In Section 3.4 we have seen how τ_w depends on height. In addition, it was shown that the viscous stress only plays a minor role in the stress balance. Neglecting the viscous stress, and application of Eq. (3.153) at height $z = z_0$, gives

$$\tau_w(z_0) + \tau_{turb}(z_0) = \tau. \tag{3.155}$$

With τ_{turb} given by Eq. (3.140) and making use of the logarithmic profile in Eq. (3.147), the turbulent stress at $z = z_0$ is found to be

$$\tau_{turb}(z_0) = \tau \left(\frac{z_0}{z_0 + z_1}\right)^2. \tag{3.156}$$

The combination of Eqs. (3.155) and (3.156) then gives for $z_2 = z_0 + z_1$

$$z_2 = \frac{z_0}{\sqrt{1-x}}, \quad x = \frac{\tau_w}{\tau}, \tag{3.157}$$

where $\tau_w = \tau_w(z_0)$. For large height L, $L \gg z_1$, the drag coefficient, defined as $C_D = (u_* / U_0(L))^2$, becomes

$$C_D(z = L) = \left(\frac{\kappa}{\log(L/z_2)} \right)^2. \tag{3.158}$$

Since the total stress $\tau = C_D(L)U_0^2(L)$, the total stress in the surface layer is obtained from an iterative solution of

$$\tau = \left[\frac{\kappa U_0(L)}{\log(L/z_2)} \right]^2. \tag{3.159}$$

Therefore, for given wave stress τ_w and wind speed $U_0(L)$ we can determine the total stress τ and the roughness length z_2. What remains is a determination of the wave-induced stress τ_w.

Using conservation of momentum, the wave-induced stress $\tau_w(z = 0)$ may be related to the rate of change of wave momentum due to wind (Janssen, 1989). Here, the wave momentum is given by

$$\mathbf{P} = \mathbf{k}N \tag{3.160}$$

and thus the wave-induced stress $\tau_w(z = 0)$ is

$$\tau_w(z = 0) = \int d\omega \, d\theta \, \mathbf{k} \frac{\partial}{\partial t} N \Big|_{\text{wind}} \tag{3.161}$$

where the rate of change of the action density due to wind is given by Eq. (3.152). Note that the frequency integral extends to infinity. A practical wave model only solves the energy balance equation in a finite frequency range. However, waves with a frequency higher than the cut-off value ω_c also contribute to the stress on the airflow. A parametrization of the unresolved part of the frequency spectrum is therefore required. Because observations seem to favour an ω^{-5} power law for high frequencies (Forristal, 1981; Birch and Ewing, 1986; Banner, 1990; Hara and Karachintsev, 2003) we use this power law for the unresolved, high-frequency part of the spectrum, where the directional distribution is determined by the spectral density at the highest resolved frequency $\omega = \omega_c$. Hence, for $\omega > \omega_c$ we have

$$N(\omega, \theta) = \left(\frac{\omega_c}{\omega} \right)^5 N(\omega_c, \theta). \tag{3.162}$$

In addition, we assume that the wave stress points in the wind direction as it is mainly determined by the high-frequency waves which respond quickly to changes in the wind direction. This assumption and the simple power law of Eq. (3.162) allow one to simplify the high-frequency contribution to the wave-induced stress.

This completes the parametrization of quasi-linear theory. It is emphasized that in this parametrization a key role is played by Eq. (3.157). It shows that the roughness length is given by a Charnock relation (Charnock, 1955). Combined with Eq. (3.146) one finds

$$z_2 = \alpha u_*^2 / g. \tag{3.163}$$

However, the dimensionless Charnock parameter α is not constant but depends on the sea state through the wave-induced stress since

$$\alpha = \hat{\alpha} \Big/ \sqrt{1 - \frac{\tau_w}{\tau}}. \tag{3.164}$$

Evidently, whenever τ_w becomes of the order of the total stress in the surface layer (this happens, for example, for young windsea which may be quite steep) a considerable enhancement of the Charnock parameter is found, giving an efficient momentum transfer from air to water. The possible consequences of this sea-state-dependent momentum transfer will be discussed in Chapter 5.

This finally leaves us with the choice of two unknowns, namely $\hat{\alpha}$ from Eq. (3.146) and β_m from Eq. (3.150). The constant $\hat{\alpha}$ was chosen in such a way that for old windsea the Charnock parameter α has the value of 0.0185, in agreement with observations collected by Wu (1982), on the drag over sea waves. It should be realized, though, that the determination of $\hat{\alpha}$ is not a trivial task, as beforehand the ratio of wave-induced stress to total stress is simply not known. It requires the running of a wave model. By trial and error the constant $\hat{\alpha}$ was found to be $\hat{\alpha} = 0.01$.

The constant β_m was chosen in such a way that the growth rate γ in Eq. (3.149) is in agreement with the numerical results obtained from Miles' theory. For $\beta_m = 1.2$ and $\alpha = 0.0144$ we have shown in Fig. 3.14 the comparison between Miles' theory and Eq. (3.149). In addition, observations as compiled by Plant (1982) are shown. Realizing that the relative growth rate γ/f varies by four orders of magnitude it is concluded that there is a fair agreement between the fit of Eq. (3.149), Miles' theory and observations.

The next issue to be considered is how well the approximation of the surface stress compares with observed surface stress at sea. Fortunately, during the HEXOS experiment (Katsaros *et al.*, 1987) wind speed at 10 m height, U_{10}, surface stress τ and the one-dimensional frequency spectrum were measured simultaneously so that the parametrization of the surface stress may be verified experimentally.

A first attempt towards verification of some of the consequences of the quasi-linear theory of wind-wave generation was made by Maat *et al.* (1991) using

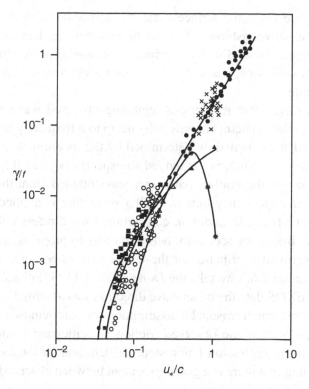

Fig. 3.14. Dimensionless growth rate γ/f as a function of u_*/c according to the observations compiled by Plant (1982). Continuous line; Miles' theory with $\alpha_{CH} = 0.0144$. ■, Eq. (3.149) with $\alpha_{CH} = 0.0144$; *, Eq. (3.149) with a 10 times larger Charnock parameter, $\alpha_{CH} = 0.144$ (from Janssen, 1991).

HEXOS data. They used measured spectra and the observed friction velocity u_* to determine the wave-age parameter $\chi = c_p/u_*$. It is only for windsea that the wave age is expected to be a good measure of the sea state. For this reason Maat *et al.* carefully selected windsea cases by only considering single-peaked spectra and by studying the appropriate weather maps to ensure that the selected cases were indeed windsea. In this fashion they were able to relate the drag coefficient C_D to the wave age. It was confirmed that, indeed, the drag of airflow over sea waves is sea-state dependent in agreement with the results from quasi-linear theory as presented in Section 3.4.3. The number of cases studied by Maat *et al.* (1991) was small, and questions on the statistical significance of the results were raised. The recent work of Oost *et al.* (2002) involved more cases by an order of magnitude and confirmed the earlier findings; see also the discussion in Section 2.6.4.

Let us now discuss a direct comparison between observations and the parametrized quasi-linear theory by using the observed wave spectrum for the determination of the wave-induced stress (Janssen, 1992). For a given observed wind speed and wave spectrum, the surface stress is obtained by solving Eq. (3.159) for the stress in an iterative fashion as the roughness length z_2 depends, in a complicated manner, on the stress. Since the surface stress was measured by means of the eddy-correlation technique, a direct comparison between observed and modelled stresses is possible.

There are, however, two restrictions regarding observed wave spectra from a wave rider. Wave-rider spectra are only reliable up to a frequency of 0.5 Hz. Since the wave-induced stress is mainly determined by the medium- to high-frequency range of the wave spectrum, we extended the spectra beyond 0.5 Hz by means of an f^{-5} tail, where the Phillips parameter was obtained from the spectral values of the last three frequency bins (a similar procedure is applied to the model spectra, see Eq. (3.162)). In addition, conventional wave riders only observe the one-dimensional frequency spectrum, hence in order to progress, assumptions regarding the directional distribution of the waves have to be made. Following the discussion in Section 2.6.3 we take the Donelan *et al.* (1985) distribution (2.130), where for the HEXOS data the mean wave direction was assumed to coincide with the local wind direction, a reasonable assumption for pure windsea cases.

In Fig. 3.15, modelled and observed friction velocities are plotted. Fitting the squares with a linear regression line a slope of 0.96 and an intercept of 0.05 are obtained indicating that there is a good agreement between observed and modelled

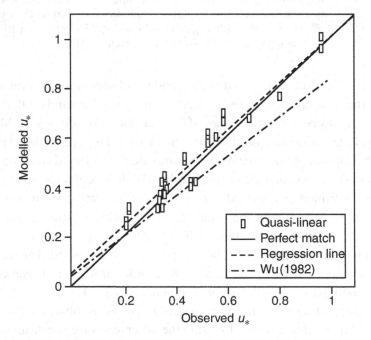

Fig. 3.15. Comparison of modelled and observed friction velocity (from Janssen, 1992).

stress. For comparison purposes we have also plotted results for friction velocity with a sea-state-independent roughness length, the Charnock relation. As suggested by Wu (1982) we took as Charnock parameter $\alpha_{CH} = 0.0185$. The resulting regression line is also given in Fig. 3.15, giving a considerable underestimation of friction velocity of about 20 per cent.

It is emphasized that Fig. 3.15 also illustrates in an elegant way the notion of a sea-state-dependent roughness. Far away from coasts and in steady circumstances a description of momentum transfer in terms of a constant Charnock parameter is likely to be appropriate. Open-ocean observations of Smith (1980) and Yelland and Taylor (1996) seem to give values for the Charnock parameter as low as 0.01. In coastal areas, such as the North Sea where HEXOS took place, or in rapidly varying circumstances on the open ocean, young windseas seem to prevail giving a much larger Charnock parameter (even up to 0.1). This large variation of the Charnock parameter is captured by quasi-linear theory in a realistic manner. It is of interest to study the consequences of the sea-state-dependent roughness on the modelling of the weather over the oceans, storm surges and the ocean circulation since all these subjects depend on an accurate description of the momentum transfer at the ocean surface. However, first we need to discuss two other important source terms of the energy balance equation, namely nonlinear transfer and dissipation of wave energy by, for example, white capping; these source terms will be examined in Chapter 4.

3.6 Summary of conclusions

In this chapter we have given an overview of the problem of wind-wave generation, with emphasis on the Miles instability. Miles' theory was extended by including wave–mean-flow interactions which results in a sea-state dependence of the momentum transfer from air to water, while the role of gusts in the process of the generation of waves by wind was also discussed. Although this point was not addressed here, Miles' theory can also be extended by including the effects of atmospheric stratification (Janssen and Komen, 1985).

We also discussed in some detail results from numerical models that include effects of small-scale turbulence. Results from these models regarding the growth rate of gravity waves agree with Miles' theory for phase speeds smaller than the wind speed, but predict in contrast to Miles' theory, wave damping when the phase speed is larger than the wind speed. As already discussed these damping rates are expected to be much smaller than the corresponding growth rates because otherwise swells from the extra-tropics would never arrive in the tropics. It was pointed out that the usual mixing-length model for turbulence may not be adequate for these long waves as the dominant eddies do not have sufficient time to transport a significant amount of momentum during a wave period. Presumably, mixing-length modelling therefore overestimates the impact of turbulence on growth and damping of the long waves. And indeed, in the context of rapid-distortion theory, Belcher and Hunt (1993) and Mastenbroek (1996) have found reduced damping rates.

Based on the above turbulence models it has been suggested that Miles' critical-layer mechanism does not play a role in the problem of wind-wave generation, even in the context of rapid distortion theory. However, the turbulent relaxation time scale may be considerably longer than suggested by the scaling arguments of Belcher and Hunt (1993), reducing the role of turbulent fluctuations. In any case, recent work with direct numerical simulations of wave growth by wind (Sullivan *et al.*, 2000) and detailed observations of Hristov *et al.* (2003) suggest that, at least for the longer waves, a critical-layer mechanism seems to operate.

The process of the generation of ocean waves by wind is an example of two-way interaction, because as soon as the waves become sufficiently steep the associated wave-induced stress gives rise to a slowing down of the airflow. Thus, this should result in a sea-state dependence of air–sea momentum transfer. The possible consequences of this sea-state dependence will be discussed in Chapter 5.

4

Nonlinear wave–wave interactions and wave dissipation

In this chapter we study the effects of nonlinearity on the evolution of deep-water gravity waves. Eventually this will result in an expression for the source function for nonlinear wave–wave interactions and dissipation (presumably by white capping), which completes the description of the energy balance equation.

We shall begin with a fairly extensive discussion of nonlinear wave–wave interactions, followed by a brief treatment of dissipation of wave energy by white capping. The latter treatment is only very schematic, however, because this process involves steep waves which only occur sporadically. At best the choice of the white-capping source function can be made plausible. It turns out that the overall dissipation rate is in agreement with observed dissipation rates. Much more is known regarding nonlinear wave–wave interactions. An important reason for this is that ocean waves may be regarded most of the time as weakly nonlinear, dispersive waves. Because of this there is a small parameter present which permits the study of the effect of nonlinearity on wave evolution by means of a perturbation expansion with starting point linear, freely propagating ocean waves. In addition, it should be pointed out that the subject of nonlinear ocean waves has conceptually much in common with nonlinear wave phenomena arising in diverse fields such as optics and plasma physics. In particular, since the beginning of the 1960s many people have contributed to a better understanding of the properties of nonlinear waves, and because of the common denominater we have seen relatively rapid progress in the field of nonlinear ocean waves.

Nonlinear waves is a vast field and it should be clear, therefore, that only a glimpse can be presented here of what has been going on in the field. Still, an attempt will be made to summarize the progress in our knowledge of nonlinear effects. In this summary a key role will be played by the Hamiltonian for water waves. Once the Hamiltonian (presented in Chapter 2) is found it is relatively straightforward to develop the theory of deep-water waves, including the effect of nonlinearity.

We start with a discussion of nonlinear effects on a single, deterministic wave train. In fact, the study of nonlinear effects on deep-water gravity waves started in the nineteenth century with the important contribution of Stokes (1847). He considered a single wave of permanent shape and was able to find the effects of finite amplitude on the dispersion relation by means of a so-called 'singular perturbation technique'. In hindsight, he may be regarded as the 'father' of the renormalization technique – nowadays so popular in particle physics – because he found the dispersion relation of finite-amplitude gravity waves by renormalization of the acceleration of gravity g. Later, Levi-Cività (1925) proved the convergence of the Stokes series solution.

In 1965, Lighthill discovered, using Whitham's variational approach, that a nonlinear, deep-water gravity-wave train is unstable to modulational perturbations, giving rise to a deeply modulated wave train, hence focussing wave energy in space and time, and as a consequence growth of sidebands in the corresponding spectrum. This instability may be regarded as a special case of a four-wave interaction process, which, as we will see, plays an important role in the physics of ocean waves. In the field of fluid dynamics this instability is nowadays known as the Benjamin–Feir instability, because Benjamin and Feir (1967) were the first to give experimental evidence of its existence. In other fields it is referred to as the modulational instability or sideband instability.

An interesting question one may then ask is what will happen to this unstable wave train. The answer is somewhat surprising and was given by Lake *et al.* (1977). It turned out that the wave train did not disintegrate but instead it recurred after a finite time (Fermi–Pasta–Ulam recurrence) as an almost uniform wave train. This feature also follows from a simplified description of wave dynamics in the limit of a narrow-band, weakly nonlinear wave train. In that event the envelope of the wave train is determined by the so-called 'nonlinear Schrödinger equation'. We will discuss some of the properties of this classical evolution equation because it arises in many different applications, although it only has a restricted validity.

Finally, to close the discussion of deterministic aspects of nonlinear waves we give results on the instability of a nonlinear wave to three-dimensional perturbations. This instability is related to a five-wave interaction process (McLean, 1982; McLean *et al.*, 1982) and there is experimental evidence for the existence of this process (Su, 1982). In the context of ocean-wave forecasting this process is usually ignored because it is of higher order in action density than four-wave interactions.

After having discussed some effects of nonlinearity on the evolution of deterministic, deep-water waves we proceed with the study of nonlinear transfer in a stochastic sea and we derive the nonlinear source function of the energy balance equation. Some consequences of nonlinear transfer are briefly discussed, in particular the role played by nonlinear interactions in shaping the high-frequency part of the spectrum and in shifting the peak of the spectrum towards lower frequencies.

Nonlinear interactions also have a pronounced impact on the probability distribution of the surface elevation; because in deep water nonlinearity results in focussing there is a tendency to have an increased probability of extreme events, signalling the increased probability of freak wave events. Finally, we also devote some attention to the relation between the statistical theory of four-wave processes and the Benjamin–Feir instability from deterministic theory.

4.1 Evolution equation for deep-water waves derived from a Hamiltonian

We recall from Chapter 2 that there are two equivalent variational formulations for the evolution of deep-water gravity waves, namely by means of a Hamiltonian (Eq. (2.7)) and by means of a Lagrangian (Eq. (2.10)). The equivalence between the two formulations was also discussed in Chapter 2. Starting from the Lagrangian formulation we obtained the basic evolution equation for waves, namely the energy balance equation (2.71) which, for a random sea, describes the rate of change of the action density as caused by advection, refraction and physical processes. In this chapter, the main goal is to derive the rate of change of the spectrum owing to nonlinear interactions. This is done in two steps. First, we obtain the deterministic evolution equation for surface gravity waves in deep water. It is called the Zakharov equation and is obtained from the Hamiltonian of water waves first found by Zakharov (1968). In the next step, equations for the moments of the probability distribution function (pdf) of the waves are obtained from the Zakharov equation. This infinite hierarchy of equations will be closed by applying the random-phase approximation to the sixth moment and the resulting system of equations is solved by means of the multiple-time-scale technique (Davidson, 1972).

Consider the potential flow of an ideal fluid of infinite depth. Coordinates are chosen in such a way that the undisturbed surface of the fluid coincides with the $x-y$ plane. The z-axis is pointed upward, and the acceleration of gravity g is pointed in the negative z-direction. Let η be the shape of the surface of the fluid, and let ϕ be the potential of the flow. Hence, the velocity of the flow follows from $\mathbf{u} = -\nabla\phi$.

By choosing as canonical variables

$$\eta, \text{ and } \psi(\mathbf{x}, t) = \phi(\mathbf{x}, z = \eta, t), \tag{4.1}$$

Zakharov (1968) showed that the total energy E of the fluid may be used as a Hamiltonian. Apart from the constant water density ρ_w the energy is

$$E = \frac{1}{2} \int d\mathbf{x} \int_{-\infty}^{\eta} dz \left((\nabla\phi)^2 + \left(\frac{\partial\phi}{\partial z}\right)^2 \right) + \frac{g}{2} \int d\mathbf{x}\, \eta^2. \tag{4.2}$$

The x-integrals extend over the total basin considered. If an infinite basin is considered, the resulting total energy is infinite, unless the wave motion is localized within a finite region. This problem may be avoided by introducing the energy per unit area by dividing Eq. (4.2) by the total surface $L \times L$, where L is the length of the basin, and taking the limit of $L \to \infty$ afterwards. As a consequence, integrals over wavenumber $\mathbf{k}$ are replaced by summations while δ-functions are replaced by Kronecker δs. For a more complete discussion see Komen *et al.* (1994). We will adopt this approach implicitly in the remainder of this text.

The boundary conditions at the surface, namely the kinematic boundary condition and Bernoulli's equation, are then equivalent to Hamilton's equations,

$$\frac{\partial \eta}{\partial t} = \frac{\delta E}{\delta \psi}, \quad \frac{\partial \psi}{\partial t} = -\frac{\delta E}{\delta \eta}, \tag{4.3}$$

where $\delta E / \delta \psi$ is the functional derivative of E with respect to ψ, etc. Inside the fluid the potential ϕ satisfies Laplace's equation,

$$\nabla^2 \phi + \frac{\partial^2 \phi}{\partial z^2} = 0$$

with boundary conditions

$$\phi(\mathbf{x}, z = \eta) = \psi$$

and

$$\frac{\partial \phi(\mathbf{x}, z)}{\partial z} \to 0, \ z \to \infty.$$

If one is able to solve the potential problem, then ϕ may be expressed in terms of the canonical variables η and ψ. Then the energy E may be evaluated in terms of the canonical variables, and the evolution in time of η and ψ follows at once from Hamilton's equations (4.3). This was done by Zakharov (1968), who obtained the deterministic evolution equations for deep-water waves by solving the potential problem in an iterative fashion for small steepness ϵ.

In order to make progress, it is most convenient to introduce the Fourier transform of ϕ, denoted by $\hat{\phi}$,

$$\phi = \int d\mathbf{k} \ \hat{\phi} \ e^{i\mathbf{k}\cdot\mathbf{x}}. \tag{4.4}$$

The potential problem for ϕ then becomes an ordinary differential equation for $\hat{\phi}$. The relevant solution which satisfies the boundary condition at $z = -\infty$ reads

$$\hat{\phi} = \hat{\phi}(t) \ e^{kz},$$

where $k = |\mathbf{k}|$. The difficult part is now to try to satisfy the boundary condition at

$z = \eta$. Explicitly we then have

$$\phi(x, z = \eta) = \psi(x, t) = \int d\mathbf{k}\, \hat{\psi}\, e^{i\mathbf{k}\cdot\mathbf{x}} \qquad (4.5)$$

and progress can only be made by means of a Taylor expansion of ϕ around $z = 0$. Thus,

$$\phi(\mathbf{x}, z = \eta) = \phi(\mathbf{x}, 0) + \eta\frac{\partial}{\partial z}\phi + \frac{1}{2}\eta^2\frac{\partial^2}{\partial z^2}\phi + \cdots = \psi \qquad (4.6)$$

and introducing the Fourier transform of η

$$\eta = \int d\mathbf{k}\, \hat{\eta}\, e^{i\mathbf{k}\cdot\mathbf{x}},$$

Eq. (4.6) is solved for ϕ by means of iteration with the result

$$\hat{\phi}_0 = \hat{\psi}_0 - \int d\mathbf{k}_{1,2}\, k_1\, \hat{\psi}_1\hat{\eta}_2\, \delta_{0-1-2}$$

$$- \int d\mathbf{k}_{1,2,3}\, k_1 D_{0,1,2,3}\, \hat{\psi}_1\hat{\eta}_2\hat{\eta}_3\, \delta_{0-1-2-3} + \cdots \qquad (4.7)$$

where we have introduced the short-hand notation $d\mathbf{k}_{1,2} = d\mathbf{k}_1 d\mathbf{k}_2$, $\hat{\psi}_1 = \hat{\psi}(\mathbf{k}_1)$, $\delta_{0-1-2} = \delta(\mathbf{k}_0 - \mathbf{k}_1 - \mathbf{k}_2)$, etc. and

$$D_{0,1,2,3} = \frac{1}{4}\,(2|\mathbf{k}_1| - |\mathbf{k}_0 - \mathbf{k}_1| - |\mathbf{k}_0 - \mathbf{k}_2| - |\mathbf{k}_1 + \mathbf{k}_2| - |\mathbf{k}_1 + \mathbf{k}_3|).$$

Equation (4.7) expresses $\hat{\phi}$ in terms of the Fourier transforms of the canonical variables η and ψ and is formally correct up to third order in amplitude (this is what is needed in order to describe four-wave processes). It should be pointed out that the solution (4.7) has a restricted validity for two reasons. First, the steepness of the dominant waves should be small. Second, there may be a restriction on the wavenumber range to which Eq. (4.7) is applied. For example, consider the case of short waves riding on a long wave. This case is not well described by the solution (4.7) because in that event the Taylor expansion around the mean sea level, as given in Eq. (4.6) may, in principle, become invalid.

A great simplification is achieved by introducing the complex variable $A(\mathbf{k})$ through

$$\hat{\eta}(\mathbf{k}) = \left(\frac{k}{2\omega}\right)^{1/2} [A(\mathbf{k}) + A^*(-\mathbf{k})],$$

$$\qquad (4.8)$$

$$\hat{\psi}(\mathbf{k}) = -i\left(\frac{\omega}{2k}\right)^{1/2} [A(\mathbf{k}) - A^*(-\mathbf{k})]$$

where ω is just given by the dispersion law for gravity waves,

$$\omega = \sqrt{gk}. \tag{4.9}$$

The advantage of the use of the action variable A will become clear in a moment. Using Eq. (4.8) and substitution of Eq. (4.7) into the energy E of the surface waves in Eq. (4.2) then gives up to fourth order in amplitude

$$E = \int d\mathbf{k}_1 \omega_1 A_1 A_1^* + \int d\mathbf{k}_{1,2,3} \delta_{1-2-3} V_{1,2,3}^{(-)} \left[A_1^* A_2 A_3 + \text{c.c.} \right]$$

$$+ \frac{1}{3} \int d\mathbf{k}_{1,2,3} \delta_{1+2+3} V_{1,2,3}^{(+)} \left[A_1 A_2 A_3 + \text{c.c.} \right]$$

$$+ \int d\mathbf{k}_{1,2,3,4} \delta_{1-2-3-4} W_{1,2,3,4}^{(1)} \left[A_1^* A_2 A_3 A_4 + \text{c.c.} \right]$$

$$+ \frac{1}{2} \int d\mathbf{k}_{1,2,3,4} \delta_{1+2-3-4} W_{1,2,3,4}^{(2)} A_1^* A_2^* A_3 A_4$$

$$+ \frac{1}{4} \int d\mathbf{k}_{1,2,3,4} \delta_{1+2+3+4} W_{1,2,3,4}^{(4)} \left[A_1^* A_2^* A_3^* A_4^* + \text{c.c.} \right]. \tag{4.10}$$

Here, V^0 and W^0 are complicated expressions in ω and $\mathbf{k}$ which are given by Krasitskii (1994). For convenience all relevant interaction coefficients are also recorded in the Appendix to this chapter.

The introduction of the action variable A has the advantage that the second-order expression for the energy,

$$E_2 = \int d\mathbf{k}_1 \omega_1 |A_1|^2 \tag{4.11}$$

assumes a very simple form, much simpler than in terms of the original variables $\hat{\eta}$ and $\hat{\psi}$. When comparing Eq. (4.11) with Eq. (2.46) it is recognized that $|A_1|^2$ is nothing but the action density. Another advantage of the introduction of the complex variable A is that the Hamilton equations (4.3) become the single equation

$$\frac{\partial A}{\partial t} = -i \frac{\delta E}{\delta A^*}. \tag{4.12}$$

In the linear approximation, i.e. using the lowest-order expression for the energy in Eq. (4.11), the evolution equation of the action variable assumes the simple form

$$\frac{\partial A}{\partial t} = -i\omega A. \tag{4.13}$$

Normally, in terms of the Fourier transform of the surface elevation, gravity waves are determined by a second-order differential equation (see Eq. (3.83)), reflecting the property that for a given wavenumber there are two modes, namely one with positive frequency and one with negative frequency. The transformation (4.8) allows one to obtain a first-order evolution equation for the action variable A which simplifies the subsequent nonlinear development considerably, but still incorporates the property that there are two modes, because A is essentially complex. For example, while because of the reality of the surface elevation its Fourier transform enjoys the property $\hat{\eta}^*(-\mathbf{k}) = \hat{\eta}(\mathbf{k})$, such a relation does not apply to the action variable A.

Evaluating now the functional derivative of the full expression for E with respect to A^* we find as evolution equation for A,

$$
\frac{\partial}{\partial t} A_1 + i\omega_1 A_1 = -i \int d\mathbf{k}_{2,3} \left\{ V_{1,2,3}^{(-)} A_2 A_3 \delta_{1-2-3} + 2 V_{2,1,3}^{(-)} A_2 A_3^* \delta_{1-2+3} \right.
$$

$$
\left. + V_{1,2,3}^{(+)} A_2^* A_3^* \delta_{1+2+3} \right\}
$$

$$
- i \int d\mathbf{k}_{2,3,4} \left\{ W_{1,2,3,4}^{(1)} A_2 A_3 A_4 \delta_{1-2-3-4} \right.
$$

$$
+ W_{1,2,3,4}^{(2)} A_2^* A_3 A_4 \delta_{1+2-3-4}
$$

$$
+ 3 W_{4,3,2,1}^{(1)} A_2^* A_3^* A_4 \delta_{1+2+3-4}
$$

$$
\left. + W_{1,2,3,4}^{(4)} A_2^* A_3^* A_4^* \delta_{1+2+3+4} \right\}. \qquad (4.14)
$$

Equation (4.14) is the basic evolution equation of weakly nonlinear gravity waves and it includes the relevant amplitude effects up to third order. Admittedly, this might not seem an easy way to obtain the fundamental evolution equation but it is the most transparent route I could find. The key is that once an approximate solution to the potential problem is found the energy may be determined to any required order. The time evolution of $A(\mathbf{k})$ then follows at once from Hamilton's equation (4.12). This elegant derivation is due to Zakharov (1968).

The nonlinear evolution equation (4.14) contains the effects of both three- and four-wave interactions. In order to see this it must be remembered that Eq. (4.14) was obtained under the assumption of small wave steepness so that in lowest order we are dealing with a linear oscillation and the right-hand side of Eq. (4.14) gives small (but nevertheless important) corrections to this linear oscillation. Consider now the quadratic terms on the right-hand side of Eq. (4.14). They oscillate with frequencies $-(\omega_2 + \omega_3)$, $-\omega_2 + \omega_3$ and $\omega_2 + \omega_3$ respectively. If they match the oscillation frequency $-\omega_1$ of the linear system, a resonant energy transfer between the modes with wavenumber $\mathbf{k}_1$, $\mathbf{k}_2$ and $\mathbf{k}_3$ is possible. In a similar vein, one of the cubic terms oscillates with frequency $\omega_2 - \omega_3 - \omega_4$ and if this frequency equals $-\omega_1$ of the linear system a resonant interaction between four modes is possible. (Note, however, that the quadratic terms may also contribute to four-wave processes, see the discussion that follows shortly.)

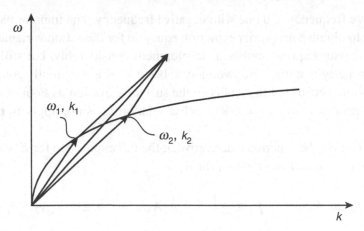

Fig. 4.1. For gravity waves resonant three-wave interactions are impossible.

To sum up, the evolution equation (4.14) allows, in principle, three- and four-wave interactions. Three-wave processes are very effective when the resonance conditions

$$\omega_1 \pm \omega_2 \pm \omega_3 = 0, \ \mathbf{k}_1 \pm \mathbf{k}_2 \pm \mathbf{k}_3 = 0 \qquad (4.15)$$

are satisfied. Similarly, four-wave processes are expected to affect the evolution of the linear waves in an appreciable manner when resonance conditions of the type

$$\omega_1 + \omega_2 = \omega_3 + \omega_4, \ \mathbf{k}_1 + \mathbf{k}_2 = \mathbf{k}_3 + \mathbf{k}_4 \qquad (4.16)$$

can be met. However, it should be emphasized that the resonance conditions of Eqs. (4.15) and (4.16) cannot always be satisfied. The occurrence of these resonant processes depends on the type of dispersion relation, $\omega = \omega(k)$. For example, for deep-water gravity waves the dispersion relation reads $\omega = \sqrt{gk}$, which as shown in Fig. 4.1 is concave. From the graphical construction, given in the figure, it is immediately concluded that resonant three-wave processes are impossible.

Phillips (1960) has shown, however, that resonant four-wave interactions are permitted by the dispersion relation for deep-water gravity waves. The possible solutions are sketched in Fig. 4.2, which has become known as 'Phillips' figure of eight'. Note, however, that gravity waves only enjoy resonant four-wave transfer for the resonance conditions, given in Eq. (4.16); for example, there are no resonant four-wave interactions of the type $\mathbf{k}_1 + \mathbf{k}_2 + \mathbf{k}_3 = \mathbf{k}_4$, or of the type $\mathbf{k}_1 + \mathbf{k}_2 + \mathbf{k}_3 + \mathbf{k}_4 = 0$.

As deep-water waves only allow resonant four-wave interactions of the type given in Eq. (4.16), a considerable simplification of the nonlinear evolution equation results. The distinction between resonant and nonresonant interactions has important consequences, namely, nonresonant interactions, resulting in, for example, bound

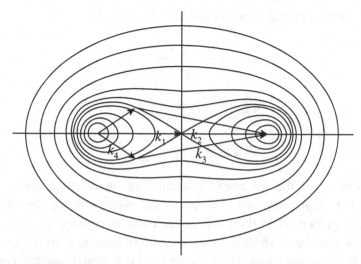

Fig. 4.2. Phillips' figure of eight: resonant four-wave interactions are possible (from Phillips, 1960).

harmonics, may be eliminated by means of a canonical transformation. Let us return now to the amplitude expansion of the wave energy given in Eq. (4.10). In fact, Krasitskii (1990, 1994) has shown that in the absence of resonant three-wave interactions there is a nonsingular, canonical transformation from the action variable A to a new variable a that allows elimination of the nonresonant third- and fourth-order contributions to the wave energy E. Note that, by definition, a canonical transformation has the property that, in terms of the new variables, the system is also Hamiltonian. The transformation $A = A(a, a^*)$ is only explicitly known in terms of an amplitude expansion. The first few terms are given by

$$
\begin{aligned}
A_1 = a_1 + \int \mathrm{d}\mathbf{k}_{2,3} &\left\{ A^{(1)}_{1,2,3} a_2 a_3 \delta_{1-2-3} + A^{(2)}_{1,2,3} a_2^* a_3 \delta_{1+2-3} \right. \\
&\left. + A^{(3)}_{1,2,3} a_2^* a_3^* \delta_{1+2+3} \right\} \\
+ \int \mathrm{d}\mathbf{k}_{2,3,4} &\left\{ B^{(1)}_{1,2,3,4} a_2 a_3 a_4 \delta_{1-2-3-4} \right. \\
&+ B^{(2)}_{1,2,3,4} a_2^* a_3 a_4 \delta_{1+2-3-4} + B^{(3)}_{1,2,3,4} a_2^* a_3^* a_4 \delta_{1+2+3-4} \\
&\left. + B^{(4)}_{1,2,3,4} a_2^* a_3^* a_4^* \delta_{1+2+3+4} \right\} + \cdots
\end{aligned}
\tag{4.17}
$$

The unknowns A^0 and B^0 are obtained by systematically removing the nonresonant third- and fourth-order contributions to the wave energy, and insisting that the form of the energy remains symmetric. These expressions are quite involved and have been given by Krasitskii (1990, 1994) for example. We only give the transfer coefficients for the quadratic terms explicitly. They read

$$A^{(1)}_{1,2,3} = -\frac{V^{(-)}_{1,2,3}}{\omega_1 - \omega_2 - \omega_3},$$

$$A^{(2)}_{1,2,3} = -2\frac{V^{(-)}_{2,1,3}}{\omega_1 + \omega_2 - \omega_3},$$

$$A^{(3)}_{1,2,3} = -\frac{V^{(+)}_{1,2,3}}{\omega_1 + \omega_2 + \omega_3}$$

and they show that in the absence of resonant three-wave interactions of the type in Eq. (4.15), the transformation $A = A(a, a^*)$ is indeed nonsingular. Nevertheless it is emphasized that when three waves are close to resonance, as may happen for shallow-water waves (because they are almost nondispersive), the canonical transformation is almost singular. This, in fact, is a signal that the perturbation expansion is not convergent, and a different approach is required which will lead to a variant of the Boussinesq equations.

Elimination of the variable A in favour of the new action variable a results in a great simplification of the wave energy E (Eq. (4.10)). It becomes

$$E = \int dk_1 \omega_1 a^*_1 a_1 + \frac{1}{2} \int dk_{1,2,3,4} T_{1,2,3,4} a^*_1 a^*_2 a_3 a_4 \delta_{1+2-3-4}, \qquad (4.18)$$

where the interaction coefficient $T_{1,2,3,4}$ is given by Krasitskii (1990, 1994) and in the Appendix to this chapter. The interaction coefficient enjoys a number of symmetry conditions, of which the most important one is $T_{1,2,3,4} = T_{3,4,1,2}$, because this condition implies that E is conserved. In terms of the new action variable a, Hamilton's equation becomes $\partial a/\partial t = \delta E/\delta a^*$, or $-i$

$$\frac{\partial a_1}{\partial t} + i\omega_1 a_1 = -i \int dk_{2,3,4} T_{1,2,3,4} a^*_2 a_3 a_4 \delta_{1+2-3-4}, \qquad (4.19)$$

which is known as the Zakharov equation. Clearly, by removing the non-resonant terms, a considerable simplification of the evolution equation describing four-wave processes has been achieved (cf. Eq. (4.14)).

It is emphasized that the transfer function $T_{1,2,3,4}$ contains two types of contributions, one directly from the cubic terms of Eq. (4.14) and a number of terms related to the interaction of the bound waves (given by the quadratic terms in the canonical transformation (4.17)), with two free waves that satisfy the dispersion relation. For this reason one may distinguish two types of resonant four-wave interactions: the first one is called a 'direct interaction' and the second one is called a 'virtual-state interaction'. Here, the direct interaction only involves interaction between free waves, while in the virtual-state interaction two free waves generate a virtual state consisting of bound waves which then decays into a different set of free waves.

In Section 4.2, we investigate a number of properties of the Zakharov equation. However, before moving on, two points must be raised. First, it is noted that once the solution to the Zakharov equation (4.19) is known for a, one still needs to apply the canonical transformation (4.17) to recover the actual action variable A. Although the difference between the two action variables is only of the order of the wave steepness, there are a number of applications where one is interested in the effects of bound waves. If one then needs to obtain the second-order corrections to the wave spectrum, knowledge of the canonical transformation up to third order in amplitude is required. This will be discussed further in Section 4.7.3 when we study the high-frequency part of the spectrum.

The second point is of a historical nature and concerns the Hamiltonian character of Eq. (4.19). As was shown by Caponi *et al.* (1982) the original Zakharov equation was not Hamiltonian; by numerical means it was found that the energy of the waves was not conserved. The reason for this is that in Zakharov (1968) a transformation was used that was not canonical. This was pointed out by Krasitskii (1990),[1] who found the proper transformation from A to a, i.e. Eq. (4.17), which up to second order in amplitude is identical to the original transformation of Zakharov (1968). However, Krasitskii found that additional terms of order a^3 are needed to preserve the Hamiltonian nature of the problem. As a result, the transfer function enjoys the symmetry

$$T_{1,2,3,4} = T_{3,4,1,2} \qquad (4.20)$$

and this condition is sufficient for conservation of the wave energy E from Eq. (4.18). Note, however, that for resonant four-wave processes (obeying the resonance conditions in Eq. (4.16)) the two forms of T are identical.

Many results regarding the stability properties of deep-water gravity waves have been obtained with the original form of the transfer function T (for a review see Yuen and Lake, 1982). Krasitskii and Kalmykov (1993) reported that for large steepness, quantitative (but no qualitative) differences between stability results were obtained with the original form of T and the symmetrical form of T. Note, however, that the use of different transformations $A = A(a, a^*)$ will affect the results for a, because a will obey different evolution equations; but, of course, the results for the original action variable A are *not* affected. In terms of A, the results reported in Yuen and Lake (1982) are, therefore, expected to be correct, even for large wave steepness. Thus, the discrepancies found by Krasitskii and Kalmykov (1993)

[1] But Zakharov was aware of this problem as well and sketched its solution in a review of the Russian translation of Yuen and Lake (1982).

are not understood. Note, however, that there are subtle differences between the two approaches. For example, stability results are usually presented as a function of the physical wave steepness values (the one based on A). Using the Zakharov transformation the steepness obtained from a is indeed identical to the physical wave steepness. But the Krasitskii transformation contains cubic terms in a, and therefore in those circumstances the steepness values based on A and a are different. This is an important difference for large values of the wave steepness.

4.2 Finite-amplitude effects on the dispersion relation and the instability of finite-amplitude deep-water waves

A finite-amplitude wave has a dispersion relation that depends on the amplitude. This was already recognized by Stokes in the nineteenth century. Although the effect of the amplitude on the frequency is quite small (for example, a steepness of 10 per cent only gives a 0.5 per cent effect on frequency), nonlinearity still has a tremendous impact on the evolution of a wave train. This follows from the work of Lighthill, Whitham, and Benjamin and Feir, who discovered that a uniform wave train may be unstable to sideband perturbations. This instability, which is called the 'Benjamin–Feir instability' (in other fields it is known as the modulational instability or sideband instability), is just an example of a four-wave interaction.

In this section and Section 4.3 we shall explore some of the consequences of nonlinearity on wave evolution in a deterministic context. This is then followed by a discussion of the effects of nonlinearity in a statistical framework, which is more appropriate for actual wave forecasting. It will be seen that the deterministic and standard, statistical approaches result in apparently conflicting conclusions. These apparent conflicts may be resolved, however, by a slightly more general statistical approach that includes both resonant and nonresonant interactions.

The nonlinear dispersion relation for deep-water gravity waves is obtained immediately from the Zakharov equation (4.19). Consider the case of a single wave, e.g.

$$a(k) = \hat{a}\, \delta(k - k_0). \tag{4.21}$$

Then, substitution of Eq. (4.21) into Eq. (4.19) gives

$$\frac{\partial}{\partial t}\hat{a} = -iT_0|\hat{a}|^2\hat{a}, \tag{4.22}$$

where it may be verified that $T_0 = T_{0,0,0,0} = k_0^3$. Equation (4.22) may be solved at once by writing

$$\hat{a} = a_0\, e^{-i\Omega t}$$

where Ω denotes the correction of the dispersion relation due to nonlinearity. It is given by

$$\Omega = T_0 |a_0|^2. \tag{4.23}$$

Therefore, the dispersion relation of a weakly nonlinear gravity wave is given by

$$\omega = \omega_0 \left(1 + \frac{1}{2}s^2\right), \quad \omega_0 = \sqrt{g|k_0|}, \tag{4.24}$$

and s is the wave steepness, defined as wavenumber times surface elevation amplitude, hence wave steepness is related to the action density variable in the following way: $s = k_0 a_0 \sqrt{2k_0/\omega_0}$. The result in Eq. (4.24) was obtained by Stokes (1847) using a singular perturbation method.

The dependence of the dispersion relation on the wave steepness will have a profound impact on the time evolution of a weakly nonlinear wave train. This will be discussed in Section 4.3, but let us first discuss the short-time behaviour of a nonlinear wave train by means of a linear stability analysis.

Therefore, let us now address the question of whether a weakly nonlinear wave train is stable or not. To test the stability of a uniform wave train we perturb it by a pair of sidebands with wavenumber $k_\pm = k_0 \pm K$ (with K the modulation wavenumber) and amplitude $A_\pm(t)$, e.g.

$$a = A_0 \delta(k - k_0) + A_+ \delta(k - k_+) + A_- \delta(k - k_-).$$

Assuming that the sideband amplitudes are small compared with the amplitude A_0 of the carrier wave and neglecting the square of small quantities, the following evolution equations for $A_\pm$ are found from the Zakharov equation (4.19),

$$i\frac{\mathrm{d}}{\mathrm{d}t}A_\pm = T_{\pm,\mp} a_0^2 A_\mp^* \exp[-i(\Delta\omega + 2T_0 a_0^2)t] + 2T_{\pm,\pm} a_0^2 A_\pm \tag{4.25}$$

where

$$T_{\pm,\pm} = T(k_0 \pm K, k_0, k_0, k_0 \pm K),$$
$$T_{\pm,\mp} = T(k_0 \pm K, k_0 \mp K, k_0, k_0),$$
$$T_0 = T(k_0, k_0, k_0, k_0),$$
$$\Delta\omega = 2\omega(k_0) - \omega(k_0 + K) - \omega(k_0 - K)$$

and a_0 is the same quantity as given in Eq. (4.23).

By means of the substitution

$$A_+ = \hat{A}_+ \exp\left[-i\left(\frac{1}{2}\Delta\omega + T_0 a_0^2\right)t - i\Omega t\right],$$
$$A_-^* = \hat{A}_-^* \exp\left[+i\left(\frac{1}{2}\Delta\omega + T_0 a_0^2\right)t - i\Omega t\right],$$

where Ω is still unknown, a set of differential equations is obtained that contains no explicit time dependence. A nontrivial solution is then found provided that Ω satisfies the dispersion relation

$$\Omega = (T_{+,+} - T_{-,-})a_0^2$$
$$\pm \left\{ -T_{+,-}T_{-,+}a_0^4 + \left[-\frac{1}{2}\Delta\omega + a_0^2(T_{+,+} + T_{-,-} - T_0) \right]^2 \right\}^{1/2}. \quad (4.26)$$

We have instability provided that the term under the square root is negative. This result, including the discussion that follows, is due to Crawford *et al.* (1981) (see also Yuen and Lake, 1982; Krasitskii and Kalmykov, 1993).

A considerable simplification of the dispersion relation is found when close sidebands are considered. In order to see this, one introduces the dimensionless perturbation wavenumber

$$\kappa = K/k_0 \quad (4.27)$$

and one considers $\kappa \ll 1$. The frequency mismatch is then approximately given by

$$\Delta\omega \simeq \frac{1}{4}\omega_0\kappa^2.$$

Expanding then the expressions for T in powers of κ and retaining only terms quadratic in the steepness s and modulation wavenumber κ, one finds

$$\Omega = \omega_0 \left(-\frac{\kappa^2}{8}s^2 + \frac{\kappa^4}{64} \right)^{1/2}. \quad (4.28)$$

This is the classical result of Benjamin and Feir (1967). Instability is found for sufficiently steep waves, or, in other words, for sufficiently long wavelength perturbations:

$$\kappa^2 \leq 8s^2. \quad (4.29)$$

Note that Lighthill (1965) only discussed very long wavelength modulations and therefore did not find the above threshold for instability.

In Fig. 4.3 the normalized growth rate $\Im(\Omega)/\frac{1}{2}\omega_0 s^2$, obtained from Eq. (4.26), is plotted as a function of the normalized sideband wavenumber $\Delta = \kappa/2s$. Here, we have used the wave steepness s as a label and the results of Benjamin and Feir correspond to the limit $s \to 0$. As can be seen from Fig. 4.3, small but finite amplitude gives considerable deviations from their classical result. For $s = 0.2$, Eq. (4.26) gives a maximum growth rate which disagrees with Eq. (4.28) by about 40 per cent. Furthermore, from Eq. (4.26) it is found that the very long waves ($\Delta \to 0$) become stable again for a steepness s which is larger than 0.39. This

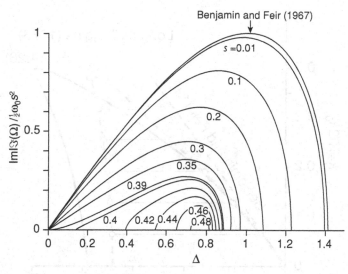

Fig. 4.3. Two-dimensional instability growth rate as a function of perturbation wavenumber for various values of wave steepness. The Benjamin–Feir result is recovered by taking the limit as wave steepness approaches zero (from Crawford *et al.*, 1981).

restabilization of the very long waves is in qualitative agreement with results from Whitham's average Lagrangian approach which yields restabilization for $s = 0.34$. The quantitative discrepancy of only 14 per cent is better than expected since the present theory is formally accurate to $\mathcal{O}(s^2)$. Finally, Fig. 4.3 shows restabilization for all modulation wavenumbers for sufficiently large steepness ($s \simeq 0.5$). This property is in qualitative agreement with numerical results of Longuet-Higgins (1978). This is shown more clearly in Fig. 4.4 where the marginal stability boundary is plotted in the κ–s plane.

The agreement between the results from the Zakharov equation and experimental results is also very encouraging. This is shown in Fig. 4.5 where normalized growth rate as a function of wave steepness is presented. The experimental results are from Benjamin (1967) and Lake *et al.* (1977). For reference, the theoretical result of Benjamin and Feir (Eq. (4.28)) is also shown.

To summarize, we have seen that a weakly nonlinear wave train is unstable to sideband perturbations. An energy transfer occurs from the basic wave to the sidebands through a four-wave interaction process. The results from the Zakharov equation compare favourably with exact computations and experiment. This theoretical approach has a surprisingly large range of validity, considering that it is formally only valid up to $\mathcal{O}(s^2)$.

The discussion so far has been confined to the case of sidebands that propagate in the same direction as the basic wave (two-dimensional modulations). The present theory is also applicable to three-dimensional modulations. Let the basic wave

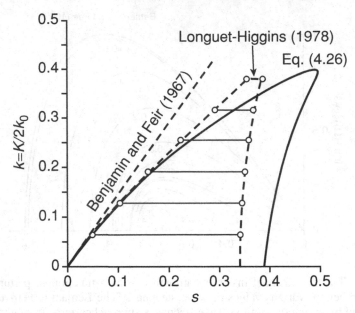

Fig. 4.4. Stability diagram for two-dimensional perturbations on a uniform wave train from the Zakharov equation and comparison with results from Longuet-Higgins (1978) (from Crawford *et al.*, 1981).

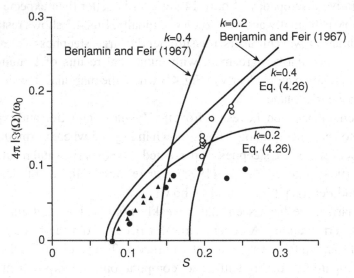

Fig. 4.5. Comparison of calculated instability growth rate with experimental results as a function of wave steepness for two values of perturbation wavenumber. $\circ$, $\kappa = 0.4$; $\bullet$, $\kappa = 0.2$ (Lake *et al.*, 1977); $\triangle$, data from Benjamin (1967) (from Crawford *et al.*, 1981).

propagate in the x-direction and let

$$\Delta_x = \kappa_x/2s, \quad \Delta_y = \kappa_y/2s,$$

where $\kappa_x = K_x/k_0$ and $\kappa_y = K_y/k_0$ are the x- and y-components of the normalized sideband wavenumber vector. Returning to Eq. (4.26) we again expand the frequency mismatch $\Delta\omega$ for very long wavelength modulation

$$\Delta\omega = \frac{1}{4}\omega_0\kappa_x^2 - \frac{1}{2}\omega_0\kappa_y^2.$$

Also, expanding T and retaining terms to $\mathcal{O}(s^2)$ and $\mathcal{O}(|\kappa|^2)$ one finds from Eq. (4.26)

$$\Omega = \left(-\frac{1}{2}\Delta\omega s^2\omega_0 + \frac{1}{4}(\Delta\omega)^2\right)^{1/2}. \tag{4.30}$$

The stability boundary is then defined by the curves $\Delta\omega = 0$ and $\Delta\omega = 2s^2\omega_0$ and they are sketched in the Δ_x–Δ_y plane in Fig. 4.6. In the long-wavelength approximation the instability region is infinite in extent and as a result one would have an energy cascade from low-modulation wavenumbers to high-modulation wavenumbers. However, the long-wavelength approximation obviously becomes invalid for large Δ_x and Δ_y. Results from the Zakharov equation (Eq. (4.26) with the full expression for $\Delta\omega$ and T) show that the instability region is finite in extent and therefore a qualitatively different picture emerges. For small steepness s, the

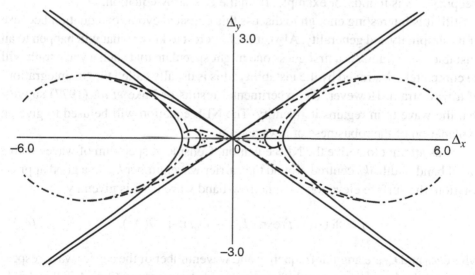

Fig. 4.6. Three-dimensional stability boundary from the Zakharov equation in strained coordinates Δ_x and Δ_y. —, $s = 0.01$; –·–, $s = 0.1$; —·—, $s = 0.4$; ····, $s = 0.48$ (from Crawford *et al.*, 1981).

instability region is very close to the Phillips figure of eight, but this should be no surprise as the figure of eight was obtained from the four-wave resonance conditions using the linear dispersion relation. For larger steepness, the wavenumbers near the outer edges of the figure of eight stabilize and the diagram resembles a pair of touching horseshoes. For even larger steepness, the longer waves also begin to stabilize and the two horseshoes separate. Just before the system stabilizes the instability is concentrated at $K = \pm 0.78\,k_0$ and is strongly two-dimensional.

4.3 Nonlinear Schrödinger equation and long-time behaviour of the Benjamin–Feir instability

According to the Zakharov equation a nonlinear wave train is unstable to sideband perturbations. It is of interest to point out that this is not just a property of gravity waves but that this instability may happen for any nonlinear dispersive wave with a nonlinear dispersion relation provided that certain conditions of nonlinear focussing are met. Starting from a general dispersion relation, a heuristic derivation is given of the evolution equation of the envelope of a narrow-band wave train: 'the nonlinear Schrödinger equation', abbreviated to the NLS equation. In the gravity-wave context the NLS equation was first derived by Zakharov (1968). This equation gives a correct description of the case of very long wavelength modulation. The linear instability results are identical to the results from Benjamin and Feir. However, the NLS equation will not give the restabilization of the sideband instability for large steepness, as is found, for example, from the Zakharov equation.

Still, it is interesting enough to discuss this classical evolution equation because of its simplicity and generality. Also, it is of interest to know what will happen to an unstable wave train. As a first guess one might speculate that such a wave train will be completely destroyed by the instability; this is usually called the 'disintegration' of a wave train. However, the experimental results of Lake *et al.* (1977) suggest that the wave train regains its identity. The NLS equation will be used to give an explanation of the robustness of a wave train.

Let us attempt to derive the NLS equation. Consider a spectrum of waves with a small band width δk_0 centred around the carrier wavenumber k_0. In a good approximation, the surface elevation for a narrow-band wave train is given by

$$\eta = \Re\left(A(x,t)\exp\mathrm{i}(k_0 x - \omega_0 t)\right) + \mathcal{O}(A^2), \qquad (4.31)$$

where ω_0 and k_0 are angular frequency and wavenumber of the carrier wave respectively, while, in agreement with the narrow-spectrum assumption, A is the slowly varying, complex envelope of the wave. For constant amplitude $|A|$ the nonlinear dispersion relation is given by

$$\omega = \omega(k, |A|^2). \tag{4.32}$$

For narrow-band, weakly nonlinear waves one may expand the dispersion relation around the carrier frequency ω_0 and carrier wavenumber k_0, and around zero amplitude. The result is

$$\omega - \omega_0 = \frac{\partial \omega}{\partial k_0}(k - k_0) + \frac{1}{2!}\frac{\partial^2 \omega}{\partial k_0^2}(k - k_0)^2 + \frac{\partial \omega}{\partial |A|^2}|A|^2 + \cdots, \tag{4.33}$$

where all partial derivatives are evaluated at $k = k_0$ and $A = 0$.

In a linear wave system there is a direct correspondence between the dispersion relation and the governing partial differential equations. This correspondence states that

$$-\mathrm{i}(\omega - \omega_0) \rightarrow \frac{\partial}{\partial t}, \quad \mathrm{i}(k - k_0) \rightarrow \frac{\partial}{\partial x}. \tag{4.34}$$

This correspondence can be generalized in a straightforward manner to weakly nonlinear systems where the nonlinear term does not depend on the modulation wavenumber $k - k_0$, e.g. as in the case of Eq. (4.33). Invoking Eq. (4.34), we obtain from Eq. (4.33) a differential operator which is to be applied to the envelope A. The result is the nonlinear Schrödinger equation

$$\mathrm{i}\left(\frac{\partial}{\partial t} + \frac{\partial \omega}{\partial k_0}\frac{\partial}{\partial x}\right)A + \frac{1}{2}\frac{\partial^2 \omega}{\partial k_0^2}\frac{\partial^2}{\partial x^2}A - \frac{\partial \omega}{\partial |A|^2}|A|^2 A = 0, \tag{4.35}$$

which governs the evolution of the envelope of a weakly nonlinear wave train in the narrow-band approximation. It is emphasized that Eq. (4.35) only holds for slowly varying envelopes, i.e.

$$\frac{\partial}{\partial t} \ll \omega_0, \quad \frac{\partial}{\partial x} \ll k_0.$$

Then, to lowest order, the envelope (and hence the wave energy) is advected by the group velocity $\partial \omega / \partial k_0$ while dispersion and nonlinearity give nontrivial corrections to the dynamics of the wave train, in particular for large times.

Clearly, this heuristic derivation suggests that the NLS equation arises in a variety of applications where nonlinear effects and dispersive waves are relevant. For this reason it has been obtained in a number of fields, notably in plasma physics, nonlinear optics and hydrodynamics. The concept of a nonlinear dispersion relation is the link between these fields. In this context it is worthwhile pointing out an important connection with Whitham's average Lagrangian method presented in Chapter 2. Whitham's approach is valid for weakly nonlinear, narrow-band wave trains (see the condition in Eq. (2.31) and the consequent mathematical development). The resulting evolution equations (2.40a)–(2.40c) may be applied immediately to the stability of a weakly nonlinear wave train and there is *always* instability when nonlinear focussing counteracts linear dispersion, independent of the modulation wavenumber (Whitham, 1974; Lighthill, 1965). Whitham (1974; pp. 522–526) has pointed out that when higher-order dispersion is allowed for, a threshold of instability arises when the modulation wavenumber is increased, in agreement with the results of Benjamin and Feir (1967), see Eq. (4.29). Therefore, if a nonlinear wave system has a Lagrangian, then Whitham's average Lagrangian method (including higher-order dispersion) results in the NLS equation (4.35).

Let us now consider the stability of a uniform wave train. For convenience we transform to a frame moving with the group velocity,

$$\tau = t, \ \xi = x - \frac{\partial \omega_0}{\partial k} \tau,$$

then Eq. (4.35) becomes

$$i\frac{\partial}{\partial \tau}A + \frac{1}{2}\omega''\frac{\partial^2}{\partial \xi^2}A - \omega_{A^2}|A|^2 A = 0, \tag{4.36}$$

where $\omega'' = \partial^2 \omega_0 / \partial k^2$ and $\omega_{A^2} = \partial \omega / \partial |A|^2$.

A special solution of Eq. (4.36) is the uniform wave train

$$A = A_0 \, e^{-i\delta\omega\tau} (A_0 \text{ and } \delta\omega \text{ are real constants}). \tag{4.37}$$

Substitution of Eq. (4.37) in Eq. (4.36) gives

$$\delta\omega = \omega_{A^2} A_0^2 \tag{4.38}$$

which represents the usual nonlinear Stokes correction to the frequency of gravity waves, hence Eq. (4.38) agrees with Eqs. (4.23) and (4.24). In order to study the stability of this solution it is most convenient to apply the transformation

$$A = \hat{A}(\xi, \tau) \exp i(-i\omega_{A^2} A_0^2 \tau) \tag{4.39}$$

with the result

$$i\frac{\partial}{\partial \tau}\hat{A} + \frac{1}{2}\omega''\frac{\partial^2}{\partial \xi^2}\hat{A} - \omega_{A^2}(|\hat{A}|^2 - A_0^2)\hat{A} = 0, \tag{4.40}$$

and the steady state is now given by $\hat{A} = A_0$. Perturbing this steady state,

$$\hat{A} = A_0 + A_1, \ A_1 \ll A_0, \tag{4.41}$$

and linearizing in A_1 we have

$$i\frac{\partial}{\partial \tau}A_1 + \frac{1}{2}\omega''\frac{\partial^2}{\partial \xi^2}A_1 - \omega_{A^2}A_0^2(A_1 + A_1^*) = 0. \tag{4.42}$$

Since A_1 is complex we write

$$A_1 = u + iv$$

to obtain for the real part of A_1

$$\frac{\partial}{\partial \tau} u + \frac{1}{2} \omega'' \frac{\partial^2}{\partial \xi^2} v = 0,$$

while the equation for the imaginary part becomes

$$\frac{\partial}{\partial \tau} v - \frac{1}{2} \omega'' \frac{\partial^2}{\partial \xi^2} u + 2\omega_{A^2} A_0^2 u = 0.$$

For normal modes ($u, v \sim \exp i(k\xi + \Omega\tau)$) one obtains

$$i\Omega u - \frac{1}{2} \omega'' k^2 v = 0,$$

$$i\Omega v + \frac{1}{2} \omega'' k^2 u + 2\omega_{A^2} A_0^2 u = 0.$$

This system has a nontrivial solution provided that Ω satisfies the dispersion relation

$$\Omega^2 = \frac{1}{4} k^2 \left[(\omega'')^2 k^2 + 4\omega'' \omega_{A^2} A_0^2 \right] \tag{4.43}$$

and instability is found when $\Omega^2 < 0$, hence a necessary condition for instability is

$$\omega'' \omega_{A^2} < 0. \tag{4.44}$$

This is indeed the case for gravity waves as $\omega_{A^2} = \omega_0 k_0^2 / 2 > 0$ while $\omega'' = -g^{1/2} k_0^{-3/2}/4 < 0$. Substitution of the expressions for ω_{A^2} and ω'' results in Eq. (4.28) which is the narrow-band limit of the dispersion relation obtained from the Zakharov equation.

Equation (4.44) shows an important result for nonlinear wave trains as we have only modulational instability provided that linear dispersion and nonlinear dispersion (ω_{A^2}) counteract each other. Note that when there is a balance between nonlinearity and dispersion, solutions of permanent shape are possible. These are called envelope solitary waves. This is very similar to the occurrence of solitary waves in the Korteweg–de Vries (KdV) equation (Whitham, 1974). These solutions play an important role in the large-time behaviour of the exact solution to the one-dimensional NLS equation for the case of vanishing boundary conditions (Zakharov and Shabat, 1972).

Let us now discuss the physical interpretation of the Benjamin–Feir instability. As already noted, this instability is an example of a four-wave interaction process. To see this in more detail, let us perturb the spectrum with peak at $k = k_0$ with two sidebands at $k_0 \pm \kappa$. Since the carrier wave has finite amplitude, there is a second harmonic with wavenumber $k_0 + k_0 = 2k_0$. Hence, there is a degenerate four-wave interaction possible between the two sidebands and the second harmonic as

$$2k_0 = k_0 + k_0 = k_0 + \kappa + k_0 - \kappa. \tag{4.45}$$

A *resonant* transfer between second harmonic and sidebands is then possible provided that the frequency mismatch $\Delta\sigma$ vanishes, where

$$\Delta\sigma = 2\sigma(k_0) - \sigma(k_0 + \kappa) - \sigma(k_0 - \kappa). \tag{4.46}$$

For definiteness, we have denoted the frequency of a wave, including nonlinear effects, by σ.

The process just described is an example of a *virtual* interaction as discussed in Section 4.1 This is, however, not the whole story as there is also a *direct* interaction possible between the carrier wave, counted twice, and the two sidebands. In the narrow-band approximation these two types of interaction, virtual and direct, have the same weight in the magnitude of the nonlinear coupling coefficient T_0 (see Eq. (4.23)).

The virtual and the direct interaction give rise to the same resonance conditions (Eqs. (4.45) and (4.46)). The frequency resonance condition, however, cannot always be satisfied. The frequency of the carrier wave is, of course, given by the well-known nonlinear dispersion relation

$$\sigma(k_0) = \omega_0 \left(1 + \frac{1}{2}s^2 \right), \quad \omega_0 = \sqrt{gk_0},$$

see Eq. (4.24). The dispersion relation of the sidebands is less easy to understand. It should be realized that the small-amplitude sidebands are riding on a finite-amplitude carrier wave. Because of the finite-amplitude carrier a current, the Stokes drift, $\mathbf{u}$ is set up. Thus, the frequency of the sidebands will get a Doppler shift equal to $(k_0 \pm \kappa)\mathbf{u}$, where $\mathbf{u} = cs^2$ and $c = \omega_0/k_0$. Therefore,

$$\sigma(k \pm \kappa) = \omega_0(k_0 \pm \kappa) + (k_0 \pm \kappa)\omega_0(k_0)s^2/k_0.$$

This result is in agreement with the evolution equation for one sideband, Eq. (4.25), in the absence of the other. Expanding the dispersion relation to second order in the modulation wavenumber κ, the frequency mismatch becomes

$$\Delta\sigma = -s^2\omega_0 - \kappa^2\omega_0''.$$

For gravity waves, the frequency mismatch vanishes for the wavenumber given by the condition $\kappa^2 = -s^2\omega_0/\omega_0''$, or using the dispersion relation,

$$\left(\frac{k}{k_0} \right)^2 = 4s^2, \tag{4.47}$$

and for this wavenumber the growth rate of the Benjamin–Feir instability attains a maximum (see Fig. 4.3 and Eq. (4.28)). This makes sense as the energy transfer is the most efficient for perfect matching of the frequencies of the waves involved. A frequency mismatch obviously reduces the efficiency of the energy transfer, until the mismatch becomes so large that no energy transfer is possible. Referring again to Fig. 4.3 this seems to occur for a dimensionless modulation wavenumber $\Delta = \kappa/(2s) = \sqrt{2}$.

We have seen that a nonlinear gravity wave is unstable to sideband perturbations. It is of interest to speculate about the subsequent evolution in time of this instability. In the past, several proposals have been made regarding the end-state of the Benjamin–Feir instability, ranging from a complete disintegration of the wave train to an end-state involving cnoidal waves (the periodic equivalent of envelope solitary waves) (see, for example, Benjamin, 1967; Hasselmann, 1967; Hasimoto and Ono, 1972). It should be noted that the proposal of an end-state consisting of envelope cnoidal waves is certainly not a wild guess. That is, Zakharov and Shabat (1972) applied the inverse scattering technique to the initial-value problem of the NLS equation where the initial state was of compact support. They found that the end-state usually consisted of a number of envelope solitons. Whitham (1974) then conjectured that the Benjamin–Feir instability (an initial-value problem with periodic boundary conditions) would have cnoidal waves as its end-state.

Surprisingly, none of the above-mentioned end-states were found in the experiment of Lake *et al.* (1977) who studied the long-time behaviour of the Benjamin–Feir instability. This experiment was performed in a wave tank with a programmable wave maker so that control of the size of the sideband perturbations on the carrier wave was possible. The resulting evolution of the wave train as a function of fetch (i.e. the distance from the wave maker) is displayed in Fig. 4.7. It is found that after the wave train reaches a strongly modulated state (which indeed resembles cnoidal envelope waves) the wave train nearly returns to its initial state, although there is clear evidence of a frequency downshift because the frequency of the wave train at the end of the wave tank is smaller than at the beginning. In this particular case the frequency downshift was thought to be caused by wave breaking. Recently, Tulin and Waseda (1999) investigated in the laboratory the relationship between the occurrence of wave breaking and the frequency downshift. In a clean experimental set-up they also showed that, in the absence of breaking, no downshifting occurred. In other words, in the absence of dissipation, the long-time evolution of the unstable wave train would show recurrence; this is known as the Fermi–Pasta–Ulam recurrence. Fermi *et al.* (1955) showed its existence when studying oscillations of an anharmonic lattice.

Indeed, numerical solution of the NLS equation showed a perfect recurrence of the Benjamin–Feir instability. This recurrence was also found by means of an approximate solution of the NLS equation in the limit of small growth rate. The Benjamin–Feir instability has, as shown in Fig. 4.8, a threshold for instability.

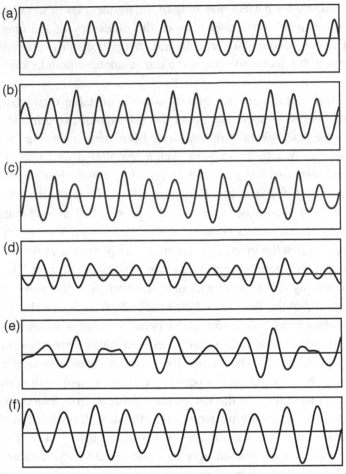

Fig. 4.7. Example of the long-time evolution of an initially uniform nonlinear wave train, showing onset of modulational instability and subsequent demodulations: (a) $x = 5$ ft (1.5 m); (b) $x = 10$ ft (3.0 m); (c) $x = 15$ ft (4.5 m); (d) $x = 20$ ft (6 m); (e) $x = 25$ ft (7.5 m); (f) $x = 30$ ft (9 m). Note that this example does not show full recurrence because of frequency downshifting (redrawn from Lake *et al.*, 1977).

Therefore, the magnitude of the growth rate may be controlled by an appropriate choice of the modulation wavenumber. In addition, if the distance Δ to the threshold is not too large, the higher harmonics are in the stable region, which simplifies the problem considerably. Assuming $\Delta \ll 1$, application of the method of multiple time scales results in the following evolution equation for the amplitude Γ of the unstable sideband (Janssen, 1981):

$$\frac{d^2}{dt^2}\Gamma = \gamma^2\Gamma - \beta^2|\Gamma|^2\Gamma, \qquad (4.48)$$

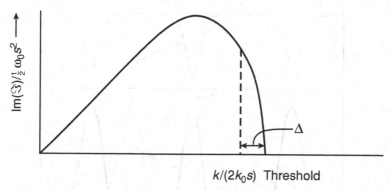

$k/(2k_0s)$ Threshold

Fig. 4.8. Growth rate of Benjamin–Feir instability and the definition of the distance Δ to the threshold for instability.

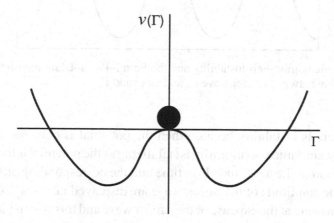

Fig. 4.9. The potential $\mathcal{V}(\Gamma)$ as function of the amplitude $|\Gamma|$.

where γ is the growth rate according to linear theory and β measures the strength of nonlinearity. Equation (4.48) has periodic solutions in time since the evolution equation is just equivalent to the motion of a particle in a potential well $\mathcal{V}$, displayed in Fig. 4.9. In order to see this equivalence, multiply Eq. (4.48) by $d\Gamma^*/dt$ and add the complex conjugate to the result. Integration with to time gives the conservation law

$$\frac{1}{2}\left|\frac{d\Gamma}{dt}\right|^2 + \mathcal{V}(\Gamma) = \text{constant}$$

where the potential is given by

$$\mathcal{V}(\Gamma) = -\frac{\gamma^2}{2}|\Gamma|^2 + \frac{\beta}{4}|\Gamma|^4.$$

It is evident from Fig. 4.9 and the expression for the potential $\mathcal{V}$ that for small

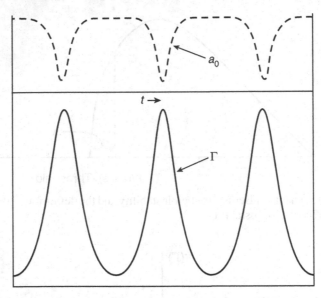

Fig. 4.10. Benjamin–Feir instability and the Fermi–Pasta–Ulam recurrence. Note the interplay between carrier wave a_0 and sideband Γ.

amplitude there is instability because then the potential is concave. On the other hand, for large amplitude nonlinearity is stabilizing as the potential is locally convex.

The evolution of the amplitude Γ in time and the corresponding effect of the instability on the amplitude of the carrier wave are displayed in Fig. 4.10. Clearly, the sideband is growing at the expense of the carrier wave and this is at the same time the main reason for quenching the energy transfer from the carrier wave to the sideband. To be more precise, while the amplitude of the carrier wave decreases, the threshold for instability will move to lower values of the dimensionless modulation wavenumber until the sideband enters the stable regime. In other words, the unstable sideband extracts energy from the carrier wave, thereby modifying the nonlinear wave in such a way that it becomes stable to the very same sideband perturbation. Finally, in Fig. 4.11 a comparison is given between the approximate solution of the NLS equation and the numerical result. It shows the maximum of the modulation amplitude as a function of the distance Δ to the threshold. For small Δ the agreement is good. The numerical result was obtained by means of a straightforward numerical solver of the NLS equation. To that end a Fourier expansion of the envelope A is performed, resulting in a coupled set of ordinary differential equations for the Fourier amplitudes. The set of ordinary differential equations is solved with a Runge–Kutta method with variable time step. For a similar approach see Yuen and Ferguson (1978).

The existence of the Fermi–Pasta–Ulam recurrence illustrates the fact that nonlinear systems may have a long memory of the initial condition. For periodic boundary

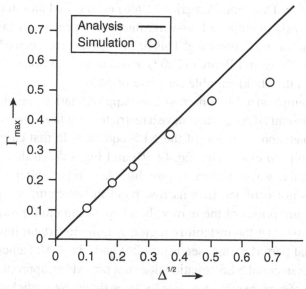

Fig. 4.11. Modulation depth as a function of $\Delta^{1/2}$ for analytical and numerical solution of the NLS equation.

conditions, the one-dimensional NLS equation (4.36) provides us with a good example of this. The NLS equation is an integrable system and there is even Fermi–Pasta–Ulam recurrence when two or more modes are unstable (Osborne *et al.*, 2000). However, it should be remarked that the simple picture of the Fermi–Pasta–Ulam recurrence is, probably, not universally valid. For more-complicated evolution equations analytical approaches fail and one has to rely on computer simulations. The first numerical solutions of the Zakharov equation were given by Caponi *et al.* (1982) (see also Krasitskii and Kalmykov, 1993), but for a rather coarse spatial resolution. No recurrence was found and, in fact, the solution appears to be rather chaotic.

In this context the interesting work of Dold and Peregrine (1986) should be mentioned. In the case of deep modulation the amplitude of the carrier wave may become so large that its steepness locally exceeds the maximum steepness of gravity waves, signalling the onset of wave breaking. Such a nonlinear process cannot be described in the context of the Zakharov equation, which follows from a weakly nonlinear expansion. Therefore, Dold and Peregrine (1986) numerically solved the problem for potential flow including the nonlinear boundary conditions (Eqs. (2.5) and (2.6)) for one-dimensional propagation. When there is Benjamin–Feir instability (see Eq. (4.29) or (4.44)) these authors found that for short times the unstable sidebands evolved according to theory. However, for large times the modulation may become so deep that the carrier wave breaks. Clearly, perfect recurrence is then no longer possible. Dold and Peregrine (1986) established an interesting connection between a weakly nonlinear four-wave interaction process and a truly nonlinear phenomenon such as wave breaking. This line of thought has been taken up recently by, for example, Song and Banner (2002) and Banner and Song (2002) who both inferred a robust threshold variable for wave breaking.

Finally, it is emphasized that narrow-band approximations resulting in the one- and two-dimensional NLS equation have a restricted validity. This is clearly evident for the two-dimensional version of the NLS equation. In that case the instability region is not finite in extent (see Eq. (4.30) and Fig. 4.6) so that energy leakage to high-modulation wavenumbers is possible. As a result, after a finite time the wave energy is not confined to a narrow band in wavenumber space, therefore violating the assumptions of the narrow-band approximation. From the Zakharov equation we know that the instability region is finite in extent, however, even for two-dimensional perturbations (see Eq. (4.26) and Fig. 4.4). Hence, this suggests that the two-dimensional NLS equation does not provide an appropriate description of the envelope of surface gravity waves for large times. Nevertheless, studies of the properties of the NLS equation have been vital in understanding nonlinear wave– wave interactions and in understanding the conditions under which freak waves, for example, may occur. For a more complete discussion on the issue of freak waves and nonlinear focussing, consult Trulsen and Dysthe (1997), Osborne *et al.* (2000) and Trulsen and Stansberg (2001).

4.4 Beyond the Zakharov equation: five-wave interactions

We have discussed in some detail four-wave interactions of deep-water gravity waves. One may wonder whether higher-order resonant interactions are possible, and whether they may be observed for nonlinear gravity waves. The answer to both these questions is in the affirmative. In the 1980s there was much interest in five-wave interactions. The resonance conditions are given by

$$\mathbf{k}_1 + \mathbf{k}_2 = \mathbf{k}_3 + \mathbf{k}_4 + \mathbf{k}_5, \quad \omega_1 + \omega_2 = \omega_3 + \omega_4 + \omega_5. \qquad (4.49)$$

A description of this interaction cannot be given in the framework of the Zakharov equation (4.19) because fourth-order terms need to be included. This means one has to go back to the Hamiltonian of water waves and one has to derive an approximate Hamiltonian correct to fifth order in amplitude. This has been accomplished by Krasitskii (1994), but details will not be discussed here because they are too involved. Stiassnie and Shemer (1984) also obtained a fourth-order evolution equation and studied five-wave interactions in detail (but their evolution equation was not in Hamiltonian form, for the same reason as for the original Zakharov equation). Five-wave interactions give rise to an interesting surface pattern, resembling a horseshoe. An interesting study on this subject was reported by Annenkov and Shrira (1999).

Earlier, Longuet-Higgins (1978), McLean (1982) and McLean *et al.* (1982) discussed the five-wave interaction process using numerical computations of the exact water-wave equations. The surface displacement of a finite-amplitude wave train of wavelength $2\pi/k$ was represented as a Stokes expansion

$$\eta = \sum_{n=0}^{\infty} A_n \cos n(x - ct),$$

with normalized x, t coordinates, with known Fourier coefficients A_n and with phase speed c which depends on wave steepness $s = ka$. This nonlinear wave is perturbed by an infinitesimal three-dimensional disturbance

$$\eta' = \exp i\,[p(x - ct) + qy - \Omega t] \sum_{n=-\infty}^{\infty} a_n \exp i n(x - ct) + \text{c.c.},$$

where $\Omega = \Omega(p, q, s)$ is an eigenvalue to be found. Instability arises when $\Im(\Omega) \neq 0$, with roots occurring in complex-conjugate pairs, because the system is Hamiltonian.

McLean's computations were accomplished by truncating the expansions at high order. He found two distinct instability regions in the p–q plane for various values of steepness s. Examples are given in Fig. 4.12.

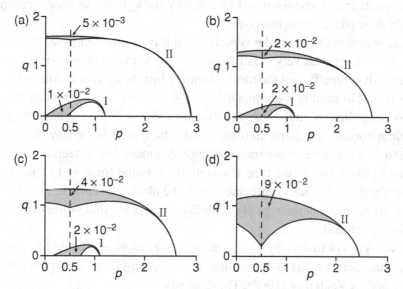

Fig. 4.12. Type I and type II instability of finite-amplitude gravity waves in deep water (from McLean *et al.*, 1982) for increasing values of steepness. Panels (a)–(d) correspond to $s = 0.064\pi$, $s = 0.095\pi$, $s = 0.111\pi$ and $s = 0.127\pi$ respectively (from McLean, 1982).

The type I instability which is found near the origin reduces to that for the Benjamin–Feir instability of the two-dimensional NLS equation when p, q and s are sufficiently small. For larger p and q values but with steepness still small it coincides with Phillips' figure of eight (note that only one-quarter of the figure of eight is shown). For larger steepness the type I instability disappears, in agreement with the results found with the Zakharov equation in Section 4.2.

The outer region of the type II instability may, for small steepness, be interpreted as a degenerate five-wave resonant interaction (Eq. (4.49)) where, in normalized form

$$\mathbf{k}_3 = \mathbf{k}_4 = \mathbf{k}_5 = (1, 0), \quad \omega_3 = \omega_4 = \omega_5 = c = 1$$

represent the fundamental wave train while

$$\mathbf{k}_1 = (1 + p, q), \quad \mathbf{k}_2 = (2 - p, -q),$$

represent the perturbation. Using the linear dispersion relation (with $g = 1$)

$$\omega = g^{1/2}(k^2 + l^2)^{1/4}, \quad \mathbf{k} = (k, l),$$

and the frequency-matching condition of Eq. (4.49) the result is

$$[(p + 1)^2 + q^2]^{1/4} + [(2 - p)^2 + q^2]^{1/4} = 3. \tag{4.50}$$

The line labelled 'II' in Fig. 4.12a corresponds to Eq. (4.50). The type II instability has maximum growth rate for $p = 1/2$, when $\mathbf{k}_1$ and $\mathbf{k}_2$ have the same k-component, $k = 3/2$; their phase speed then equals c.

Experimental evidence of the type II instability is reported by Melville (1982) and Su (1982). At relatively small slopes, $s < 0.3$, the Benjamin–Feir instability is stronger than type II, and the most dominant instability is one-dimensional. But oblique type II instability is dominant for $0.3 < s < 0.44$. The latter instability type leads to three-dimensional waves as shown in Fig. 4.13, from Su (1982), which resemble a horseshoe pattern. In other words, the type II instability dominates the behaviour of a nonlinear wave train for large steepness. For a steepness of 0.33 the maximum growth rate of the type II instability is found for $p = 1/2$ and $q = 1.2$. As these length scales are in agreement with the observed scales in Fig. 4.13, there is reason to believe that the type II instability triggers the three-dimensional waves found in experiment.

The two types of instability discussed so far are probably the first two members of an infinite class of higher-order interactions among three distinct wave modes, as suggested by Zakharov (1968). These satisfy

$$\mathbf{k}_1 + \mathbf{k}_2 = N\mathbf{k}_0, \quad \omega_1 + \omega_2 = N\omega_0$$

where $\mathbf{k}_0 = (1, 0)$, $\omega_0 = 1$ denotes the fundamental wave train and $N > 1$. Since

Fig. 4.13. Three-dimensional wave configuration resulting from oblique type II instability, for $s = 0.33$ (from Su, 1982).

the characteristic growth rates of the $\mathbf{k}_1$ and the $\mathbf{k}_2$ perturbation are $\mathcal{O}(s^N)$, these higher-order resonances are only expected to be relevant for very steep waves. It may very well be that they play a role in the wave breaking process. On the other hand, it should be emphasized that for small wave steepness ($s < 0.25$) the four-wave interaction process is most dominant.

4.5 Statistical approach to nonlinear interactions

In the previous sections we have discussed in some detail the effects of nonlinearity on the evolution of deep-water waves. The key result was that a nonlinear gravity wave is unstable to sideband perturbations, in agreement with experimental results from the laboratory. It was emphasized that this instability was a special case of a four-wave interaction process. Clearly, thus far we have been engaged with a deterministic description of the water surface.

Let us ask ourselves now how our insights may be used in a practical application, such as wave prediction on a global scale. In that event, as already discussed in Chapter 2, one is interested in a statistical description of the sea surface, which means one is concerned with the evolution of the energy of an ensemble of waves. Although for extreme events, such as occur in the presence of freak waves, there is a need for information on the phases of the waves it is noted that the prediction of the phase of the individual waves is a hopeless venture. First, we have no observations of the initial phases of the waves, and, second, long-time integrations of the Zakharov equation exhibit features of chaotic behaviour (Annenkov and Shrira, 2001). In other words, at finite time the phases of the waves will show a sensitive dependence on the initial conditions, and therefore in practice they are not predictable.

Therefore, at best one can hope to predict average quantities such as the second moment

$$\langle a_1 a_2^* \rangle, \tag{4.51}$$

where the angle brackets denote an ensemble average. In most practical applications this turns out to be sufficient.

Here, we sketch the derivation of the evolution equation for the second moment from the Zakharov equation (Eq. (4.19)), assuming a zero mean value of the amplitude a_1, $\langle a_1 \rangle = 0$. It is known, however, that because of nonlinearity, the evolution of the second moment is determined by the fourth moment, and so on, resulting in an infinite hierarchy of equations (Davidson, 1972). The question then is how to obtain a meaningful truncation of this hierarchy. Historically, this truncation of the hierarchy is obtained by making two assumptions, namely that the ensemble of waves is spatially homogeneous and stationary, and that the probability distribution for the complex amplitude of the waves a_1 is *close* to a *Gaussian*. Later, it will be discussed that the assumption of spatial homogeneity is not really necessary (Alber, 1978) but for sufficiently broad spectra it turns out to be a valid assumption.

Let us discuss these assumptions in more detail first. A wave field is considered to be homogeneous if the two-point correlation function $\langle \eta(\mathbf{x}_1)\eta(\mathbf{x}_2) \rangle$ depends only on the distance $\mathbf{x}_1 - \mathbf{x}_2$. Using the expression for the surface elevation, Eq. (4.8), it is then straightforward to verify that a wave field is homogeneous provided that the second moment satisfies

$$\langle a_i a_j^* \rangle = N_i \delta(\mathbf{k}_i - \mathbf{k}_j), \tag{4.52}$$

where N_i is the spectral action density, which is equivalent to a number density because $\omega_i N_i$ is the spectral energy density, while $\mathbf{k}_i N_i$ is the spectral momentum density (apart from a factor ρ_w).

The complex amplitude a_1 is regarded as a stochastic variable with a certain probability distribution. One of the main problems is now to determine the pdf. It is common practice (see Cook, 1974) to introduce the characteristic functional of the probability distribution. It is defined as

$$G(\{\mu_k\}, t) = \langle \exp i \sum \mu_k a_k \rangle \tag{4.53}$$

where $\{\mu_k\}$ denotes all the variables μ_k. G contains all the statistical information, e.g. the moments of the distribution function are related to derivatives of G with respect to μ_k. Hence,

$$\left.\frac{\partial G}{\partial \mu_k}\right|_{\{\mu_k=0\}} = i\langle a_k \rangle = 0, \qquad \left.\frac{\partial^2 G}{\partial \mu_k \partial \mu_l}\right|_0 = -\langle a_k a_j \rangle, \dots \qquad (4.54)$$

Therefore, the moments are related to the coefficients of the Taylor expansion of G about the origin. This expansion is, however, not very useful because it does not bring out the significance of a special characteristic function – that of a Gaussian distribution – which is expected to be of great importance, as explained shortly. Here, the Gaussian characteristic function is given by

$$G_0 = \exp -\frac{1}{2}\sum \mu_k \mu_j B_{k,j} \qquad (4.55)$$

and by making an expansion around G_0 the so-called 'cumulants' of the distribution function are introduced. These are the coefficients of the Taylor expansion of the logarithm of G,

$$G = \exp\left[i\sum \mu_k A_k - \frac{1}{2!}\sum \mu_k \mu_j B_{k,j} - i\frac{1}{3!}\sum \mu_k \mu_j \mu_l C_{k,j,l}\right.$$
$$\left. + \frac{1}{4!}\sum \mu_k \mu_j \mu_l \mu_m D_{k,j,l,m} + \cdots\right]. \qquad (4.56)$$

By differentiation of G the following relation between the moments and the cumulants is found:

$$\begin{aligned}
\langle a_k \rangle &= A_k = 0, \\
\langle a_k a_j \rangle &= B_{k,j}, \\
\langle a_k a_j a_l \rangle &= C_{k,j,l}, \\
\langle a_k a_j a_l a_m \rangle &= B_{k,j} B_{l,m} + B_{k,l} B_{j,m} + B_{k,m} B_{j,l} + D_{k,j,l,m}, \text{ etc.}
\end{aligned} \qquad (4.57)$$

Evidently, for a Gaussian distribution with zero mean all cumulants except the second-order one, $B_{k,j}$, vanish. The finiteness of the other cumulants is therefore a measure for the deviation from normality. From Eq. (4.57) it is clear, however, that for a Gaussian the moments do not vanish.

In the statistical theory of the evolution of a random wave field the Gaussian distribution plays a central role. Mathematically, this follows from the central limit theorem, which tells us that if the amplitudes have random and independent phase then the probability distribution is Gaussian (for this reason one frequently uses the term 'random-phase approximation'). If the waves are noninteracting then the phases remain uncorrelated. However, nonlinear interactions tend to create correlations (higher cumulants) but if the waves have a small steepness this effect is expected to be weak. Hence, for weakly nonlinear waves one may expect that the wave field is near normality so that the rate of change of the wave spectrum (or second moment) is expected to be small; for large times this small effect may have, however, significant consequences regarding the evolution of the wave spectrum.

Let us now sketch the derivation of the evolution equation for the second moment $\langle a_i a_j^* \rangle$ from the Zakharov equation (4.19). To that end, we multiply Eq. (4.19) for a_i by a_j^*, add the complex conjugate with i and j interchanged, and take the ensemble average:

$$\left[\frac{\partial}{\partial t} + i(\omega_i - \omega_j) \right] \langle a_i a_j^* \rangle$$

$$= -i \int d\mathbf{k}_{2,3,4} [T_{i,2,3,4} \langle a_j^* a_2^* a_3 a_4 \rangle \delta_{i+2-3-4} - \text{c.c.} \ (i \leftrightarrow j)], \quad (4.58)$$

where c.c. denotes complex conjugate, and $i \leftrightarrow j$ denotes the operation of interchanging indices i and j in the previous term. Because of nonlinearity the equation for the second moment involves the fourth moment. Similarly, the equation for the fourth moment involves the sixth moment. It becomes

$$\left[\frac{\partial}{\partial t} + i(\omega_i + \omega_j - \omega_k - \omega_l) \right] \langle a_i a_j a_k^* a_l^* \rangle$$

$$= -i \int d\mathbf{k}_{2,3,4} [T_{i,2,3,4} \langle a_2^* a_k^* a_l^* a_3 a_4 a_j \rangle \delta_{i+2-3-4} + (i \leftrightarrow j)]$$

$$+ i \int d\mathbf{k}_{2,3,4} [T_{k,2,3,4} \langle a_3^* a_4^* a_l^* a_2 a_i a_j \rangle \delta_{k+2-3-4} + (k \leftrightarrow l)]. \quad (4.59)$$

We see from this that an infinite hierarchy of equations is found, known as the 'BBGKY hierarchy' (after Bogoliuboff, Born, Green, Kirkwood and Yvon). To close this hierarchy we will use the assumptions of a near-Gaussian, homogeneous wave field.

Specifically, we proceed as follows:

(1) For weakly nonlinear waves we assume that the cumulants, denoted from now on by g_n ($n \geq 2$), obey the following ordering:

$$g_n = \mathcal{O}(\epsilon^{n-1}), \quad (4.60)$$

where ϵ is a small parameter, reflecting our assumption that we are dealing with the case of weakly nonlinear waves. Here, in the spirit of Eq. (4.57), we write

$$g_2(i, j) = \langle a_i a_j^* \rangle, \quad (4.61)$$

$$\langle a_i a_j a_k^* a_l^* \rangle = g_2(i, k) g_2(j, l) + g_2(i, l) g_2(j, k) + g_4(i, j, k, l),$$

and (Crawford *et al.*, 1980),

$$\langle a_i a_j a_k a_l^* a_m^* a_n^* \rangle = 2 g_2(i, l) g_2(j, m) g_2(k, n) + 2 g_2(i, m) g_2(j, n) g_2(k, l)$$

$$+ 2 g_2(i, m) g_2(j, l) g_2(k, n) + R_6(i, j, k, l, m, n).$$

Here, the remainder term R_6 contains the sixth cumulant g_6 and products of the fourth and the second cumulant and, therefore, applying the ordering in Eq. (4.60) the remainder term is $\mathcal{O}(\epsilon^4)$ which is small compared with the terms involving the second cumulant. Hence, the ordering in Eq. (4.60) implies that the wave field is close to the Gaussian state. In addition, g_2 varies slowly in time for a homogeneous sea, according to Eq. (4.58).

(2) At first sight one would think that a meaningful truncation of this BBGKY hierarchy may be achieved by neglecting the fourth cumulant in the evolution equation (4.58) for g_2. The combination with homogeneity of the wave field results, however, in constancy of g_2 so that one needs to go to higher order, i.e. g_4 needs to be determined. It turns out then that neglecting the sixth cumulant is a meaningful closure hypothesis.

(3) As already noted, for homogeneous sea g_2 is a slowly varying function of time. Apparently, there are several time scales in the problem as, for example, g_4 does vary on a faster time scale. It is therefore assumed that the cumulants depend on many time scales, e.g.

$$g_2 = g_2(\tau_0, \tau_1, \tau_2, \ldots), \quad \tau_n = \epsilon^n t,$$

so that

$$\frac{\partial}{\partial t} g_2 = \frac{\partial}{\partial \tau_0} g_2 + \epsilon \frac{\partial}{\partial \tau_1} g_2 + \epsilon^2 \frac{\partial}{\partial \tau_2} g_2 + \cdots \tag{4.62}$$

Likewise, the cumulants are expanded themselves in the small parameter ϵ, i.e.

$$g_2 = \epsilon g_2^{(1)} + \epsilon^2 g_2^{(2)} + \cdots,$$
$$g_4 = \epsilon^3 g_4^{(3)} + \epsilon^4 g_4^{(4)} + \cdots \tag{4.63}$$

Therefore, we obtain an approximate solution from the BBGKY hierarchy of Eqs. (4.58) and (4.59), utilizing the assumptions of near-Gaussianity and homogeneity and using the multiple-time-scale method.

Intermezzo Let us illustrate the usefulness of the multiple-time-scale method by applying it to a simple problem, namely the solution of the ordinary differential equation

$$\frac{d^2 x}{dt^2} + 2\epsilon \frac{dx}{dt} + x = 0, \quad \epsilon \ll 1,$$

which models a slightly damped oscillation. The exact solution is

$$x = a\, e^{-\epsilon t} \sin\left[t(1 - \epsilon^2)^{1/2} + \phi\right],$$

where amplitude a and phase ϕ are constants.

Using a naive perturbation technique,

$$x = x_0 + \epsilon x_1 + \epsilon^2 x_2 + \cdots,$$

one finds

$$x = a\left[\sin(t + \phi) - \epsilon t\, \sin(t + \phi) + \mathcal{O}(\epsilon t)^2\right],$$

which is an inadequate approximation of the exact solution for large times. The second term is called a 'secular' term.

A uniform valid solution, valid up to $t = \mathcal{O}(1/\epsilon)$, may be obtained with the multiple-time-scale method. As may be inferred from the exact solution and from the differential equation there are several time scales in the problem. We therefore write

$$x = x(\tau_0, \tau_1, \ldots), \quad \tau_0 = t, \quad \tau_1 = \epsilon t, \ldots,$$

where the time scales are considered to be independent. Thus

$$\frac{\mathrm{d}}{\mathrm{d}t} x = \frac{\partial x}{\partial \tau_0} \frac{\mathrm{d}\tau_0}{\mathrm{d}t} + \frac{\partial x}{\partial \tau_1} \frac{\mathrm{d}\tau_1}{\mathrm{d}t} + \cdots = \left(\frac{\partial}{\partial \tau_0} + \epsilon \frac{\partial}{\partial \tau_1} + \cdots \right) x.$$

Likewise,

$$\frac{\mathrm{d}^2}{\mathrm{d}t^2} = \frac{\partial^2}{\partial \tau_0^2} + 2\epsilon \frac{\partial^2}{\partial \tau_0 \partial \tau_1} + \mathcal{O}(\epsilon^2).$$

In addition, we expand x in a power series of ϵ,

$$x = x_0 + \epsilon x_1 + \cdots,$$

to obtain

$$\left[\frac{\partial^2}{\partial \tau_0^2} + 2\epsilon \frac{\partial^2}{\partial \tau_0 \partial \tau_1} + 2\epsilon \left(\frac{\partial}{\partial \tau_0} + \epsilon \frac{\partial}{\partial \tau_1} \right) + 1 \right] (x_0 + \epsilon x_1) = \mathcal{O}(\epsilon^2).$$

Equating like powers of ϵ we obtain a hierarchy of equations. In lowest order we find

$$\mathcal{L}x_0 \doteq \left(\frac{\partial^2}{\partial \tau_0^2} + 1 \right) x_0 = 0$$

with general solution

$$x_0 = a(\tau_1) \sin(\tau_0 + \phi(\tau_1)),$$

where the integration constants a and ϕ are independent of the fast time scale but are allowed to depend on the slow time scale τ_1. In first order we find

$$\mathcal{L}x_1 = -2\frac{\partial x_0}{\partial \tau_0} - 2\frac{\partial^2 x_0}{\partial \tau_0 \partial \tau_1} = -2\left\{ a \cos(\tau_0 + \phi) + \frac{\partial}{\partial \tau_1} [a \cos(\tau_0 + \phi)] \right\}.$$

The source term in this equation oscillates with the eigenfrequency of the homogeneous equation and therefore produces resonance. The particular solution

$$x_1 = -\tau_0 \left[\left(a + \frac{\partial a}{\partial \tau_1} \right) \sin(\tau_0 + \phi) + \frac{\partial \phi}{\partial \tau_1} \cos(\tau_0 + \phi) \right]$$

shows that secular behaviour occurs. However, the introduction of the slow time scale τ_1 gives us the freedom to prevent this undesirable behaviour (as we insist on having $x_1 \ll x_0$ for all time). Removal of secular behaviour thus results in

$$\frac{\partial \phi}{\partial \tau_1} = 0 \text{ and } \frac{\partial a}{\partial \tau_1} + a = 0.$$

Hence

$$x = a_0 \, e^{-\epsilon t} \sin(t + \phi_0) + \mathcal{O}(\epsilon^2),$$

giving a uniform valid approximation to the solution of the ordinary differential equation.

The multiple-time-scale method proves to be very powerful and it has been applied to a variety of singular perturbation problems. For a more complete account of this method see Davidson (1972). In the context of systems having a Lagrangian, an elegant and systematic approach to two-timing has been presented by Whitham (1974). His approach has been presented in some detail in Chapter 2 when we discussed the dynamics of wave groups. The problem of the Fermi–Pasta–Ulam recurrence in solutions of the NLS equation (see Section 4.3) was treated by means of the multiple-time-scale method by Janssen (1981).

We have now introduced all the tools needed for giving an approximate solution to the BBGKY hierarchy of Eqs. (4.58) and (4.59). Thus, we use the expansion equations (4.62) and (4.63) in Eq. (4.58) to obtain in lowest order

$$\frac{\partial}{\partial \tau_0} g_2^{(1)}(i, j) = -i \left(\omega_i - \omega_j \right) g_2^{(1)}(i, j)$$

and for homogeneous sea

$$g_2^{(1)}(i, j) = N_i \delta(i - j), \tag{4.64}$$

we find the usual result that the action density N_i does not depend on the fast time scale τ_0. Note that strictly speaking we should use the notation $N_i^{(1)}$ for the action density to lowest significant order, but as long as there is no confusion we simply write N_i. In second order we find

$$\left[\frac{\partial}{\partial \tau_0} + i \left(\omega_i - \omega_j \right) \right] g_2^{(2)}(i, j) = -\frac{\partial}{\partial \tau_1} g_2^{(1)}$$

$$- i \int \mathrm{dk}_{2,3,4} T_{i,2,3,4} \delta_{i+2-3-4} N_3 N_4 \left[\delta_{3-j} \delta_{4-2} + \delta_{3-2} \delta_{4-j} \right] + \text{c.c.} \ (i \leftrightarrow j). \tag{4.65}$$

For homogeneous sea the nonlinear term vanishes so that removal of secularity results in the condition $\partial g_2^{(1)} / \partial \tau_1 = 0$. In third order we then find

$$\left[\frac{\partial}{\partial \tau_0} + i \left(\omega_i - \omega_j \right) \right] g_2^{(3)}(i, j) = -\frac{\partial}{\partial \tau_2} g_2^{(1)}$$

$$- i \int \mathrm{dk}_{2,3,4} T_{i,2,3,4} \delta_{i+2-3-4} \, g_4^{(3)}(3, 4, j, 2) + \text{c.c.} \ (i \leftrightarrow j). \tag{4.66}$$

hence we need to know the fourth cumulant g_4. Its evolution follows from Eq. (4.59). Using the assumption that the pdf of the surface elevation is close to a Gaussian, hence, neglecting the effect of the sixth cumulant g_6, we obtain in lowest order from Eq. (4.59)

$$\left[\frac{\partial}{\partial \tau_0} + i\left(\omega_i + \omega_j - \omega_k - \omega_l\right)\right] g_4^{(3)}(i, j, k, l)$$
$$= 2 i T_{i,j,k,l}\delta_{i+j-k-l}\left[N_i N_j(N_k + N_l) - (N_i + N_j)N_k N_l\right]. \quad (4.67)$$

In order to obtain this result extensive use has been made of the symmetry properties of the nonlinear transfer coefficient T, in particular the Hamiltonian symmetry. This equation may be solved by means of Laplace transformation with respect to the fast time scale τ_0, where it is noted that N may be regarded as a constant as it only depends on the slow time scale τ_2. For the initial condition of vanishing fourth cumulant, i.e. $g_4^{(3)}(t = 0) = 0$, the solution becomes

$$g_4^{(3)}(i, j, k, l) = 2T_{i,j,k,l}\delta_{i+j-k-l}G(\Delta\omega, t)\left[N_i N_j(N_k + N_l)\right.$$
$$\left. -(N_i + N_j)N_k N_l\right] \quad (4.68)$$

where $\Delta\omega$ is shorthand for $\omega_i + \omega_j - \omega_k - \omega_l$. The function G is defined as

$$G(\Delta\omega, t) = i \int_0^t d\tau \, e^{i\Delta\omega(\tau - t)} = R_r(\Delta\omega, t) + i R_i(\Delta\omega, t), \quad (4.69)$$

where R_r and R_i are the real and imaginary parts of the resonance function R, defined as

$$R_r(\Delta\omega, t) = \frac{1 - \cos(\Delta\omega t)}{\Delta\omega}, \quad (4.70)$$

and

$$R_i(\Delta\omega, t) = \frac{\sin(\Delta\omega t)}{\Delta\omega}. \quad (4.71)$$

For large time t, G develops into the usual generalized functions $P/\Delta\omega$ (where P is the Cauchy principal value), and $\delta(\Delta\omega)$, since

$$\lim_{t \to \infty} G(\Delta\omega, t) = \frac{P}{\Delta\omega} + \pi i\delta(\Delta\omega). \quad (4.72)$$

The limit in Eq. (4.72) is a limit in the sense of generalized functions and is, strictly speaking, only meaningful inside integrals over wavenumber when multiplied by a smooth function.

Equation (4.68) shows that starting from an ensemble of waves that has initially Gaussian statistics, hence $g_4^{(3)}(t = 0) = 0$, nonlinearity will give rise to deviations from normality. Although for large times the deviations are the largest for resonant four-wave interactions – as is evident from the δ-function with argument $\Delta\omega$ in Eq. (4.72) – in general both resonant and nonresonant interactions will contribute to deviations of the probability distribution from the Gaussian distribution.

Finally, substitution of Eq. (4.68) into Eq. (4.66) gives

$$
\left[\frac{\partial}{\partial \tau_0} + i\left(\omega_i - \omega_j\right)\right] g_2^{(3)}(i, j) = -\frac{\partial}{\partial \tau_2} g_2^{(1)} + 4\delta(i - j) \int d\mathbf{k}_{2,3,4} |T_{i,2,3,4}|^2
$$

$$
\times \delta_{i+2-3-4}\, R_i(\Delta\omega, t)\,[N_3 N_4(N_i + N_2) - N_i N_2(N_3 + N_4)]. \tag{4.73}
$$

In order to remove secular behaviour, the normal procedure is now to study the right-hand side of Eq. (4.73) for large times τ_0. Using the limiting behaviour of $G = R_r + iR_i$ given in Eq. (4.72) it is immediately evident that the integral on the right-hand side of Eq. (4.73) becomes independent of τ_0 for large times and will therefore produce secular behaviour of $g_2^{(3)}$ on the time scale τ_0. Because of the introduction of many time scales, secularity from $g_2^{(3)}$ may be removed by demanding that for large times τ_0 the right-hand side of Eq. (4.73) vanishes. The result is that the action density evolves on the slow time scale τ_2 according to

$$
\frac{\partial}{\partial \tau_2} N_1 = 4\pi \int d\mathbf{k}_{2,3,4} |T_{1,2,3,4}|^2 \delta_{1+2-3-4}\delta(\omega_1 + \omega_2 - \omega_3 - \omega_4)
$$

$$
\times [N_3 N_4(N_1 + N_2) - N_1 N_2(N_3 + N_4)]. \tag{4.74}
$$

For gravity waves this evolution equation was first obtained by Hasselmann (1962). Equation (4.74) tells us that the action density evolves on the long τ_2 time scale owing to *resonant* four-wave interactions only.

In practical application, such as in wave prediction, resonant interactions dominate the nonlinear evolution of gravity waves. However, the result in Eq. (4.74), obtained from the condition that the series expansion for the second moment g_2 is regular, does not imply that nonresonant interactions are *not* relevant for wave evolution. On the contrary. Equation (4.73) also tells us that the second-order correction to the action density $g_2^{(3)}$ evolves on the short time scale. Using Eq. (4.73), the evolution of the action density on the short and the long time scale can be combined by imposing the condition of a homogeneous wave field and by writing

$$
\frac{\partial}{\partial t} N_i = \epsilon^3 \left(\frac{\partial}{\partial \tau_2} N_i^{(1)} + \frac{\partial}{\partial \tau_0} N_i^{(3)}\right).
$$

Then, replacing $N_i^{(1)}$ in the nonlinear term by N_i and putting $\epsilon = 1$ one finds

$$
\frac{\partial}{\partial t} N_1 = 4 \int d\mathbf{k}_{2,3,4} |T_{1,2,3,4}|^2 \delta_{1+2-3-4} R_i(\Delta\omega, t)
$$

$$
\times [N_3 N_4(N_1 + N_2) - N_1 N_2(N_3 + N_4)]. \tag{4.75}
$$

where now $\Delta\omega = \omega_1 + \omega_2 - \omega_3 - \omega_4$. This evolution equation will be called the Boltzmann equation and was first obtained by Janssen (2003). It is emphasized that Eq. (4.75) is slightly more general than the Hasselmann equation (4.74) because both resonant and nonresonant wave–wave interactions are included. Although the inclusion of the nonresonant interactions is probably not important for practical applications, it is nevertheless helpful to include them here in order to try to resolve a number of objections that have been raised against the idea of nonlinear transfer in the past.

In order to stress that Eq. (4.75) contains both nonresonant and resonant transfer, the resonance function $R_i(\Delta\omega, t)$ is studied. Two limits of R_i are of interest to mention. For small times we have

$$\lim_{t \to 0} R_i(\Delta\omega, t) = t \tag{4.76}$$

while for large times we have

$$\lim_{t \to \infty} R_i(\Delta\omega, t) = \pi\delta(\Delta\omega). \tag{4.77}$$

Hence, according to Eq. (4.75), for short times the evolution of the action density N is caused by both resonant and nonresonant four-wave interactions, while for large times, when the resonance function evolves towards a δ-function, only resonant interactions contribute to spectral change.

In the standard treatment of resonant wave–wave interactions (see, for example Hasselmann, 1962; Davidson, 1972) it is implicitly argued that the resonance function $R_i(\Delta\omega, t)$ may be replaced by its time-asymptotic value (Eq. (4.77)), because the action density spectrum is a slowly varying function of time. However, the time required for the resonance function to evolve towards a δ-function may be so large that in the meantime considerable changes in the action density may have occurred. For this reason the full expression for the resonance function is kept.

An important consequence of this choice concerns the estimation of a typical time scale T_{nl} for the nonlinear wave–wave interactions in a homogeneous wave field. With ϵ the square of a typical wave steepness and ω_0 a typical angular frequency of the wave field, one finds from the Boltzmann equation (4.75) that for short times $T_{nl} = \mathcal{O}(1/\epsilon\omega_0)$, while for large times $T_{nl} = \mathcal{O}(1/\epsilon^2\omega_0)$. Hence, although the standard nonlinear transfer, which uses Eq. (4.77) as a resonance function, does not capture the physics of the modulational instability (which operates on the fast time scale $1/\epsilon\omega_0$), the full resonance function does not suffer from this defect.

It is also important to note that according to the standard theory there is only nonlinear transfer for two-dimensional wave propagation. In the one-dimensional case there is no nonlinear transfer in a homogeneous wave field. The reason for this has to do with some special properties of the resonance conditions and the nonlinear transfer function T. In one dimension there are two different solutions to the resonance conditions, $\mathbf{k}_1 + \mathbf{k}_2 = \mathbf{k}_3 + \mathbf{k}_4$ and $\omega_1 + \omega_2 = \omega_3 + \omega_4$, namely a trivial one and a nontrivial one. The trivial one occurs for the combinations $\mathbf{k}_1 = \mathbf{k}_3, \mathbf{k}_2 = \mathbf{k}_4$ or $\mathbf{k}_1 = \mathbf{k}_4, \mathbf{k}_2 = \mathbf{k}_3$. Then, the rate of change of the action density, as given by Eq. (4.74), vanishes identically because of the symmetry properties of the term involving the action densities. The nontrivial one has the special property that for those waves the nonlinear transfer coefficient T vanishes (Dyachenko and Zakharov, 1994). Hence, in one dimension there is no resonant transfer; this is in sharp contrast with the Benjamin–Feir instability which has its largest growth rates for waves in one dimension. On the other hand, using the complete expression for the resonance function, there is always an irreversible nonlinear transfer even in the case of one-dimensional propagation.

Before we discuss the limitations of the present approach, followed by a study of the properties of four-wave nonlinear transfer, we make an important remark regarding deviations from the Gaussian distribution. We have studied the statistical aspects of random, weakly nonlinear waves in the context of the Zakharov equation. In particular, the relation between the deviations from the Gaussian distribution and four-wave interactions is then of interest. Because of the symmetries of the Zakharov equation, the first moment of interest is then the fourth moment and the related kurtosis. The third moment and its related skewness vanish: information on the odd moments can only be obtained by making explicit use of Krasitskii's (1994) canonical transformation. Now, the fourth moment $\langle \eta^4 \rangle$ may be obtained in a straightforward manner from Eq. (4.61) and the expression for the fourth cumulant Eq. (4.68) as

$$\langle \eta^4 \rangle = \frac{3}{4g^2} \int dk_{1,2,3,4} (\omega_1 \omega_2 \omega_3 \omega_4)^{1/2} \langle a_1 a_2 a_3^* a_4^* \rangle + \text{c.c.} \qquad (4.78)$$

Denoting the second moment $\langle \eta^2 \rangle$ by m_0, deviations from normality are then most conveniently established by calculating the kurtosis

$$C_4 = \langle \eta^4 \rangle / 3m_0^2 - 1,$$

since for a Gaussian pdf C_4 vanishes. The result for C_4 is

$$C_4 = \frac{4}{g^2 m_0^2} \int dk_{1,2,3,4} T_{1,2,3,4} \delta_{1+2-3-4} (\omega_1 \omega_2 \omega_3 \omega_4)^{1/2}$$
$$\times R_r(\Delta \omega, t) N_1 N_2 N_3, \qquad (4.79)$$

where R_r is defined by Eq. (4.70), while the integral should be interpreted as a principal-value integral. For large times, unlike the evolution of the action density, the kurtosis does not involve a Dirac δ-function but rather depends on $P/\Delta \omega$. Therefore, the kurtosis is determined by the resonant and nonresonant interactions. It is instructive to apply Eq. (4.79) to the case of a narrow-band wave spectrum in one dimension. Hence, performing the usual Taylor expansions around the carrier wavenumber k_0 to lowest significant order, one finds for large times

$$C_4 = \frac{8\omega_0^2}{g^2 m_0^2} \frac{T_0}{\omega_0''} \int dp_{1,2,3,4} \frac{\delta_{1+2-3-4}}{p_1^2 + p_2^2 - p_3^2 - p_4^2} N_1 N_2 N_3, \qquad (4.80)$$

where $p = k - k_0$ is the wavenumber with respect to the carrier. It is seen that the sign of the kurtosis is determined by the ratio T_0/ω_0'', which is the same parameter as the one that determines whether a wave train is stable or not to sideband perturbations (see Eq. (4.44)). Note that numerically the integral is found to be negative, at least for bell-shaped spectra. Hence, from Eq. (4.80) it is immediately plausible that for an unstable wave system that has negative T_0/ω_0'', the kurtosis will be positive and thus will result in an increased probability of extreme events. On the other hand, for a stable wave system there will be a reduction in the probability of extreme events.

Finally, a further simplification of the expression for the kurtosis may be achieved if it is assumed that the wavenumber spectrum $F(p) = \omega_0 N(p)/g$ only depends on two parameters, namely the variance m_0 and the spectral width σ_k. Introduce the scaled wavenumber $x = p/\sigma_k$ and the correspondingly scaled spectrum $m_0 H(x) \, dx = F(p) \, dp$. Then, using the deep-water dispersion relation and $T_0 = k_0^3$, Eq. (4.80) becomes

$$C_4 = -\frac{8k_0^2 m_0}{\sigma_\omega'^2} J, \qquad (4.81)$$

where $k_0 m_0^{1/2}$ is the significant steepness while σ_ω' is the relative width in angular frequency space $\sigma_\omega/\omega_0 = 0.5\sigma_k/k_0$. The parameter J is given by the expression

$$J = \int dx_{1,2,3,4} \frac{\delta_{1+2-3-4}}{x_1^2 + x_2^2 - x_3^2 - x_4^2} H_1 H_2 H_3,$$

and is independent of the spectral parameters m_0 and σ_k. Therefore, Eq. (4.81) suggests a simple dependence of the kurtosis on spectral parameters. In fact, the kurtosis depends on the same parameter as the one that determines the stability of a uniform wave train. For a spectrum of waves, Janssen (2003) has introduced the so-called 'Benjamin–Feir index' (BFI) which compares effects of nonlinearity and linear dispersion. It is defined as

$$\text{BFI} = \sqrt{2k_0^2 m_0}/\sigma_\omega'. \qquad (4.82)$$

The BFI turns out to be very useful in ordering the theoretical and numerical results presented in the following sections. For simple initial wave spectra (defined in terms of the modulation wavenumber $\mathbf{p}$) that only depend on the variance and on the spectral width, it can be shown that for the NLS equation the large-time solution is completely characterized by the BFI. For the Zakharov equation this is not the case, but the BFI is still expected to be a useful parameter for narrow-band wave trains. The BFI plays a key role in the inhomogeneous theory of wave–wave interactions (Alber, 1978), while a similar parameter has been introduced and discussed in the context of freak waves in random sea states by Onorato *et al*. (2001).

In summary, for narrow-band waves the kurtosis depends on the square of the BFI, while the sign of the kurtosis parameter (note that with our definition the kurtosis vanishes for a Gaussian sea state) is determined by the stability properties of a uniform wave train. In the case of Benjamin–Feir instability there will be an increased probability of extreme events, while in the case of stability there is, compared with the normal distribution, a reduced probability of extreme events.

4.6 Discussion of the assumptions underlying the statistical approach

Before we discuss the important consequences of the Boltzman equation (4.75) for the evolution of ocean waves it is relevant to explore the limitations of the approach that has been followed. This implies a discussion of the consequences of the assumption of the homogeneity of the ensemble of waves and the assumption of near-Gaussian statistics (frequently referred to as the random-phase approximation). Related to these assumptions it is fair to ask the question: what is the relation between the statistical theory of four-wave interactions and the Benjamin–Feir instability from deterministic theory? Furthermore, the starting point of the statistical analysis is the Zakharov equation. This equation is reversible because it is invariant with respect to reversals of the time and the horizontal coordinate axis. However, for large times, Eq. (4.75) is not reversible. Clearly, some explanation of this apparent contradiction is required.

Apart from the qualitative arguments on the range of validity of the statistical approach, one may wonder whether it is not possible to give direct evidence of range of validity of the four-wave interaction approach. The idea is to start from a realistic set of deterministic evolution equations for the amplitude and phase of deep-water gravity waves and to do a Monte Carlo simulation for an ensemble of ocean waves. The time evolution of the ensemble mean of the wave spectrum is then compared with the predicted evolution by the Boltzmann equation and, if there is sufficiently close agreement between the two, a direct proof for the existence of resonant and nonresonant four-wave interactions is given. At the same time this then provides information on the statistical approach. Following Janssen (2003), a first attempt for the case of one-dimensional wave propagation will be presented. Surprisingly, for deep-water waves the homogeneity condition and the random-phase approximation are not very restrictive, and a good agreement between results from the Monte Carlo simulations and from the Boltzmann equation is found, even for narrow-band, strongly nonlinear waves.

4.6.1 Homogeneity

Let us first discuss in some detail the assumption of the homogeneity of the ensemble of ocean waves, which at the same time will shed some light on the relation between the statistical approach and the Benjamin–Feir instability.

Alber and Saffman (1978) (see also Alber, 1978; Janssen, 1983) have studied some aspects of the evolution of an inhomogeneous, random wave field. These authors concentrated on the case of narrow-band waves and therefore the starting point was the NLS equation (4.35). In order to simplify the discussion which follows, we introduce a frame moving with the group velocity $\partial\omega/\partial k_0$ and dimensionless units $\tilde{t} = \omega_0 t/2$, $\tilde{x} = 2k_0 x$ and $\tilde{A} = k_0 A$ are introduced. Dropping the tildes, the equation for A (which is twice the steepness for a uniform wave train) reads

$$i\frac{\partial A}{\partial t} - \frac{\partial^2 A}{\partial x^2} - |A|^2 A = 0. \tag{4.83}$$

In order to study effects of inhomogeneity of the wave field, one introduces the two-point correlation function $\rho(x_1, x_2, t)$ as

$$\rho(x, r, t) = \langle A(x_1, t)A^*(x_2, t)\rangle, \tag{4.84}$$

where the average coordinate x is defined by $x = (x_1 + x_2)/2$ while the separation coordinate is $r = x_2 - x_1$. For an inhomogeneous wave field the correlation function ρ depends on both r and x. Multiplying Eq. (4.83) by $A^*(x_2, t)$, adding the complex-conjugate expression with x_1 and x_2 interchanged and averaging gives the following transport equation for $\rho(x, r, t)$:

$$i\frac{\partial}{\partial t}\rho - \left(\frac{\partial^2}{\partial x_1^2} - \frac{\partial^2}{\partial x_2^2}\right)\rho - \langle A^2(x_1)A^*(x_1)A^*(x_2)\rangle$$
$$+ \langle A^2(x_2)A^*(x_2)A^*(x_1)\rangle = 0, \tag{4.85}$$

and, as usual, the rate of change of the two-point correlation function is related to the four-point correlation function. Closure is now achieved by assuming the random-phase approximation, e.g.

$$\langle A^2(x_1)A^*(x_1)A^*(x_2)\rangle \simeq 2\langle A(x_1)A^*(x_1)\rangle\langle A(x_1)A^*(x_2)\rangle. \tag{4.86}$$

In addition, performing a transformation to the averaged coordinate x and the separation coordinate r results in the following difference–differential equation for ρ:

$$i\frac{\partial}{\partial t}\rho(x,r) - 2\frac{\partial^2}{\partial x \partial r}\rho(x,r) - 2\rho(x,r)\left[\rho\left(x+\frac{1}{2}r,0\right) - \rho\left(x-\frac{1}{2}r,0\right)\right] = 0.$$

(4.87)

Equation (4.87) describes the evolution of an inhomogeneous ensemble of narrow-band wave trains. The time scale for nonlinear transfer can easily be estimated from Eq. (4.87) with the result

$$\frac{\partial}{\partial t}\rho = \mathcal{O}(\rho) = \mathcal{O}(\epsilon),$$

(4.88)

(with ϵ again the square of the significant steepness), which coincides with the Benjamin–Feir time scale. In agreement with the discussion of the previous section, however, energy transfer is only possible for an inhomogeneous system. In other words, assuming spatial homogeneity, hence ignoring the x dependence in Eq. (4.87), it is seen that ρ does not change with time. For a homogeneous system energy transfer only exists if one includes deviations from Gaussian statistics. These are generated because nonlinearity gives rise to correlations between the different components of the wave spectrum. Crawford *et al.* (1980) have shown that deviations from normality associated with resonant nonlinear interactions result in a higher-order effect because the energy transfer occurs on the much longer time scale $\tau = \mathcal{O}(\epsilon^{-2})$. In fact, for homogeneous sea one then rediscovers the resonant irreversible energy transfer found by Hasselmann (1962), which was discussed in Section 4.5. However, we have also seen that nonresonant interactions cause deviations from normality, and these interactions act on the same time scale (see Eq. (4.88)) as the effects of inhomogeneity. Therefore, strictly speaking deviations resulting from nonresonant interactions should also be included in the present treatment; for the moment, however, we shall ignore this complication.

Using the evolution equation (4.87), Alber and Saffman (1978) discovered the random version of the Benjamin–Feir instability. The key result was that a homogeneous ensemble of waves is unstable to long wavelength perturbations provided the width of the spectrum is sufficiently small. In other words, introduce the wave spectrum $W(x, p)$ according to

$$W(x,p) = \frac{1}{2\pi}\int dr\, e^{ipr}\rho(x,r),$$

(4.89)

where, in general, W depends on the averaged coordinate x while p may be regarded as the usual wavenumber. Then, Alber and Saffman (1978) found the remarkable result that a homogeneous wave field, described by a homogeneous spectrum

$$W(x,p) = W_0(p),$$

is unstable to modulations in x-space provided that the spectrum $W_0(p)$ is sufficiently narrow. The resulting dispersion relation is hard to analyse, but for the Lorentz spectrum

$$W_0(p) = \frac{\langle A_0^2 \rangle \sigma}{\pi(p^2 + \sigma^2)}, \tag{4.90}$$

where σ denotes the width of the spectrum and $\langle A_0^2 \rangle$ is twice the mean-square slope, Crawford *et al.* (1980) found that the perturbations with wavenumber k and angular frequency ω obey the simple dispersion relation

$$\omega = 2k \left[-i\sigma \pm \left(\frac{1}{4}k^2 - \langle A_0^2 \rangle \right)^{1/2} \right]. \tag{4.91}$$

There is instability for $\Im(\omega) > 0$. Note that in the limit of vanishing bandwidth σ the growth rate (Eq. (4.91)) reduces to the result of Benjamin and Feir (see Eq. (4.43)) if one makes the identification $2\langle A_0^2 \rangle$ (Gaussian and random) $= A_0^2$ (deterministic). We also note that finite bandwidth gives a reduction of the growth rate and that when

$$\sigma \geq \langle A_0^2 \rangle^{1/2} \tag{4.92}$$

the instability disappears. Returning to dimensional quantities the stability condition (4.92) boils down to

$$\text{BFI} \leq 1, \tag{4.93}$$

where BFI is the Benjamin–Feir index introduced in Eq. (4.82).

The question of the nature of the stabilizing effect of finite bandwidth is important, but it is not easy to provide a simple explanation of this important result. Loosely speaking it may perhaps be argued that the waves in question have random phase which upsets the coherence between the four interacting waves thereby reducing the efficiency of the (resonant) energy transfer. A more precise explanation was provided by Janssen (1983) who studied the stability of a homogeneous wave spectrum in terms of normal modes of the linear problem. Although each mode is undamped, a smooth perturbation will always excite a continuum of these normal modes. In the course of time the perturbation to the wave spectrum evolves into an erratic function (or has a nonresolvable fine structure) because of phase mixing in such a way that all its moments vanish rapidly with time (van Kampen, 1955).

The inhomogeneous theory of wave evolution has provided us, therefore, with an important restriction on the validity of the Boltzman equation (4.75). Apart from the already mentioned condition that the wave steepness should be sufficiently small that the near-Gaussian assumption may be justified, there is an additional restriction on the width of the spectrum. The wave spectrum should be sufficiently broad (BFI $\leq$ 1, or Eq. (4.92)) that effects of inhomogeneities in the wave field may be disregarded. In the opposite case of narrow-band waves (BFI > 1), the Benjamin–Feir instability would occur, resulting in a rapid growth of sidebands, at the expense of the main peak of the spectrum. Because of the energy transfer from the main peak of the spectrum to the sidebands there is a considerable broadening of the spectral shape, hence the growth rate of the instability reduces until, for a sufficiently broad spectrum (of the order given in Eq. (4.92)), the random version of the Benjamin–Feir instability is quenched. This broadening of the spectrum is an irreversible process because of phase mixing (Janssen, 1983). According to the present approach the broadening of the spectrum is associated with the generation of inhomogeneities in the wave field.

In order to summarize the present discussion, we remark that the central role of the BFI is immediately evident in the context of the lowest-order inhomogeneous theory of wave–wave interactions. According to the stability criterion of Eq. (4.92) there is change of stability for BFI = 1. In other words, BFI is a bifurcation parameter: on the short time scale, spectra will be stable and therefore do not change if BFI < 1 while in the opposite case inhomogeneities will be generated giving rise to a broadening of the spectrum. However, this prediction follows from an approximate theory that neglects deviations from normality. In general, considerable deviations from normality are to be expected, in particular in case of Benjamin–Feir instability. It is therefore of interest to explore the consequences of non-normality. This will be done in Section 4.6.3 by means of a numerical simulation of an ensemble of surface gravity waves.

4.6.2 Gaussian assumption, nonresonant interactions and irreversibility

In the past there has been a considerable debate about the question of the irreversibility of the evolution equation for wave–wave interactions. However, it should be emphasized that the Boltzmann equation (4.75), which accounts for both resonant and nonresonant interactions, has the time-reversal symmetry of the original Zakharov equation, since the resonance function changes sign when time t changes sign. Also, as R_i vanishes for $t = 0$, the time derivative of the action density spectrum is continuous around $t = 0$ and does not show a cusp (see Komen *et al.*, 1994). Nevertheless, despite the fact that there is time reversal, Eq. (4.75) has the irreversibility property: the memory of the initial conditions gets lost in the course of time owing to phase mixing.

Therefore, by retaining the effects of the nonresonant interactions, the evolution equation for four-wave interactions shares the time-reversal property of the Zakharov equation. In contrast, by taking the limit of large times in the resonance function R_i from the outset one obtains the Hasselmann equation (4.74) and this equation clearly no longer has the time-reversal symmetry; the left-hand side of the equation changes sign with a change in the sign of time, while the right-hand side remains unchanged provided that the spectrum is not modified. Komen *et al.* (1994) concluded that an additional assumption must have been introduced that destroys this symmetry. And this assumption is the Gaussian hypothesis. It is to be emphasized, however, that in obtaining Eq. (4.74) we have made one additional step, namely, we have only used the solution for the fourth cumulant $g_4^{(3)}$ in the limit of large times. In the limit of large times it may be argued (see Davidson, 1972) that all the transients phase mix to zero and therefore can be ignored when one is interested in the large-time behaviour of the action density. This type of argument has been used widely with success in many applications, but it is evident that it implies an 'arrow of time'. In other words, when invoking the argument of phase mixing, thereby ignoring transient effects for large positive times, it should be clear that it is not formally allowed to consider the limit of small times, let alone to reverse time.

In this context it is noted that the breaking of time reversal occurs frequently in asymptotic analysis. A well-known example is discussed in detail by Whitham (1974). He considered the problem of the large-time behaviour of a continuous spectrum of linear (gravity) waves. The exact solution to this problem is readily written down; however, to investigate its large-time behaviour an asymptotic analysis is required. Fortunately, because of the dispersive nature of the gravity waves one may apply the method of stationary phase to obtain the well-known result that wave energy propagates with the group velocity and that the amplitude of the waves decays as $t^{-1/2}$. Clearly, the answer carries all the properties of an irreversible solution, but it is of course not allowed to take the limit of small time.

Regarding the role of the nonresonant interactions we conclude with the following remark. The extended version of the homogeneous four-wave theory has two time scales, a fast one on which the nonresonant interactions take place and a long time scale on which the resonant interactions occur. The nonresonant interactions play a similar role to the transients in the solution of an initial-value problem. They are simply generated because initially there is a mismatch between the choice of the probability distribution of the waves, a Gaussian, and the initial choice of the wave spectrum, representing a sea state with narrow-band, steep nonlinear waves. For example, if one could choose a pdf that is in equilibrium with the nonlinear sea state (theoretically one can, by the way), then nonresonant interactions would not contribute. Only resonant wave–wave interactions will then give rise to nonlinear transfer. In the general case for which there is a finite mismatch between pdf and wave spectrum, the nonresonant contribution will die out very quickly owing to phase mixing, but will, nevertheless, as we will see in Section 4.6.3, result in considerable changes in the wave spectrum. The question therefore is whether there is a need to include effects of nonresonant interactions. This depends on the application. In wave-tank experiments, where one can program a wave maker to produce the initial conditions used here, it seems that effects of nonresonant transfer need to be taken into account. For the open ocean case this is not clear. The point is that in nature the combination of steep waves and a strictly Gaussian distribution probably does not occur. Changes in nature are expected to be more gradual so that the mismatch between pdf and wave spectrum is small. Only when a wind starts blowing suddenly, hence for short fetches and duration, are the effects of nonresonant interactions expected to be relevant. More research in this direction is, however, required.

4.6.3 Numerical simulation of an ensemble of waves

It is important to determine the range of validity of the theory of resonant and nonresonant four-wave interactions. For example this theory assumes that the wave steepness is sufficiently small and the pdf of the surface elevation is close to a Gaussian. In order to address these questions we simulate the evolution of an ensemble of waves by running a deterministic model with random initial conditions. Only wave propagation in one dimension will be considered from now on.

For given wavenumber spectrum $F(k)$, which is related to the action density spectrum through $F = \omega N/g$, initial conditions for the amplitude and phase of the waves are drawn from a Gaussian probability distribution of the surface elevation. The phase of the wave components is then random between 0 and 2π while the amplitude is taken as deterministic; hence,

$$a(k) = \sqrt{N(k)\Delta k}\, e^{i\theta(k)}, \tag{4.94}$$

where $\theta(k)$ is a random phase $= 2\pi x_r$, x_r is a random number between 0 and 1, and Δk the resolution in wavenumber space.

Each member of the ensemble is integrated for a long enough time to reach equilibrium conditions, typically of the order of 60 dominant wave periods. At every time step of interest the ensemble average of quantities such as the two-point correlation function g_2, the pdf of the surface elevation, and integral parameters such as wave height, spectral width and kurtosis is taken. Typically, the size of the ensemble N_{ens} is 500 members. This choice was made to ensure that quantities such as the wave spectrum were sufficiently smooth and that the statistical scatter in the spectra, which is inversely proportional to $\sqrt{N_{ens}}$, is small enough to give statistically significant results. In Janssen (2003) the Monte Carlo approach was applied both to the NLS equation and to the Zakharov equation. Here, only the results obtained with the Zakharov equation are briefly reported.

As a starting point we choose the Zakharov equation (4.19). The action variable is written as a sum of δ-functions,

$$a(k) = \sum_{i=-N}^{N} a_i \, \delta(k - i \Delta k), \tag{4.95}$$

where Δk is the resolution in wavenumber space and $2N + 1$ is the total number of modes. Substitution of Eq. (4.95) into Eq. (4.19) gives the following set of ordinary differential equations for the amplitude a_1:

$$\frac{d}{dt} a_1 + i\omega_1 a_1 = -i \sum_{1+2-3-4=0} T_{1,2,3,4} \, a_2^* a_3 a_4. \tag{4.96}$$

We have solved this set of differential equations with a Runge–Kutta method with variable time step. Relative and absolute error of the solution have been chosen in such a way that conserved quantities such as action, wave momentum and wave energy are conserved to at least five significant digits. Amplitude and phase needed for the initial condition for Eq. (4.96) are generated by Eq. (4.94) where the wavenumber spectrum is given by a Gaussian shape,

$$F(p) = \frac{\langle \eta^2 \rangle}{\sigma_k \sqrt{2\pi}} \exp\left(-\frac{p^2}{2\sigma_k^2}\right), \tag{4.97}$$

where $p = k - k_0$, k_0 is the peak wavenumber and σ_k is the width of the wavenumber spectrum.

The nonlinear transfer coefficient was from Krasitskii (1994), while the exact dispersion relation for deep-water gravity waves was taken. Because the Zakharov equation contains all higher-order terms in the modulation wavenumber $p = k - k_0$ it is not possible to prove that the large-time solution of the initial-value problem is determined solely by the BFI, but in good approximation the BFI can still be used for this purpose as long as the spectra are narrow-banded.

In Janssen (2003) an extensive set of experiments has been performed by increasing the initial value of the BFI in small steps from 0 to 2. Even for small values of the BFI the wave spectrum changes in time in such a way that the spectrum broadens and as a consequence the final time value of the BFI is smaller than its initial value. In fact, the final time value is of the order of 1 at the most. It is remarkable that these experiments show that the solution depends in a continuous manner on the initial value of the BFI (see also Dysthe *et al.*, 2003). There is no evidence of a bifurcation, which according to the inhomogeneous approach of Alber (1978) should occur at a finite value of the BFI (for the NLS equation one finds change of stability of a uniform wave spectrum at BFI = 1). Furthermore, Janssen (2003) measured from the Monte Carlo simulations the inhomogeneity of the wave field. Although, in agreement with Alber (1978), inhomogeneity grows rapidly in the

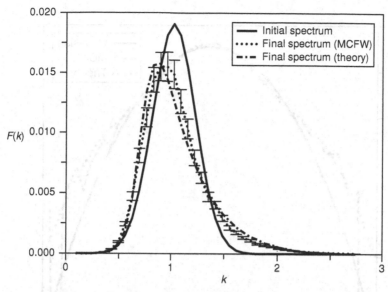

Fig. 4.14. Initial- and final-time wavenumber spectrum according to Monte Carlo forecasting of waves (MCFW) using the Zakharov equation. Error bars give 95 per cent confidence limits. Results from theory are also shown. Spectrum for BFI $= 1.4$ (from Janssen, 2003).

course of time, the level of inhomogeneity saturates at an extremely small level and it cannot explain the large changes in the wavenumber spectrum seen in the numerical simulations. These spectral changes are probably related to deviations from the normal distribution and not by effects of inhomogeneity. Indeed, nonresonant interactions act on the same fast time scale as inhomogeneities and they result in deviations from normality.

In Fig. 4.14 we have plotted the ensemble-averaged wavenumber spectrum for BFI $= 1.4$, which may be regarded as an extreme case. It shows a clear downshift of the peak of the spectrum while also considerable amounts of energy have been pumped into the high-wavenumber part of the spectrum. The wavenumber downshift is caused by the asymmetries in the nonlinear transfer coefficient and to the same extent by the asymmetries in the angular frequency with respect to the carrier wavenumber. This was checked by running Eq. (4.96) with constant nonlinear transfer coefficient, and similar looking ensemble mean spectra, but with half the wavenumber downshift, were obtained. There is also a noticeable broadening of the spectrum.

The most important results are shown in Figs. 4.15 and 4.16. In Fig. 4.15 a comparison of the simulated pdf for the surface elevation and the Gaussian distribution is given for BFI $= 1.4$. Note that the Zakharov equation describes the evolution of the free waves and therefore the pdf is symmetrical. Skewness is only

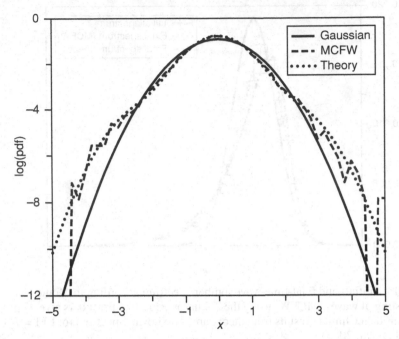

Fig. 4.15. Probability distribution function for surface elevation as a function of normalized height $\eta/\sqrt{m_0}$. Results from numerical simulations with the Zakharov equation and homogeneous theory in the case of focussing (BFI of 1.4). For reference, the Gaussian distribution is also shown. Freak waves correspond to a normalized height of 4.4 or larger.

introduced when employing Krasitskii's canonical transformation. Considerable deviations from normality are seen, both for intermediate and extreme wave height. Figure 4.16 shows a summary of our results on the deviations from normality by plotting the final-time value of the kurtosis $C_4 = \langle \eta^4 \rangle / 3m_0^2 - 1$ as function of the final-time BF. Here, the fourth moment $\langle \eta^4 \rangle$ was determined from the pdf of the surface elevation which was obtained by sampling the second half of the time series for the surface elevation at an arbitrarily chosen location. Results from the Zakharov equation are in qualitative agreement with the ones from the NLS equation, which are also shown for comparison purposes in Fig. 4.16. For small nonlinearity one would expect a normal distribution and hence a vanishing kurtosis, but the simulation underestimates the kurtosis by a small amount of 2 per cent. This underestimation is caused by the relatively low number of modes used in the simulations (typically, 41 modes were taken). The kurtosis depends almost quadratically on the BFI up to a value close to 1. Near BFI = 1, on the other hand, the kurtosis behaves in a more singular fashion, as explained in Janssen (2003).

To summarize, we have discussed results from the Monte Carlo simulation of the Zakharov equation. These results show that, on average, there is a considerable

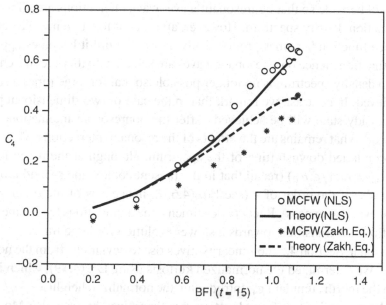

Fig. 4.16. Normalized kurtosis as a function of BFI. Results show focussing from simulations with the nonlinear Schrödinger (NLS) equation and with the Zakharov (Zakh.) equation. The corresponding theoretical results are shown as well (from Janssen, 2003).

broadening and downshift of the wave spectrum, while nonlinearity gives rise to large deviations from Gaussian statistics. The question now is whether the average of the Monte Carlo results may be obtained in the framework of a simple theoretical description, namely by the standard theory of wave–wave interactions, extended with the effects of nonresonant four-wave interactions. This approach assumes a homogeneous wave field but allows for deviations from the Gaussian sea state.

Therefore, we used the Boltzmann equation (4.75) to evolve the action density $N(k)$ for the same cases. The differential equation was solved with a Runge–Kutta method with variable time step, and the continuous problem was discretized in the same way as was done in the case of the solution of the Zakharov equation. Run times using the homogeneous theory are typically two orders of magnitude faster than when following the ensemble approach.

An example of the comparison between theoretical and simulated spectrum is given in Fig. 4.14. There is a fair agreement between the two. However, it should be mentioned that typically the simulated spectrum shows a somewhat smaller downshift in peak wavenumber than the theoretical one. Although this is not shown in detail here (see for this Janssen, 2003) there is a clear trend towards a steady state in the case of one-dimensional propagation. This can be understood as follows. First, it should be noted that for one-dimensional propagation there is no nonlinear transfer due to resonant nonlinear interactions. Now, initially the resonance function

$R_i(\Delta\omega, t)$ will be wide so that nonresonant wave–wave interactions are allowed to modify the action density spectrum. However, after about five to ten wave periods the resonance function becomes progressively narrower until it becomes approximately a δ-function, hence only resonant waves are selected. In that event, a change of the action density spectrum is no longer possible so that for large times a steady state is achieved. In contrast, we remark that in the case of two-dimensional propagation no steady state will be achieved. After the nonresonant interactions have done their work what remains are the effects of the resonant interactions. These will result in a continued downshifting of the spectrum, although at the much longer time scale $T_{nl} = \mathcal{O}(1/\epsilon^2\omega_0)$ (recall that in the present context the second moment is thought to be of the order of ϵ (see Eq. (4.63)), hence ϵ is of the order of the square of the wave steepness). Recent two-dimensional simulations by Dysthe *et al.* (2003) confirm this transition towards a slower evolution for large times.

As discussed in Section 4.5, nonlinearity gives rise to deviations from the normal distribution. We determined the normalized kurtosis using Eq. (4.79), which is obtained from the fourth cumulant g_4. Introducing the normalized height $x = \eta/\sqrt{m_0}$, the pdf of the normalized surface elevation x is then given by (see also Mori and Yasuda, 2002)

$$p(x) = \left(1 + \frac{1}{8}C_4\frac{d^4}{dx^4}\right) f_0, \qquad (4.98)$$

where f_0 is given by the normal distribution

$$f_0(x) = \frac{1}{\sqrt{2\pi}} \exp\left(-\frac{x^2}{2}\right). \qquad (4.99)$$

Equation (4.98) follows from an expansion of the pdf p in terms of orthogonal functions $(d/dx)^n f_0$. Here, n is even because of the symmetry of the Zakharov equation. The expansion coefficients are then obtained by determining the first, second and fourth moments. For the range of BFI studied here it was verified that higher moments only gave a small contribution to the shape of the pdf $p(x)$. The pdf according to theory is compared in Fig. 4.15 with the simulated one, and a good agreement is obtained, even for extreme sea-state conditions. Clearly in the case of nonlinear focussing, the probability of extreme states is, as expected, larger when compared to the normal distribution. Finally, in Fig. 4.16 theoretical and simulated final-time kurtosis is plotted as function of the final-time BFI. A good agreement between the two results is obtained even close to the limiting value of the final-time BFI. For BFI < 1 both simulated and theoretical kurtosis depend in an almost quadratic fashion on BFI, in agreement with the simple estimates of C_4 given in Section 4.5 (see Eq. (4.81)).

4.7 Consequences of four-wave interactions

In this section we discuss some important consequences of the four-wave interactions on the evolution of a random, homogeneous wave field and we discuss the key role played by these nonlinear interactions in modern wave prediction models. The need for parametrization of the nonlinear transfer is elucidated and some well-known examples of parametrization are presented.

In Section 4.6, we have already seen one important property of nonlinear transfer, namely it gives rise to a downshift of the wavenumber spectrum while, in order to conserve energy and action, considerable amounts of energy are transferred from the region just beyond the location of the spectral peak to the high-wavenumber part of the spectrum. As a consequence, the rate of change of the wavenumber spectrum due to nonlinear interactions shows the typical three-lobe structure, being positive for low wavenumbers, negative in the intermediate range, and positive for high wavenumbers. Because of the downshift towards lower wavenumbers the phase speed of the waves near the peak of the *equilibrium* windsea spectrum will be 20 to 30 per cent larger than the wind speed at 10 m height.

In the case of one-dimensional propagation, the predictions from the Boltzmann equation (4.75) were compared with Monte Carlo simulations using the Zakharov equation and a good agreement was obtained, even for steep waves with a narrow spectrum. For one-dimensional propagation, resonant four-wave interactions are absent and only nonresonant interactions will give rise to spectral change, and to deviations from the normal distribution. The time scale on which these nonresonant interactions operate is very short, typically of the order of 10 to 20 wave periods. For practical applications such as wave prediction on a global scale there is only interest in the slow time evolution of the wave spectrum. Therefore, one only deals in practice with the effects of resonant four-wave interactions because for large times the resonance function R_i has evolved into a δ-function (see Eq. (4.77)). In the remainder of this chapter we therefore only consider the Boltzmann equation (4.75) in the limit of large times, and we refer to the large-time limit of this equation as the Hasselmann equation.

4.7.1 Properties

The Hasselmann equation has the desirable property that if the action density $N(k)$ is positive initially then this feature is preserved during the course of time. In other words if $N_1 \geq 0$ at $t = 0$ then N_1 does not become negative at subsequent times. The following argument demonstrates this to be the case. Assume that the action density becomes negative at a certain instant and that this occurs for a wavenumber $k_1 = k_0$. Thus at that instant $N(k_0) = 0$. Consequently, at this instant Eq. (4.75) becomes

$$\frac{\partial}{\partial t} N_0 = 4\pi \int \mathrm{d}\mathbf{k}_{2,3,4} |T_{0,2,3,4}|^2 \delta_{0+2-3-4} \delta(\Delta\omega) N_2 N_3 N_4 \geq 0.$$

Since the right-hand side of the above equation is nonnegative, this contradicts our original hypothesis of N_1 becoming negative.

The property of the action density being always positive is very reassuring as previous statistical treatments of wave–wave interactions failed at this point; the result was that the action density became negative at a certain instant, which of course is not a desirable property.

The Hasselmann equation permits three conservation laws. These are:

(1) Conservation of action, i.e.

$$\frac{\mathrm{d}}{\mathrm{d}t} \int \mathrm{d}\mathbf{k}\, N(\mathbf{k}) = 0. \tag{4.100}$$

(2) Conservation of momentum, i.e.

$$\frac{\mathrm{d}}{\mathrm{d}t} \int \mathrm{d}\mathbf{k}\, \mathbf{k} N(\mathbf{k}) = 0. \tag{4.101}$$

(3) Conservation of wave energy,

$$\frac{\mathrm{d}}{\mathrm{d}t} \int \mathrm{d}\mathbf{k}\, \omega N(\mathbf{k}) = 0. \tag{4.102}$$

The conservation of energy and momentum applies for all resonant wave–wave interaction processes and follows generally from the invariance of the Lagrangian or the Hamiltonian with respect to time and space translations. Conservation of action, however, only holds for interaction processes in which an equal number of waves is created or annihilated. It follows from the invariance of the Lagrangian with respect to phase shifts (see Eq. (2.40b)) of the wave components. This is valid for four-wave interactions but not, for example, for three-wave interactions.

The conservation of two scalar quantities, energy and action, implies an important property of the nonlinear energy transfer. A similar relation holds for the energy transfer in a two-dimensional turbulence spectrum in the atmosphere, which also conserves two scalar quantities, namely energy and enstrophy. The argument for this is obtained from Komen *et al.* (1994) and is given below.

Consider the one-dimensional energy transfer $\mathrm{d}N(\omega)/\mathrm{d}t$ obtained by integrating the Hasselmann equation over wave propagating direction. The one-dimensional transfer must have at least three lobes of different sign (see Fig. 4.17). It cannot have a two-lobe structure representing, for example, an energy cascade from low to high frequencies, as in three-dimensional turbulence. The ratio energy/action $= \omega$ increases monotonically with frequency. Thus, if the net action lost in the negative low-frequency lobe balances the action gained in the high-frequency lobe, the

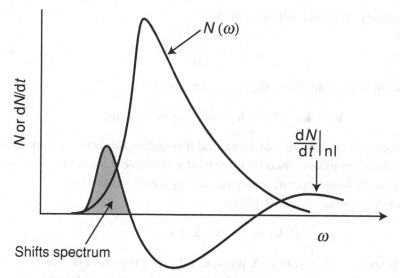

Fig. 4.17. One-dimensional energy transfer caused by resonant four-wave interactions.

energy loss in the low-frequency lobe must necessarily be smaller than the energy gained in the high-frequency lobe, therefore energy is not conserved. Numerical computations of the resonant energy transfer (and also the Monte Carlo simulations of Section 4.6) indeed give the three-lobe structure. Again, it is pointed out that the positive lobe at low frequencies has important consequences for the evolution of the wave spectrum as it shifts the spectrum towards lower frequencies.

Without proof it is mentioned here that action, momentum and energy are conserved for each resonant interaction quadruplet separately. This is plausible because the integral conservation laws hold for an arbitrary spectrum (for a more detailed discussion see Komen *et al.*, 1994). There is, therefore, a detailed balance. The principle of detailed balance is very useful for numerical computations. It was applied by Hasselmann and Hasselmann (1985) and Snyder *et al.* (1993) in developing a symmetrized method for integration of the Hasselmann equation, leading to improved conservation properties and a considerable saving in computer time.

The final important property of nonlinear transfer concerns the nonequilibrium entropy. The appropriate entropy associated with the long-time version of Eq. (4.75) is

$$S = \int d\mathbf{k} \, \ln N(\mathbf{k}). \tag{4.103}$$

From the evolution equation (4.75) it follows that $S(t)$ is a monotonic increasing function of time,

$$\frac{d}{dt} S \geq 0, \tag{4.104}$$

with the equality sign only holding when

$$\frac{1}{N_1} + \frac{1}{N_2} = \frac{1}{N_3} + \frac{1}{N_4} \tag{4.105}$$

for each set of waves satisfying the resonance conditions

$$\mathbf{k}_1 + \mathbf{k}_2 = \mathbf{k}_3 + \mathbf{k}_4, \quad \omega_1 + \omega_2 = \omega_3 + \omega_4.$$

In other words, in the absence of input and dissipation the Hasselmann equation tells us that a homogeneous, random wave field will evolve in an irreversible manner towards the equilibrium spectrum determined by condition (4.105).

The general solution of Eq. (4.105) is

$$N(\mathbf{k}) = (a + \mathbf{b} \cdot \mathbf{k} + c\omega)^{-1}$$

where a, $\mathbf{b}$ and c are constants. A positive $N(\mathbf{k})$ requires $\mathbf{b} = 0$. For $a = 0$, one obtains a uniform energy spectrum ωN in wavenumber space, while for $c = 0$ one obtains a constant number density N, which is a well-known equilibrium solution of the Liouville equation for a system governed by Hamiltonian equations.

The search for thermal equilibrium solutions and the study of spectral evolution based on nonlinear interactions alone (i.e. Eq. (4.75)) are only relevant for closed systems in which the energy is conserved. They are not applicable to numerical wave prediction because ocean waves are not a closed system, since they gain energy from the atmospheric flow and lose energy to the underlying ocean. For growing wind waves the resulting spectra are typically nonequilibrium spectra, and only in the absence of wind may ocean waves be regarded as an approximately closed system. However, in that case the (swell) waves have such a small steepness that the nonlinear transfer time scale is large compared with the travel time of the waves across a typical ocean.

4.7.2 The equilibrium energy cascade

An equilibrium solution to the Hasselmann equation is in practice not relevant because it concerns the solution of a closed system in which for large times there are no energy fluxes within the system. A statistical equilibrium in which a constant flux is maintained through the system comes closer to the real situation. This was first proposed by Zakharov and Filonenko (1967), using arguments similar to the familiar Kolmogorov inertial energy cascade concept of isotropic turbulence. It was shown that for a wind input to the waves that is mainly concentrated at low wavenumbers and an isotropic dissipation term at high wavenumbers, the nonlinear interactions would adjust the spectrum $F(\mathbf{k})$ in the region between the spectral source and sink to a constant-flux equilibrium form $\sim k^{-7/2}$. In practice, wind input and dissipation are not sufficiently separated in the wavenumber domain to apply the Zakharov and Filonenko argument rigorously, and an isotropic spectrum is also not observed. Nevertheless, numerical experiments using the exact nonlinear transfer function (Komen *et al.*, 1984) suggest that the spectra tend to adjust in such a way that the directionally averaged spectra are close to $k^{-5/2}$ (corresponding to a $k^{-7/2}$ two-dimensional spectrum), with a rather constant energy flux towards higher wavenumbers. However, the directional distribution remains anisotropic. In addition, it is found that the nonlinear flux is generally rather strong so that small changes from the $k^{-5/2}$ form are sufficient to generate divergent fluxes which can balance nonzero input and dissipation source functions in the energy cascade region.

Thus, although the conditions for a constant nonlinear energy flux are not satisfied formally, the concept appears quite useful as a first approximation and it explains why the one-dimensional spectra tend to be fairly close to a $k^{-5/2}$ power law.

Strictly speaking, the original Zakharov and Filonenko argument can only be applied to the angular average of the wavenumber spectrum. Recently, Zakharov and Zaslavskii (1982) have extended the approach and they have found the following general form of the frequency spectrum:

$$F(\omega, \theta) = \frac{g^{4/3} P^{1/3}}{\omega^4} \mathcal{F}\left(\frac{\omega Q}{P}, \frac{M}{\omega P}, \cos\theta\right)$$

where P and M are energy and momentum flux to the high-frequency domain, while Q is the action flux to small wavenumbers. Here, information on the general function $\mathcal{F}$ can only be obtained by numerical means. In other words, in order to obtain the angular distribution of the waves one needs to determine numerically the evolution of the wave spectrum according to the Hasselmann equation. In this context it is important to note that Banner and Young (1994) found that for high frequencies the nonlinear transfer has a *bimodal* distribution. This finding makes perfect sense when we return to the stability diagram of Fig. 4.6 which shows that the nonlinear transfer indeed has the tendency to transport wave energy at an angle from the main propagation direction.

As remarked in Section 2.6.3 the observed angular distribution of the high-frequency part of the windsea spectrum shows a pronounced bimodality. Banner and Young (1994) have suggested that this bimodality is caused by the action of the nonlinear transfer. However, as pointed out by Morland (1996), the critical-layer mechanism results, for sufficiently high wind speed, in a bimodal distribution of the growth rate by wind. Alves and Banner (2003) have confirmed Morland's (1996) conjecture that not only nonlinear transfer but also the wind input term is important in determining the bimodal angular distribution of the windsea spectrum.

Another illuminating application of the ideas of Kolmogorov was given by Kitaigorodskii (1983). He used a local approximation (in wavenumber space) of the nonlinear transfer in Eq. (4.75) and he found that the nonlinear energy flux Φ scales with the action density (hence, the wavenumber spectrum) to the third power, or

$$\Phi = \alpha_4 \rho_w c^3 B^3, \tag{4.106}$$

where $B = k^4 F$ is the so-called 'angular average of the degree of saturation' (Phillips, 1985), F is the angular average of the two-dimensional spectrum, c is the phase speed, α_4 a constant and ρ_w the water density. The form (4.106) may be obtained using scaling arguments. The rate of change of the energy of the waves, as follows from multiplication of Eq. (4.75) by $\rho_w \omega$, is equal to the divergence of the energy flux in wavenumber space, or

$$\frac{1}{k} \frac{\partial}{\partial k} \Phi(k) = \rho_w \omega \left. \frac{\partial}{\partial t} N \right|_{\text{nl}}. \tag{4.107}$$

Introducing now the characteristic time τ_{nl} of the nonlinear transfer one finds (note: $\omega N = gF$)

$$\Phi(k) \sim \rho_w g \frac{k^2 F}{\tau_{\text{nl}}} \tag{4.108}$$

and the time scale τ_{nl} may be obtained from Eq. (4.75) as

$$\frac{1}{\tau_{\text{nl}}} \sim g^{1/2} k^{17/2} F^2 \tag{4.109}$$

where we used the scaling behaviour of the interaction coefficient $T_0^2 \sim k^6$. Combination of Eqs. (4.108) and (4.109) then results in Eq. (4.106).

The inertial subrange is now defined by that wavenumber range where the nonlinear interactions dominate. As a consequence, the divergence of the energy flux Φ vanishes and, hence, Φ is a constant $= \Phi_0$. Therefore, in the inertial subrange the wave spectrum becomes

$$F(k) = \left(\frac{\Phi_0}{\alpha_4 \rho_w} \right)^{1/3} g^{-1/2} k^{-7/2}. \tag{4.110}$$

Kitaigorodskii (1983) argued that the constant Φ_0 is determined by the energy flux per unit area from wind to waves, denoted by Φ_w, which he estimated to be proportional to $\rho_a u_*^3$. As a consequence, F would scale with $(\rho_a / \rho_w)^{1/3}$. However, the energy flux to the waves depends on the sea state and nowadays a somewhat sharper estimate of Φ_w may be provided using the wind input source function of the energy balance equation (2.86). From Section 3.5, the rate of change of wave energy by wind is given by

$$\rho_w g \left. \frac{\partial}{\partial t} F \right|_{\text{wind}} = \rho_w g \gamma F \tag{4.111}$$

with $\gamma = \epsilon\omega\beta(u_*/c)^2$ and $\epsilon = \rho_a/\rho_w$. The energy flux Φ_w then becomes

$$\Phi_w = \rho_w g \int k \, dk \, \gamma F \tag{4.112}$$

and, in agreement with Kitaigorodskii (1983), the condition $\Phi_0 = \Phi_w$ is imposed. Note that, as expected, Φ_0 depends on the wave spectrum. Using the scaling law of Eq. (4.110) in Eq. (4.112) one then finds

$$\Phi_0 \sim \epsilon^{1/2} \rho_a u_*^3$$

and, as a consequence, the inertial subrange spectrum becomes

$$F(k) = A\epsilon^{1/2} u_* g^{-1/2} k^{-7/2}. \tag{4.113}$$

Therefore, the equilibrium spectrum in the inertial subrange depends on the friction velocity and the square root of the air–water density ratio. In the angular frequency domain one then obtains, using $E(\omega) \, d\omega = F(k)k \, dk$,

$$E(\omega) = \alpha_T u_* g \omega^{-4}, \tag{4.114}$$

where $\alpha_T = 2A\epsilon^{1/2}$ is called 'Toba's constant', because Toba (1973) was one of the first who strongly advocated the ω^{-4} power law. He obtained Eq. (4.114) from observations and on dimensional grounds, arguing that there should be a strong coupling between wind and waves.

Nowadays, there is ample evidence for the ω^{-4} power law. However, there is still debate regarding the dependence on wind-speed scale, and what wind-speed scale to take. However, following the discussion in Section 2.6.2 and the theoretical treatment in Chapter 3 (in particular Section 3.2) it is concluded that there is a preference to choose as wind scale the friction velocity u_* and that the equilibrium spectrum scales linearly with u_*.

Observations of the ω^{-4} power law have been typically between $1.3 \times \omega_p$ and $3 \times \omega_p$ where ω_p is the peak frequency of the waves. For frequencies above $3 \times \omega_p$ there may still be little direct evidence of the wavenumber spectrum or the frequency spectrum but it cannot continue with a slope of $-7/2$ or -4 respectively. For example, the mean-square slope would greatly exceed optical (Cox and Munck, 1954) and radar estimates of mean-square slope (Jackson *et al.*, 1992). In fact use of Eq. (4.113) would give rise to an unbounded mean-square slope. In order to see this it should be noted that the wave slope is basically $\partial\eta/\partial x$ and therefore the slope spectrum S becomes

$$S = k^2 F,$$

hence the mean-square slope (mss), which is the integral over S,

$$\text{mss} = \int k \, dk \, S = \int dk \, k^3 F,$$

is seen to become unbounded when Eq. (4.113) is used since the integrand behaves as $k^{-1/2}$.

Clearly, the wavenumber spectrum cannot continue with a $-7/2$ power law for increasing wavenumber. Kitaigorodskii (1983) has postulated such a break in the spectral slope from $k^{-7/2}$ to k^{-4} (fully saturated). He argues that for increasing wavenumber the downward acceleration will increase and according to Phillips (1958) wave breaking occurs when the acceleration of the fluid exceeds the acceleration of gravity g. The wavenumber where wave breaking starts to occur scales as

$$k_g \sim g/(\Phi_0/\rho_w)^{2/3}.$$

For wavenumbers much larger than k_g the spectral shape is determined by wave breaking, which, to first approximation, is a process that is completely determined by the dynamics of waves. Therefore, only acceleration of gravity g, angular frequency ω or wavenumber k are the relevant parameters that specify the spectral shape. This was first suggested by Phillips (1958) who found that

$$F(k) = Bk^{-4} \tag{4.115}$$

while

$$E(\omega) = \alpha_p g^2 \omega^{-5}, \, \alpha_p = 2B, \tag{4.116}$$

where α_p is the Phillips parameter. Nevertheless, even the power law of Eq. (4.115) does not prevent the divergence of the mean-square slope. Hence, the high-wavenumber catastrophe is not completely resolved by Kitaigorodskii's argument. Experimental evidence for the transition of Toba's to Phillips' spectrum has been discussed already in Section 2.6.2. Although the amount of evidence is fairly limited there are indications that for the frequency spectrum such a transition indeed occurs at about 3 times the peak frequency.

It is remarked that there are alternative explanations for the transition from an ω^{-4} to an ω^{-5} power law. Hara and Belcher (2002) have elaborated the idea that when waves become steep, wind input to the waves gets reduced (these authors call this the 'sheltering effect' and its effect is similar to the quasi-linear effect discussed in Chapter 3). Therefore rather than using the process of wave breaking, as Kitaigorodskii did, Hara and Belcher use the idea of sheltering. They argue that Toba's spectrum (4.114) is based on a balance of wind input and nonlinear transfer. If Toba's law were to hold, the waves would get steeper and steeper for increasing angular frequency and wavenumber, and sheltering would become more important. For wavenumbers larger than a sheltering wavenumber k_s, the wind input to the waves is so much reduced that it starts to affect the balance with the nonlinear transfer, hence a different spectral shape emerges. However, in practice, sheltering only occurs for very high wavenumbers, and as a consequence the sheltering wavenumber k_s gives a break in spectral slope at much higher frequencies than observed in the field.

4.7.3 Concluding remarks on the short-wave spectrum

We conclude our discussion of the equilibrium energy cascade by pointing out a number of important corrections to be made to the present equilibrium ideas. Let us return to the expressions for the short-wave spectrum, Eqs. (4.115) and (4.116). As pointed out, the shape of the high wavenumber spectrum still not satisfactory, because the mean-square slope diverges. It is well known, however, that the dispersion relation for the short waves is affected by the orbital motion of the long waves (see for recent experimental evidence, Donelan *et al.*, 1999). Rather than the usual linear dispersion relation, it is more appropriate to use for the short waves

$$\omega(k) = \mathbf{k} \cdot \mathbf{u}_d + \sqrt{g|\mathbf{k}|}, \tag{4.117}$$

where $\mathbf{u}_d$ is a surface drift seen by the short waves, which is caused by the long waves. This surface drift is typically of the order of the friction velocity u_*. In order to show the effect of the Doppler shift on spectral shape, it is assumed that the frequency spectrum is invariant. An argument in favour of the invariance of the frequency spectrum is that for a Hamiltonian system the total frequency ω is conserved when following a wave group (see Eq. (2.78)), hence any function of ω is invariant. Thus, the spectral form of Eq. (4.114) is taken for $\omega \le 3\omega_p$ while for $\omega > 3\omega_p$ the Phillips spectrum of Eq. (4.116) is chosen. The two spectral shapes are joined together at $\omega = 3\omega_p$ by choosing the Phillips parameter appropriately, while the Toba parameter is taken to be $\alpha_T = 0.13$. The friction velocity was chosen to be $u_* = 0.6$ m/s. In Fig. 4.18 the impact of the surface drift on the wavenumber spectrum $F(k) = v_g E(\omega)/k$ is shown. While in the absence of the drift the wavenumber spectrum obeys for high wavenumbers the usual k^{-4} power law, in the presence of such a drift spectral levels are much reduced. In fact, as shown in Fig. 4.18 one finds in that case an approximate k^{-5} power law, in agreement with the observations of Donelan *et al.* (1999). On the other hand, for low wavenumbers, the spectrum is hardly affected by the presence of the surface drift, because these waves propagate much faster than the surface drift. Interestingly, with a k^{-5} power law the mean-square slope remains finite. Evidently, the surface drift plays an important role in shaping the high-wavenumber part of the spectrum. However, the consequences for air–sea interaction, for example, have not been studied yet.

The final remark concerns the convergence of the perturbation expansion we have employed in this chapter and possible consequences for the high-wavenumber part

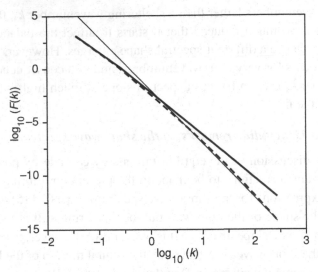

Fig. 4.18. Impact of surface drift on the wavenumber spectrum $F(k)$. Continuous line, no surface drift; dashed line, surface drift $u_d = u_*$; the thin line corresponds to the empirical relation $F(k) = 0.01k^{-5}$ of Donelan *et al.* (1999).

of the spectrum. Recall that in order to simplify the Hamiltonian of water waves, nonresonant interactions were eliminated as much as possible by means of the canonical transformation $A = A(a, a^*)$, Eq. (4.17). This transformation introduces the reduced action variable a which is used to define the action density spectrum $N(\mathbf{k})$ of the Boltzmann equation (4.75). Clearly N is just an approximation to the action density spectrum, and in order to obtain the total action density spectrum, say n, one has to employ Krasitskii's transformation (4.17) for the case of a random wave field. This is fairly straightforward to do and as a result one finds (Zakharov, 1992; Krasitskii, 1994)

$$
\begin{aligned}
n_0 = N_0 + 2 \int d\mathbf{k}_{1,2} &\left[\frac{|V^{(-)}_{0,1,2}|^2}{(\omega_0 - \omega_1 - \omega_2)^2} (N_1 N_2 - N_0 N_1 - N_0 N_2)\, \delta_{0-1-2} \right. \\
&+ \frac{|V^{(-)}_{1,0,2}|^2}{(\omega_0 - \omega_1 + \omega_2)^2} (N_1 N_2 + N_0 N_1 - N_0 N_2)\, \delta_{0-1+2} \\
&+ \frac{|V^{(-)}_{2,0,1}|^2}{(\omega_0 + \omega_1 - \omega_2)^2} (N_1 N_2 - N_0 N_1 + N_0 N_2)\, \delta_{0+1-2} \\
&\left. + \frac{|V^{(+)}_{0,1,2}|^2}{(\omega_0 + \omega_1 + \omega_2)^2} (N_1 N_2 + N_0 N_1 + N_0 N_2)\, \delta_{0+1+2} \right] \\
&+ \mathcal{O}(N^3).
\end{aligned}
$$

Note that the above expression for the complete action density contains two groups of terms, namely terms of the type N_1N_2 which give a fully nonlinear correction to the lowest-order expression for the action density, N_0, and terms involving N_0 which are termed quasi-linear because they are proportional to N_0 itself. The quasi-linear terms stem from the third-order term of the canonical transformation (4.17) and their effect is usually neglected in discussions on the so-called 'second-order wave spectrum' (Barrick and Weber, 1977). They give nevertheless an important contribution to the second-order correction of the surface-wave spectrum.

In order to illustrate the importance of the quasi-linear term we have evaluated for one-dimensional propagation the correction to the wave-height spectrum. The result for the wavenumber spectrum $F(k)$ is (see also Creamer *et al.*, 1989)

$$F(k) = F_1(k) + \frac{1}{2}k^2 \int_{k/2}^{\infty} \mathrm{d}k' \, F_1(k')F_1(|k - k'|) - k^2 F_1(k) \int_{0}^{\infty} \mathrm{d}k' \, F_1(k'),$$

$$(4.118)$$

where $F_1(k)$ is the first-order spectrum and the last two terms represent the second-order spectrum. The full nonlinear term was already found by Barrick and Weber (1977), but the last, quasi-linear term is missing. This was explained by Creamer *et al.* (1989) who noted that Barrick and Weber (1977) evaluated the wave spectrum from the surface elevation correct to second order only, but to be consistent third-order corrections are required as well. As shown in Fig. 4.19, which gives the first-order spectrum $F_1(k) \sim k^{-3}$ and the contribution of the second-order spectrum, the

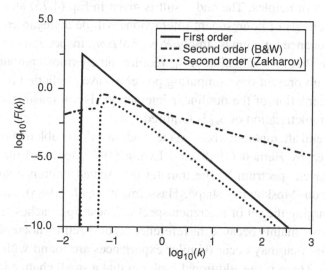

Fig. 4.19. Second-order effects on the surface wave-height spectrum, illustrating the importance of the quasi-linear term. Results are from Barrick and Weber (1977) (B&W) and Zakharov (1992).

quasi-linear term plays a vital role. When only the full nonlinear term of Barrick and Weber (1977) is taken into account, the second-order spectrum is, as expected, small for low wavenumbers, but for high wavenumbers it is far more important than the first-order contribution to the spectrum. In fact, for a k^{-3} power law it can be shown analytically that the second-order spectrum behaves like k^{-1}. This would suggest that perturbation theory is divergent. However, it can be shown analytically that the quasi-linear term cancels the singular behaviour of the full nonlinear term, and as a result the second-order spectrum only gives a small correction to the first-order spectrum. It is therefore of vital importance to include the quasi-linear term because one may now conclude that the perturbation approach of this chapter appears to be convergent. This conclusion is confirmed by recent direct numerical simulations of the Euler equations by Onorato *et al.* (2002). The total one-dimensional spectrum $F(k)$ was found to follow closely a $k^{-5/2}$ law, suggesting that the contribution of the second-order spectrum is relatively small.

Finally, we remark that for low wavenumbers too the quasi-linear term has a marked impact on the result for the second-order spectrum. This may have consequences for instruments that measure the second-order spectrum using high-frequency (HF) radar and that invert Eq. (4.118) to obtain the first-order spectrum. At present, only the full nonlinear term is used in the inversion (Wyatt, 2000).

4.8 Parametrization of nonlinear transfer

In Section 4.5 we derived the source function S_{nl}, describing the nonlinear energy transfer, from first principles. The end result is given in Eq. (4.75) and again it is pointed out that effects of nonresonant interactions will be disregarded from now on so that the resonance function is simply $R_i = \pi \delta(\Delta\omega)$. Inspection of the form of Eq. (4.75) reveals that the evaluation of S_{nl} requires an enormous amount of computation. Even with present-day computing power a wave prediction model based on the exact presentation of the nonlinear interactions is not feasible. Therefore, some form of parametrization of S_{nl} is required.

In the past, several attempts have been made to address this problem. For example, the approach used by Barnett (1968) and Ewing (1971) replaced the nonlinear transfer of any given spectrum by the transfer of a scaled reference spectrum of prescribed (Pierson–Moskowitz) shape. Hasselmann *et al.* (1985) extended this approach by enlarging the set of reference spectra. These approaches are not very successful, however, mainly because the reference set of spectra is not representative for all possibilities that may occur (similar experiences are found with the use of neural networks). There is the additional problem that a small change in the wave spectrum results in a significant change in the nonlinear transfer. In the initial stages of wave growth, the spectral shape differs from the Pierson–Moskowitz shape which holds for old windsea. Typically, the spectral peak is more narrow and enhanced compared with the one for old windsea. As a consequence, for young windsea the nonlinear transfer is an order of magnitude larger than for old windsea. Therefore, alternative parametrizations of the nonlinear transfer had to be developed, which have the same number of degrees of freedom as the spectrum itself.

Accordingly, one needs to consider approaches in which the nonlinear transfer is approximated by general nonlinear operator expressions. In the first operator parametrization, which is called the 'diffusion approximation' or 'local interaction approximation', the nonlinear transfer is given by a fourth-order, cubic diffusion operator (Hasselmann and Hasselmann, 1981). In this parametrization the nonlinear interaction coefficient $|T|^2$ is assumed to be strongly peaked near the central interaction point $k_1 = k_2 = k_3 = k_4 = k$ and by comparison the action density spectrum is slowly varying. Nevertheless, the resulting expression for the nonlinear transfer is fairly cumbersome, and Zakharov and Pushkarev (1999) have recently introduced a simplified nonlinear differential operator. The differential operator was constructed in such a way that the basic conservation laws and Boltzmann's H theorem were satisfied. The diffusion operator gives a good approximation to the contributions in the vicinity of the central interaction point, but contributions further away from the central interaction point cannot be neglected. In fact, significant contributions to the nonlinear transfer come from the intermediate quadruplets whose separations are probably slightly too large to be adequately represented by a local approximation.

Hence, Hasselmann *et al.* (1985) introduced a second operator parametrization, called the 'discrete interaction approximation' (DIA), which is similar in spirit to the diffusion approximation but overcomes the shortcomings of this approach. A nonlinear interaction operator was constructed by considering only a small number of interaction configurations consisting of neighbouring and finite-distance interactions. In fact, it was found that the nonlinear transfer could be well simulated by just one mirror-pair of intermediate-range interaction configurations. In each configuration, two wavenumbers were taken as identical $k_1 = k_2 = k_0$. The wavenumbers k_3 and k_4 are of different magnitude and lie at an angle to the wavenumber k_0, as required by the resonance conditions. The second configuration is obtained from the first by reflecting the wavenumbers k_3 and k_4 with respect to the k_0-axis. The scale and direction of the reference wavenumber are allowed to vary continuously in wavenumber space.

The simplified nonlinear operator is computed by applying the same symmetrical integration method as is used to integrate the exact transfer integral (see also Hasselmann and Hasselmann, 1985). However, DIA is much more economical because the integration is taken over a two-dimensional continuum and two discrete interactions instead of five-dimensional interaction phase space. Just as in the exact case, DIA conserves energy, momentum and action. For the configurations

$$\omega_1 = \omega_2 = \omega_0,$$
$$\omega_3 = \omega_0(1 + \lambda) = \omega_+,$$
$$\omega_4 = \omega_0(1 - \lambda) = \omega_-, \tag{4.119}$$

where $\lambda = 0.25$, satisfactory agreement with the exact computations was achieved. From the resonance conditions the angles θ_3, θ_4 of the wavenumbers $\mathbf{k}_3(\mathbf{k}_+)$ and $\mathbf{k}_4(\mathbf{k}_-)$ relative to $\mathbf{k}_0$ are found to be $\theta_3 = 11.5°$, $\theta_4 = -33.6°$.

In agreement with the principle of detailed balance (see Section 4.7.1), one finds

$$\frac{\partial}{\partial t}\begin{pmatrix} N_0 \\ N_+ \\ N_- \end{pmatrix} = \begin{pmatrix} -2 \\ +1 \\ +1 \end{pmatrix} Cg^{-8} f_0^{19}[N_0^2(N_+ + N_-) - 2N_0 N_+ N_-]J_s,$$

$$\tag{4.120}$$

where $\partial N_0/\partial t$, $\partial N_+/\partial t$, $\partial N_-/\partial t$ are the rates of change in action density at wavenumbers $\mathbf{k}_0$, $\mathbf{k}_+$, $\mathbf{k}_-$ owing to the discrete interactions within the infinitesimal interaction phase-space element $\Delta\mathbf{k}_0$, C is a numerical constant, and $f_0 = \omega_0/2\pi$. Furthermore, the Jacobian $J_s = |\Delta\mathbf{k}_0/\Delta\mathbf{k}_s|$ (with s having the values 0, + or −) only differs from 1 when the wavenumber increment depends on the wavenumber itself. The net source function S_{nl} is obtained by summing Eq. (4.120) over all wavenumbers, directions and interaction configurations.

As discussed in Hasselmann and Hasselmann (1985) the usefulness of DIA follows in particular from its correct reproduction of the growth curves for growing windsea. However, the DIA has a tendency to give, compared with the exact calculations, a stronger negative lobe, with the consequence that the high-frequency part of the spectrum is somewhat lower. However, the big advantage of DIA is its satisfactory reproduction of the low-frequency positive lobe which controls the downshift of the spectral peak. Both aspects of DIA are illustrated in Fig. 4.20 for young windsea.

Despite the success of the DIA a number of authors (Polnikov, 1991; Zakharov and Pushkarev, 1999; Lin and Perry, 1999; van Vledder *et al.*, 2000) have felt the need for improvement because the accuracy of the DIA was not regarded as sufficient for the JONSWAP spectrum. For example, Hashimoto and Kawagushi (2001) and van Vledder (2001) have tried to improve on the DIA by introducing additional quadruplets. However, these efforts have not significantly changed the situation. Recently, Polnikov and Farina (2002) have performed a 'beauty' contest between different approximations of the nonlinear transfer. The quality of any approximation was judged by a combination of a measure for relative error and efficiency. In this context it turned out that DIA performed very well, which was quite surprising considering that DIA was developed more than 15 years ago.

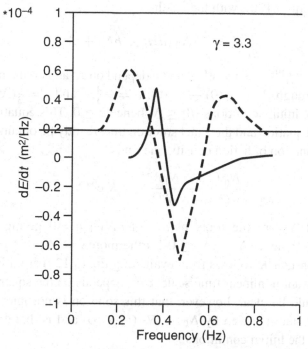

Fig. 4.20. Comparison of directionally averaged nonlinear transfer of wave variance density obtained from exact calculation (continuous line) and the DIA approximation (dashed line) for a JONSWAP spectrum with overshoot parameter $\gamma = 3.3$ (from Hasselmann *et al.*, 1985).

Before we close this section it is mentioned for completeness that for one quadruplet in isolation an exact solution to Eq. (4.120) may be obtained. We give here the solution for a fixed increment $\Delta \mathbf{k}$, hence $J_s = 1$. It is readily seen that by construction the action of the quadruplet, $\mathcal{N} = N_0 + N_+ + N_-$, is conserved. Also, because of the resonance conditions $2\omega_0 = \omega_+ + \omega_-$ and $2\mathbf{k}_0 = \mathbf{k}_+ + \mathbf{k}_-$, energy $\mathcal{E} = \omega_0 N_0 + \omega_+ N_+ + \omega_- N_-$ and momentum $\mathcal{P} = \mathbf{k}_0 N_0 + \mathbf{k}_+ N_+ + \mathbf{k}_- N_-$ of the quadruplet are conserved as well. Since there are so many conserved quantities in this problem it should not come as a big surprise that an exact solution to the time evolution of one quadruplet exists. From Eq. (4.120) one has

$$\frac{\partial}{\partial t} N_+ = \frac{\partial}{\partial t} N_-,$$

and

$$\frac{\partial}{\partial t} N_0 = -2 \frac{\partial}{\partial t} N_+.$$

Integration of these two equations allows elimination of N_+ and N_- from the first component of Eq. (4.120), with the result

$$\frac{\partial}{\partial t} N_0 = 3\alpha N_0 \left(a N_0^2 + b N_0 + c \right),$$

(4.121)

where $\alpha = C g^{-8} f_0^{19}$, $a = 1$ while b and c depend on $\mathcal{N}$ and on the initial values of N_+ and N_- through $\beta = N_+(0) - N_-(0)$; $b = -\frac{4}{3}\mathcal{N}$ and $c = \frac{1}{3}(\mathcal{N}^2 - \beta^2)$. Note that for positive inital conditions $|\beta| \leq \mathcal{N}$ hence $c \geq 0$. This equation is a special case of Abel's equation and the exact solution may readily be obtained. By solving for time as a function of action density one finds

$$\frac{N_0^2 (N_0 - N_1)^{s-1}}{(N_0 - N_2)^{s+1}} = K \, e^{6\alpha c t}$$

(4.122)

where N_1 and N_2 are the roots of $a N_0^2 + b N_0 + c = 0$ giving $N_{1,2} = \frac{2}{3}\mathcal{N} \pm (\frac{1}{3})(\mathcal{N}^2 + 3\beta^2)^{1/2}$, hence $N_1 > N_2 \geq 0$. Furthermore $s = (N_1 + N_2)/(N_1 - N_2) > 0$, while the constant K follows from evaluating Eq. (4.122) at initial time. As expected the relevant nonlinear time scale, $\alpha c t$, depends on the square of the action density. It should be clear, however, that this solution is not interesting because the asymptotic state is given by $N_0 \to N_2$ ($t \to \infty$) and is therefore completely determined by the initial conditions.

However, as already mentioned, ocean waves are not a closed system, hence it would be more relevant to try to seek a solution of the system (4.120) by adding wind input and dissipation. Denoting the sum of growth and damping by $\gamma_0 N_0$, the evolution equation for N_0 becomes

$$\frac{\partial}{\partial t} N_0 = \gamma_0 N_0 - 2\alpha N_0 \left[N_0(N_+ + N_-) - 2N_+ N_- \right]$$

while for N_+ we have

$$\frac{\partial}{\partial t} N_+ = \gamma_+ N_+ + \alpha N_0 \left[N_0(N_+ + N_-) - 2N_+ N_- \right]$$

and a similar equation for N_- (obtained by interchanging the subscripts $+$ and $-$). Although the case of equal growth rates $\gamma = \gamma_+ = \gamma_-$ may be solved exactly, this is, as far as I know, not possible for the more interesting case of $\gamma > 0$ and $\gamma_+, \gamma_- < 0$, which mimics an energy cascade. Also, the more realistic case should deal with a large number of gravity waves and therefore a numerical solution of the energy balance equation is required. This will be discussed in Chapter 5.

4.9 Wave dissipation

The least-understood aspect of the physics of wave evolution is the dissipation source function. Waves may lose energy continuously by viscous dissipation and by the highly intermittent process of wave breaking. In addition, the (small-scale) breaking waves generate eddies in the surface layer of the ocean. These eddies may give rise to damping of the longer ocean waves. Furthermore, ocean waves may lose energy because of the generation of organized motions in the ocean such as Langmuir circulations and in the presence of a vertical shear in the current. Dissipation of wave energy by molecular viscosity is well understood and can easily be calculated. However, it is only an important sink for the short gravity–capillary waves, which have wavelengths of the order of 1 cm. For the longer gravity waves viscous dissipation is not important. Dissipation of long gravity waves by ocean-surface turbulence may also be determined (Jenkins, 1987) and we briefly discuss the consequences of this approach. Finally, we are only beginning to understand the energy loss by means of the generation of Langmuir circulation; it will therefore not be discussed in this context (but see the interesting work of Teixeira and Belcher (2002)).

Understanding and modelling of the wave breaking process are of importance in achieving an accurate representation of the principal sink in the action balance equation. Unfortunately, even nowadays there has not been much progress in obtaining a convincing model of dissipation caused by wave breaking or white capping. This should not come as a surprise, because wave breaking is a truly nonlinear phenomenon that cannot be captured by the perturbation techniques that have been discussed in this chapter. A straightforward theoretical approach does not, therefore, seem possible. Although there has been considerable progress in the numerical modelling of breaking of individual waves, it is hard to see how these results may be extended to the general case of a random wave field.

Nevertheless, there has been progress in modelling dissipation by white caps in a semi-empirical manner. Basically, three approaches may be distinguished. The first approach assumes that white capping is the main cause for the dissipation process, which is regarded as local in space; this approach has been put forward by Hasselmann (1974). In the second approach, Phillips (1985) started from the completely opposite assumption that wave dissipation is local in wavenumber space. The third approach, proposed by Jenkins (1987), advocated the picture that breaking waves will generate ocean eddies which in turn will damp the waves.

4.9.1 White-cap model

The assumptions behind Hasselmann's white-cap model are that the white caps can be treated as a random distribution of perturbations, which are formally equivalent to pressure pulses, and that the scales of the white caps in space and time are small compared with the scales of the associated wave. The theoretical development consists of two steps. First, it is shown that all processes that are weak in the mean, even if they are strongly nonlinear locally, yield source functions that are quasi-linear in lowest approximation. The source function consists of the spectrum at the wavenumber considered multiplied by a factor that is a functional of the entire wave spectrum. A derivation in the general context of wave processes is given in Komen *et al.* (1994, Section I.2.8). The second step consists of the determination of the factor for the special case of white capping. This involves a number of fairly complex approximations based on the assumed scales of the white caps. The details are discussed by Hasselmann (1974).

Introducing the mean frequency $\langle\omega\rangle$ by means of the inverse mean frequency,

$$\langle\omega\rangle = \int d\mathbf{k}\ F(\mathbf{k}) \Big/ \int d\mathbf{k}\ \frac{F(\mathbf{k})}{\omega} \qquad (4.123)$$

with F the wavenumber spectrum, and a similar relation for the mean wavenumber $\langle k\rangle$, the final result is the following dissipation source function:

$$S_{ds} = -\gamma_d N, \qquad (4.124)$$

with

$$\gamma_d = \beta\langle\omega\rangle \left(\langle k\rangle^2 m_0\right)^m \left[\frac{k}{\langle k\rangle} + a\left(\frac{k}{\langle k\rangle}\right)^2 + \cdots\right]. \qquad (4.125)$$

Here, β, a and m are constants which still need to be determined. It is remarked that in the original work of Hasselmann (1974) the second term in the square bracket is absent. The reason for this is that Hasselmann assumed a large separation between the length scale of the waves and the white caps, giving a power 1 for the wavenumber in the dissipation term. For the high-frequency part of the wave spectrum, however, such a large gap between waves and white caps may not exist, therefore allowing the possibility of a different dependence of dissipation on wavenumber.

The first rational attempt to determine the unknown coefficients in the dissipation source function was reported by Komen *et al.* (1984). In passing, note that this work showed for the first time that a numerical solution of the energy balance equation gave a realistic equilibrium solution for the wave spectrum, thereby paving the way for the development of numerical wave prediction models that solve the energy balance equation from first principles.

Komen *et al.* (1984) started from the empirical expression for wind input of Snyder *et al.* (1981), which was adapted to accommodate friction-velocity scaling, while the exact form of Hasselmann's nonlinear transfer was taken. For a constant wind speed, the energy balance equation was integrated until stationary conditions were reached, and the unknown coefficients m and β were chosen in such a way that the equilibrium spectrum resembled the Pierson–Moskowitz (1964) spectrum as closely as possible (note that in their work a was put to zero from the outset). The power m was found to be equal to 2 while the coefficient β was of the order of 3.

4.9.2 Quasi-saturated model

Phillips (1985) and Donelan and Pierson (1987) have followed a different approach. While formally their model is still basically consistent with the quasi-linear form (Eq. (4.124)) for a process that is weak in the mean, the damping factor γ_d is estimated differently from Hasselmann (1974). Rather than assuming that the space and time scales of the white-capping process are small compared with the characteristic wavelength and periods of the waves for which the dissipation is being determined, i.e. white caps are highly local in space and time, they make the complementary assumption that white capping is essentially local in wavenumber space.

The wave breaking process is pictured as highly nonlinear in wave steepness, having no effect until some limiting steepness is reached when the wave becomes unstable and spills or plunges forward, producing white caps at large scales or a micro-breaker at small scales. At the end of the breaking event a substantial energy loss may occur. This is illustrated by Rapp and Melville (1990) who studied the interaction and breaking of waves in a wave packet where losses of up to 30 per cent of the momentum per breaker were found. Wave breaking is thought to occur mainly in wave groups and the onset of breaking is caused by nonlinear focussing. As a result of focussing, a uniform wave train becomes strongly modulated (Benjamin–Feir instability) and it becomes much more likely that the biggest wave in the group exceeds the limiting steepness, and hence breaks. This approach has been elaborated recently by, for example, Song and Banner (2002) and Banner and Song (2002) who inferred a robust threshold variable for wave breaking.

Phillips (1985) has argued that in the equilibrium range the wind input, nonlinear transfer and dissipation are roughly of equal importance, and as there is no internal wavenumber scale, the ratio of the three terms must be constant. Balancing dissipation with the nonlinear transfer and assuming that the four-wave interactions may be estimated by means of a local-in-wavenumber expression he finds that the dissipation rate is cubic in the degree of saturation $B(\mathbf{k})$, thus

$$S_{ds} = -C\omega k^{-4} B^3(\mathbf{k}) \tag{4.126}$$

where (see Eq. (4.106)) $B = k^4 F(\mathbf{k})$ and C is a constant. In order to compare with Hasselmann's white-cap model we write the dissipation source function in the form of Eq. (4.124) where for the Phillips model γ_d is given by

$$\gamma_d = C\omega B^2(\mathbf{k}). \tag{4.127}$$

Therefore, the damping rate in Phillips' model is determined by the local spectrum and not, as seen in the white-cap model, by the integral properties of the wave field.

Note   There are some interesting differences between the approach taken by Kitaigorodskii (1983) (see Section 4.7.2) and Phillips (1985) to understand the equilibrium range of the spectrum. Kitaigorodskii (1983) writes the nonlinear transfer as a divergence of an energy flux, and obtains the equilibrium range from the condition that nonlinear transfer dominates, hence the divergence of the energy flux vanishes. Phillips obtains the same expression for the equilibrium range by balancing Plant's (1982) wind-input term with the nonlinear dissipation source function (4.126). However, numerical simulations with the energy balance equation suggest, in agreement with Kitaigorodskii (1983) and Zakharov and Filonenko (1967), that nonlinear transfer dominates in the equilibrium range.

Recently, there have been some interesting new developments in this context. By assuming a balance between wind input and dissipation, Donelan (2001) determined from observations of spectral shape and wind input an optimal form of Eq. (4.127). However, because the state of the short gravity waves is affected by the longer waves, there is a need to introduce a dependence on the mean-square slope of the long waves. The problem with this approach is that the important effects of nonlinear four-wave interactions have not been accounted for. Furthermore, following the work of Song and Banner (2002) and Banner and Song (2002), Alves and Banner (2003) have studied the advantages of a relative of the parametrization (4.127), by replacing the degree of saturation B by its angular average. The dependence on the angular degree of saturation is only switched on when the degree of saturation exceeds an empirically determined threshold. The result is a fairly complicated expression for dissipation but simulations with the exact nonlinear transfer and various forms of wind input look reasonable.

4.9.3 Eddy-viscosity model

When waves are breaking, considerable amounts of energy and momentum are transferred to the ocean column thereby feeding the large-scale ocean motions and generating ocean turbulence. The eddies will give rise to energy and momentum transport, and Jenkins (1987) assumed that the eddy transport can be modelled in a similar way as the momentum transport by molecular viscosity, except that the eddy viscosity ν may depend on depth. It is then straightforward to obtain the wave dissipation source function. It has the same form as Eq. (4.124), where γ_d is given by

$$\gamma_d = 8k^3 \int_{-\infty}^{0} dz \, \nu(z) \, e^{2kz}.$$ (4.128)

Hence, the wavenumber dependence of the dissipation source function depends on the depth dependence of the eddy viscosity ν. For a constant eddy viscosity, $\nu = \nu_0$, the damping rate becomes $\gamma_d = 4\nu_0 k^2$ which corresponds to the familiar expression for wave damping owing to viscosity (Lamb, 1932). However, if ν increases linearly with depth the damping rate is found to increase linearly with wavenumber; but, near the ocean surface the eddy viscosity cannot vanish and a more appropriate model is (Terray *et al.*, 1999)

$$\nu = \kappa w_*(|z| + z_0)$$

where w_* is the friction velocity in water and the roughness z_0 is of the order of the significant wave height (reflecting the vigorous action of wave breaking on the ocean surface layer). Such a model for eddy viscosity is qualitatively in close agreement with Hasselmann's white-cap model (Eq. (4.125)), because the wave damping depends on a linear combination of a k term and a k^2 term.

Unfortunately, although this model gives attractive results, in practice damping rates are larger than growth rates by wind (Belcher *et al.*, 1994). The reason is simply that the eddy-turnover times in the ocean are much longer than the wave periods, hence most of the eddies do not have sufficient time to transport wave momentum. A straightforward application of rapid-distortion ideas (see Section 3.3.1) reveals that wave dissipation would scale in a similar manner as the wind-input term, namely as $\omega \, (w_*/c)^2$.

4.9.4 Discussion

Whether the wave dissipation process is local in ordinary space or in wavenumber space is hard to decide, also from observations. For example, from the work of Pierson *et al.* (1992), who examined the spectral changes of highly nonlinear gravity waves in a wave tank, one may find support for the contention that the wave-breaking process is local in wavenumber space. This is so because most of the energy loss is seen to occur at the peak of the spectrum. However, this experimental evidence is not entirely convincing because the quasi-linear white-cap model will also give rise to an energy loss that is mostly concentrated near the peak of the spectrum. Furthermore, experiments such as those reported in Pierson *et al.* (1992) involve steep waves, and therefore four-wave interactions are expected to be important as well, and they also will give rise to similar changes in the spectrum as observed.

An argument in favour of a quasi-linear form of wave breaking follows from the work of Donelan *et al.* (1972) who observed that wave breaking occurs as a wave passes through the peak of a wave group, suggesting that wave breaking depends not only on the state of the individual wave but rather on the occurrence of the wave group and nonlinear focussing. As wave groups are formed by the superposition of at least two waves with different wavenumbers, this suggests that wave damping depends on some integral properties of the wave field.

Another argument in favour of the quasi-linear formulation concerns the behaviour in the limit of a narrow spectrum. Thus,

$$F = E_0 \delta(\mathbf{k} - \mathbf{k}_p)$$

and the damping rate from the quasilinear model (4.125) becomes

$$\gamma_d = \beta \omega_p \left(k_p^2 E_0 \right)^m (1 + a)$$

which is well behaved. Accordingly, the rate of change of the energy E_0 of a single wave due to white capping depends, as expected, nonlinearly on its energy. However, for the quasi-saturated model (Eq. (4.127)) the limit for a narrow spectrum is not defined.

The above two arguments give a slight preference for the white-cap model of Eq. (4.125) which in the remainder of this book will be adopted as the model for dissipation of wave energy. Although the above arguments are to a large extent of a formal nature, this is probably the best that can be done under the present circumstances. Observational studies have, so far, not been able to discriminate between the two proposed models for wave dissipation. For example, Felizardo and Melville (1995) correlated ambient noise as produced by breaking waves with the total dissipation from Hasselmann's and Phillips' models but similar (high) correlations were found. Hence, at present, one can only utilize formal arguments to distinguish between the two. However, the recent work by Melville and Matusov (2002) may shed further light on this issue because they obtained the frequency distribution of dissipation by studying white-cap coverage from video pictures of the sea surface.

4.10 Summary of conclusions

In this chapter we have studied effects of nonlinearity on the evolution of deep-water gravity waves. Our starting point was the Hamiltonian for water waves which plays a central role in the development of the nonlinear theory. Using Hamilton's equations, the complex amplitude equations for finite amplitude may be obtained in a systematic way.

As a first example we have discussed effects of finite amplitude on the dispersion relation of a single gravity wave. Finite-amplitude gravity waves are, however, unstable to sideband perturbations. This instability is known as the Benjamin–Feir instability or the modulational instability and may be regarded as an example of a four-wave interaction process. In the context of the evolution of a random wave field, four-wave interactions are of prime importance in understanding the development of the wave spectrum with time.

We have therefore discussed in great detail the derivation of the nonlinear source function for a continuous spectrum of waves. In particular we have stressed the underlying assumptions, such as a finite but small wave steepness with the consequence that the probability distribution is close to a Gaussian, and the assumption of a homogeneous wave field. In addition, the assumption of a limited wavenumber range has also been made (the problem of short waves riding on a long wave).

It is important to ask the question of how restrictive these assumptions are. In other words, can one safely apply the Boltzmann equation (4.75), which describes both resonant and nonresonant four-wave processes, to the practical case of wave prediction? Monte Carlo simulations of the Zakharov equation for the case of one-dimensional propagation suggest that even for steep deep-water waves and narrow spectra the Boltzmann equation is giving a fair description of spectral evolution and also of the probability distribution (e.g. kurtosis) of the surface elevation. Nonresonant interactions do occur on a very short time scale but they are probably not relevant for practical wave prediction because the main interest is in large-time evolution. The nonresonant interactions play a role similar to that of transients in the solution of an initial-value problem. They die out very quickly because of phase mixing and as a consequence for large times only resonant interactions contribute to spectral change. We also discussed the consequences of the assumption of a homogeneous wave field. If inhomogeneities in the wave field are allowed for, a rapid nonlinear energy transfer is found, provided that the wave spectrum is sufficiently narrow. Energy transfer is caused by nonresonant four-wave interactions associated with the random version of the Benjamin–Feir instability. This rapid energy transfer tends to broaden the spectrum, thereby removing the cause of the instability. After a while what is left is the relatively slow nonlinear energy transfer of a homogeneous wave field. Therefore, for large-scale applications the assumption of a homogeneous wave field is well satisfied.

From the Monte Carlo simulations we have also learned that the statistical approach gives accurate estimates of the deviations of the probability distribution from the normal distribution. Parameters such as the kurtosis provide valuable information on extreme wave conditions which occur, for example, when freak waves are present. However, validation of the results of the present approach with observations of extremes collected over a long period is clearly desirable.

It may be argued that wave dissipation has been treated as the Cinderella of the source functions. To a large extent this is true because this process is really nonlinear and straightforward perturbation techniques are inadequate. Nevertheless, Hasselmann (1974) was able to impose a number of constraints on the general form of the dissipation source function, assuming that the main source of dissipation is white capping. By insisting that in equilibrium conditions the wave spectrum resembles the Pierson–Moskowitz spectrum as closely as possible, Komen *et al.* (1984) were able to determine a number of unknown constants in the dissipation term. A crucial step in this approach was the use of the empirical wind-input term of Snyder *et al.* (1981). Since the nonlinear interactions conserve momentum and energy, this implies that the overall level of dissipation is fairly well established because it almost balances the gain by wind (effects of advection are normally quite small). However, at present, uncertainty remains regarding the spectral distribution of the dissipation, but it is expected that, in the near future, observations from field campaigns will shed further light on this issue.

4.11 Appendix: nonlinear transfer coefficients

The correct version of the interaction coefficient $T(\mathbf{k}_1, \mathbf{k}_2, \mathbf{k}_3, \mathbf{k}_4)$ was first given by Krasitskii (1990, 1994) and Zakharov (1992). It is recorded here for reference. Adopting the simplified notation introduced in Section 4.1 we write

$$T_{1,2,3,4} = T(\mathbf{k}_1, \mathbf{k}_2, \mathbf{k}_3, \mathbf{k}_4).$$

We have

$$
\begin{aligned}
T_{1,2,3,4} = W_{1,2,3,4} \\
&- V^{(-)}_{1,3,1-3} V^{(-)}_{4,2,4-2} \left[\frac{1}{\omega_3 + \omega_{1-3} - \omega_1} + \frac{1}{\omega_2 + \omega_{4-2} - \omega_4} \right] \\
&- V^{(-)}_{2,3,2-3} V^{(-)}_{4,1,4-1} \left[\frac{1}{\omega_3 + \omega_{2-3} - \omega_2} + \frac{1}{\omega_1 + \omega_{4-1} - \omega_4} \right] \\
&- V^{(-)}_{1,4,1-4} V^{(-)}_{3,2,3-2} \left[\frac{1}{\omega_4 + \omega_{1-4} - \omega_1} + \frac{1}{\omega_2 + \omega_{3-2} - \omega_3} \right] \\
&- V^{(-)}_{2,4,2-4} V^{(-)}_{3,1,3-1} \left[\frac{1}{\omega_4 + \omega_{2-4} - \omega_2} + \frac{1}{\omega_1 + \omega_{3-1} - \omega_3} \right]
\end{aligned}
$$

$$-V^{(-)}_{1+2,1,2}V^{(-)}_{3+4,3,4}\left[\frac{1}{\omega_{1+2}-\omega_1-\omega_2}+\frac{1}{\omega_{3+4}-\omega_3-\omega_4}\right]$$

$$-V^{(+)}_{-1-2,1,2}V^{(+)}_{-3-4,3,4}\left[\frac{1}{\omega_{1+2}+\omega_1+\omega_2}+\frac{1}{\omega_{3+4}+\omega_3+\omega_4}\right]$$

where the second-order coefficients $V^{(\pm)}$ are defined as

$$V^{(\pm)}_{1,2,3}=\frac{1}{4\sqrt{2}}\left\{[\mathbf{k}_1\cdot\mathbf{k}_2\pm k_1k_2]\left(\frac{\omega_1\omega_2}{\omega_3}\frac{k_3}{k_1k_2}\right)^{1/2}\right.$$

$$+[\mathbf{k}_1\cdot\mathbf{k}_3\pm k_1k_3]\left(\frac{\omega_1\omega_3}{\omega_2}\frac{k_2}{k_1k_3}\right)^{1/2}$$

$$\left.+[\mathbf{k}_2\cdot\mathbf{k}_3+k_2k_3]\left(\frac{\omega_2\omega_3}{\omega_1}\frac{k_1}{k_2k_3}\right)^{1/2}\right\}$$

with $k_i=|\mathbf{k}_i|$, $\omega_i=\omega(k_i)$. The third-order coefficient is defined as

$$W_{1,2,3,4}=U_{-1,-2,3,4}+U_{3,4,-1,-2}-U_{3,-2,-1,4}-U_{-1,3,-2,4}$$
$$-U_{-1,4,3,-2}-U_{4,-2,3,-1}$$

with

$$U_{1,2,3,4}=\frac{1}{16}\left(\frac{\omega_1\omega_2}{\omega_3\omega_4}k_1k_2k_3k_4\right)^{1/2}$$

$$\times[2(k_1+k_2)-k_{2+4}-k_{2+3}-k_{1+4}-k_{1+3}],$$

and $k_{i\pm j}=|\mathbf{k}_i\pm\mathbf{k}_j|$, $\omega_{i\pm j}=\omega(k_{i\pm j})$. Note that I have used a Fourier transform without the factor of 2π. As a consequence, compared with Krasitskii (1994) and Zakharov (1992) the second-order coefficient is larger by a factor 2π, while the third-order coefficient is larger by a factor of $4\pi^2$. The advantage of this is that the narrow-band limit of $T_{1,2,3,4}$ simply becomes $T_{1,1,1,1}=k_1^3$ while the second moment simply becomes $\langle a_i a_j^*\rangle=N_i\delta(\mathbf{k}_i-\mathbf{k}_j)$.

Finally, in order to evaluate the second-order correction to the action density spectrum the coefficient $B^{(2)}_{1,2,3,4}$ of the Krasitskii transformation (4.17) is required. It reads

$$B^{(2)}_{1,2,3,4}=\Gamma^{(-)}_{2,3,2-3}\Gamma^{(-)}_{4,1,4-1}+\Gamma^{(-)}_{2,4,2-4}\Gamma^{(-)}_{3,1,3-1}-\Gamma^{(-)}_{1,3,1-3}\Gamma^{(-)}_{4,2,4-2}$$

$$-\Gamma^{(-)}_{1,4,1-4}\Gamma^{(-)}_{3,2,3-2}-\Gamma^{(-)}_{1+2,1,2}\Gamma^{(-)}_{3+4,3,4}+\Gamma^{(+)}_{-1-2,1,2}\Gamma^{(+)}_{-3-4,3,4}$$

where

$$\Gamma_{1,2,3}^{(-)} = -\frac{V_{1,2,3}^{(-)}}{\omega_1 - \omega_2 - \omega_3},$$

while

$$\Gamma_{1,2,3}^{(+)} = -\frac{V_{1,2,3}^{(+)}}{\omega_1 + \omega_2 + \omega_3}.$$

The coefficients $\Gamma^{(-)}$ and $\Gamma^{(+)}$ are identical to the coefficients $A^{(1)}$ and $A^{(3)}$ of the canonical transformation (4.17).

5

Wave forecasting and wind–wave interaction

This chapter is devoted to a synthesis of what we have previously learned about the physics of wind–wave interaction, culminating in a numerical wave prediction system. After a brief discussion of the numerics of such a model, we illustrate the combined effects of wind input, nonlinear interactions and dissipation on the simple case of the growth of surface gravity waves by wind for an infinite ocean. The resulting growth laws for parameters such as wave height, peak frequency and the Phillips parameter are compared with empirical growth laws. Also, the simulated wave-age dependence of ocean roughness is compared with a number of empirical fits. In addition, we discuss effects of gustiness on wave evolution and roughness of the sea surface which gives an indication of the well-known sensitive dependence of wave results on the forcing wind field.

The resulting wave prediction system turns out to be a promising tool for forecasting purposes and in the remainder of this chapter we discuss a number of applications of the wave forecasting system as it has been implemented at the European Centre for Medium-Range Weather Forecasts (ECMWF). Historically, ECMWF played an important role in the development of the third-generation wave prediction system, called WAveModel (WAM). They provided the necessary infrastructure (such as supercomputers and data-handling facilities), high-quality surface wind fields and support. The combination with the promise of the wealth of global wave-height and spectral data from the ERS-1 and ERS-2 satellites – launched by the European Space Agency (ESA) – resulted in a rapid development of the WAM model. The global version of the WAM model became operational at ECMWF in July 1992. Although the WAM model code was a beautiful and efficient piece of software, the ever-changing operational environment required further developments. For example, the original version of the WAM model was running on one processor. Demands such as increases in spatial spectral resolution meant that the model had to be able to run on memory-shared and distributed-memory multiple processors. Furthermore, there were numerical improvements in the advection scheme, in the time-integration scheme and in interpolation of the wave stress. Because of these differences I will therefore call the ECMWF version of the WAM model 'ECWAM'. It is emphasized, however, that the heart of ECWAM, namely the physical source functions, is identical to WAM (except for the introduction of the effects of gustiness and variable air–sea density ratio).

Traditionally, many weather centres have applied their weather forecasts to activities associated with the marine environment. Shipping, fisheries, offshore operations and coastal protection are all strongly dependent on weather and require marine weather forecasting extending to the limit of the medium-range forecasting period (typically, in the range of 5 to 10 days). An important component of the marine weather forecast is the sea state and, therefore, wave forecasting using forecast low-level winds from an atmospheric model is of prime importance. Hence, it is of paramount interest to discuss the quality of present-day wave forecasts. As an example, progress in wave forecasting at ECMWF during the period 1995 to 2000 is discussed. Here, the quality of the forecast will be judged by a comparison with buoy observations and global altimeter wave-height data and by a comparison with the verifying analysis. If these verification results are put into a historical context, it is seen that the quality of the wave forecasts has improved considerably. Hence, the useful forecast range has increased. In addition, the verification results suggest that an important error source in forecast wave height is the wind-speed error, in agreement with the practical experience of many wave modellers and forecasters (of course, this is immediately evident from the scaling relations for fully developed windsea, see Eq. (2.93)). It is therefore understandable that people from the wave-modelling community have been critical of the quality of the surface wind field, sometimes resulting in suggestions for improvement.

Because of the sensitive dependence of wave results on the quality of the surface wind field, ocean-wave information can give benefits for atmospheric modelling and data assimilation. Wave results have already been used to diagnose planetary boundary problems and overactivity of the atmospheric model. This has resulted in an improved integration scheme for vertical diffusion (Janssen *et al.*, 1992), while the diagnosis of overactivity, reflected by a rapid systematic error growth in forecast wave height, provided ECMWF with a guideline as to what extent overactivity near the surface needed to be reduced (Janssen *et al.*, 2000).

However, there are additional benefits from wave forecasting. In agreement with the arguments presented in this book, wave results may be used to obtain a consistent momentum balance at the ocean surface and they may be of help in the interpretation of satellite data. First, the momentum exchange between atmosphere and the ocean surface could be treated more accurately if the wave-induced drag were taken into account in a *two-way interaction* (Janssen, 1989); at present most atmospheric models assume that the drag over the sea is a function of instantaneous wind speed only. Second, the utilization of satellite data from the scatterometer, synthetic-aperture radar (SAR) and altimeter (e.g. winds and mean sea level) will benefit from the use of sea-state information for an optimal assimilation (Stoffelen and Anderson, 1995; Janssen *et al.*, 1998; Hasselmann and Hasselmann, 1991; Janssen, 2000). In fact, any instrument that involves specular reflection at the sea surface would benefit from knowledge of the sea state (Cox and Munk, 1954). Furthermore, because of the strong interaction between wind and waves, observed wave information may be beneficial for the atmospheric state when assimilated into a coupled ocean-wave–atmosphere model.

Finally, two-way interaction of wind and waves may also be relevant for a more accurate prediction of storm surges and even the ocean circulation. In particular, the sea-state-dependent drag may affect the wind-driven ocean circulation, and the surface current. However, clearly there is a mutual interaction as currents in turn may affect to some extent the evolution of the wave field. On the other hand, waves and winds enjoy a mutual interaction as well, which suggests that the evolution of winds, waves and the wind-driven ocean circulation should be studied in the context of a coupled atmosphere–ocean circulation system where the ocean waves are the agent that transfers energy and momentum across the air–sea interface in accordance with the energy balance equation.

Ocean waves play an important role in the interaction of the atmosphere and ocean. On the one hand, ocean waves receive energy and momentum from the atmosphere through wind input (hence, the ocean waves control to a large extent the drag of airflow over the oceans); on the other hand, through the process of white capping, the ocean waves transfer energy and momentum to the ocean, thereby feeding the turbulent and large-scale motions of the oceans. The energy-conserving nonlinear transfer plays no direct role in this interaction process, although it determines to a large extent the shape of the wave spectrum, and therefore controls energy and momentum fluxes in an indirect way. In equilibrium conditions, the fluxes received by the ocean waves from the atmosphere through the wind-input term would balance the fluxes from ocean waves to ocean via wave breaking. However, ocean waves are in general not in an equilibrium state determined by the balance of the three source functions, because advection and unsteadiness are important as well. As a rule of thumb, of the amount of energy gained by wind, about 95 per cent is lost locally to the ocean by wave breaking, while the remaining 5 per cent is either advected away or spent in local growth. Therefore, for young windseas a considerable imbalance may exist, in particular for the low-frequency waves. On the other hand, when wind waves leave a storm area the magnitude of the wind-input source function decreases dramatically, while the waves are still sufficiently steep that white capping is still important. Since dissipation dominates, wave energy will decay and as a consequence momentum and energy flux to the ocean may be larger than the amounts received by the waves from the atmosphere.

It would be of considerable interest to develop a coupled atmosphere–ocean circulation system where the ocean waves are the agent that transfers energy and momentum across the air–sea interface. A first attempt towards this goal was reported by Bao *et al.* (2000) who studied, in the context of a fully coupled model, the development of hurricanes in the Gulf of Mexico. Here, we mainly concentrate on one aspect of the overall problem, namely the mutual interaction between wind and waves. The reason for this is that in the last decade a considerable amount of research has been devoted to this subject, and there is now ample evidence that two-way interaction between wind and waves matters for prediction of both wind and waves. The study of the impact of surface waves on the ocean circulation is only beginning, and presently there is discussion on the proper primitive equations for ocean circulation including the effects of ocean waves. Nevertheless, the two-dimensional equations, obtained by an integration of the momentum equations over the vertical, are well established (Phillips, 1977) and have been applied to the modelling of storm surges in shallow water. Therefore, first results on the sensitivity of storm surges to a sea-state-dependent drag will be reported.

This chapter is organized as follows. In Section 5.1 we discuss details of the numerical implementation of the ECWAM prediction system. This is followed in Section 5.2 by a presentation of results from the single-grid-point version of the model. In Section 5.3 we discuss our experience with two-way interaction, when applied to medium-range forecasting and forecasting on the seasonal time scale, while in Section 5.4 the impact of sea-state-dependent drag on storm surges is treated. In Section 5.5 an overview of applications of ECWAM at ECMWF is given, and the verification results of analysis and forecast are presented and discussed.

5.1 Numerical implementation of the energy balance equation

In this section numerical aspects of the action balance equation are discussed as they are implemented in the ECWAM model.

Although from a theoretical point of view it is most natural to discuss the evolution of gravity waves in terms of the action density, in practice one deals with the surface elevation spectrum $F(f, \theta)$. The relation between action density and frequency spectrum is straightforward, since

$$F(f, \theta) = \sigma N(f, \theta)/g, \qquad (5.1)$$

with σ the intrinsic frequency (see Eq. (2.68)) and the action density $N(f, \theta) =$

$2\pi N(\omega, \theta)$, where the evolution of $N(\omega, \theta)$ follows from Eq. (2.137). Therefore, wave models such as the ECWAM model are formulated in terms of the frequency–direction spectrum $F(f, \theta)$.

In a numerical model, the continuous spectrum is approximated by step functions that are constant in a frequency–direction bin. It is important to make a distinction between a prognostic part and a diagnostic part of the spectrum. The prognostic part of the spectrum has *NANG* directions and *NFRE* frequencies. These frequencies are on a logarithmic scale, with $\Delta f/f = 0.1$, spanning a frequency range $f_{\max}/f_{\min} = (1.1)^{NFRE-1}$. The logarithmic scale has been chosen to have a uniform relative resolution so that the relevant low frequencies are well represented. Also, the nonlinear transfer scales with frequency. The starting frequency may be selected arbitrarily. A decade ago, most global studies used a starting frequency f_0 of 0.042 Hz, the number of frequencies *NFRE* was 25 and the number of directions *NANG* was 12 (30° resolution). However, by comparing modelled WAM spectra with observations from SAR it became evident that the directional distribution of the modelled spectra was too broad. This was caused by an angular resolution that was too crude. In addition, low-frequency swells in the Pacific were not always well captured because the starting frequency was too high. Therefore, nowadays most third-generation wave prediction models have 24 directions (15° resolution) while at ECMWF the starting frequency is 0.031 Hz with 30 frequencies. For closed basins, such as the Mediterranean Sea or the Baltic, where low-frequency swell is absent, a choice of a starting frequency f_0 of 0.05 Hz is sufficient.

Beyond the high-frequency limit f_{hf} of the prognostic region of the spectrum an f^{-5} tail is added with the same directional distribution as the last band of the prognostic region. This choice was made because observations seem to favour an f^{-5} power law (see Section 2.6.2). The diagnostic part of the spectrum is therefore given by

$$F(f, \theta) = F(f_{\text{hf}}, \theta) \left(\frac{f}{f_{\text{hf}}} \right)^{-5} \quad \text{for } f \geq f_{\text{hf}}. \tag{5.2}$$

The high-frequency limit is set as

$$f_{\text{hf}} = \min\{f_{\max}, 2.5\langle f \rangle\}, \tag{5.3}$$

where $\langle f \rangle$ is the mean frequency. A dynamic high-frequency cut-off f_{hf} rather than a fixed one at $f_{\max}$ is necessary to avoid excessive disparities in the response time scales within the spectrum. In the original WAM model the lower limit $\max\{2.5\langle f \rangle, 4f_{\text{PM}}\}$, with f_{PM} the Pierson–Moskowitz frequency, was used instead of $2.5\langle f \rangle$. This choice was made in order to ensure that even for low wind speed the spectrum was determined by the energy balance and was not treated as a diagnostic tail. In the ECWAM model the latter choice was abandoned because it resulted in unrealistically high mean-square slopes in cases of a sudden decrease in wind speed.

A diagnostic tail is added for $f \geq f_{hf}$ in order to compute the nonlinear energy transfer in the high-frequency part of the prognostic range and also to compute a number of integral quantities such as occur in the dissipation and wind-input source functions. Note that the contribution from the diagnostic tail to the total energy is normally small, but quantities such as the mean-square slope and the wave-induced stress depend in a sensitive manner on the spectral tail.

The prognostic part of the spectrum is now obtained by numerically solving the energy balance equation. Let us now discuss the different numerical schemes that are used to integrate the source functions and the advective terms of the transport equation.

5.1.1 Implicit scheme for the source functions

The energy balance equation for the wave spectrum is evaluated in detail up to a high-frequency cut-off f_{hf}. The high-frequency relaxation time scales are considerably shorter than the time scales of the energy-containing low-frequency waves, which are the main interest in practical applications. Hence, in the high-frequency region it is sufficient to determine the equilibrium level to which the spectrum adjusts in response to the slowly changing low-frequency waves. Implicit integration schemes with time steps matched to the evolution of the low-frequency waves meet this requirement automatically: for the low-frequency waves the integration scheme yields the same results as a simple forward-integration scheme, while for the high frequencies the method gives the slowly varying equilibrium spectrum (WAMDI, 1988; Komen *et al.*, 1994).

The implicit-difference equations (leaving out the advection terms) are given by

$$F_{n+1} = F_n + \Delta t \left[\alpha S_{n+1} + (1 - \alpha) S_n \right] \tag{5.4}$$

where Δt is the time step and the index n refers to the time level. The parameter α is a constant. If the source terms depend linearly on F then for a stable numerical scheme α should be at least $1/2$. For a linear source term it is straightforward to solve Eq. (5.4) directly for the spectrum F_{n+1}.

However, none of the source terms are linear. Therefore, a Taylor expansion

$$S_{n+1} = S_n + \frac{\partial S_n}{\partial F} \Delta F \tag{5.5}$$

is introduced. The functional derivative in Eq. (5.5) can be divided into a diagonal matrix Λ_n and a nondiagonal remainder R_n,

$$\frac{\partial S_n}{\partial F} = \Lambda_n + R_n. \tag{5.6}$$

Trial computations (see WAMDI, 1988; Komen *et al.*, 1994) indicated that the off-diagonal contributions were generally small for time steps that were not too large. Disregarding these contributions and substitution of Eqs. (5.6) and (5.5) into Eq. (5.4) gives, realizing that S may depend on the friction velocity at time level $n + 1$,

$$\left[1 - \alpha \Delta t \Lambda_n(u_*^{n+1})\right] \Delta F = \Delta t \left[(1 - \alpha)S_n(u_*^n) + \alpha S_n(u_*^{n+1})\right], \tag{5.7}$$

where $\Delta F = F_{n+1} - F_n$ is the increment in the spectrum owing to the physics. As a consequence, the increment ΔF becomes

$$\Delta F = \frac{\Delta t \left[(1 - \alpha)S_n(u_*^n) + \alpha S_n(u_*^{n+1})\right]}{\left[1 - \alpha \Delta t \Lambda_n(u_*^{n+1})\right]}. \tag{5.8}$$

The original WAM model had $\alpha = 1/2$; however, Hersbach and Janssen (1999) noted the occurrence of numerical noise and therefore suggested the use of a fully implicit scheme with $\alpha = 1$. The generation of noise should not come as a surprise, since even in the case of a linear source term the choice of $\alpha = 1/2$ is only marginally stable. Although numerical noise is not amplified for $\alpha = 1/2$, it is also not damped and therefore it is safer to choose $\alpha > 1/2$. For $\alpha = 1$, Eq. (5.8) becomes

$$\Delta F = \frac{\Delta t S_n(u_*^{n+1})}{1 - \Delta t \Lambda_n(u_*^{n+1})}, \tag{5.9}$$

which gives a considerable simplification because only the source term with $u_* = u_*^{n+1}$ needs to be evaluated, in addition to the diagonal part of its functional derivative.

Nevertheless, in practice, numerical instability is found in the early stages of wave growth. This is caused either by the neglect of the off-diagonal contributions of the functional derivative or more likely by the fact that the solution may not always be close to the attractor of the complete source function. Therefore, a limitation on spectral change needs to be imposed. In ECWAM one of the variants of the growth limiter of Hersbach and Janssen (1999) is used; the maximum increment in the spectrum, $|F|_{max}$, is given by

$$|F|_{max} = 5 \times 10^{-7} g u_* f^{-4} \langle f \rangle \Delta t. \tag{5.10}$$

The limiter expresses that the change in the spectrum should not exceed a certain fraction of Toba's law (Eq. (4.114)), which is the spectral shape that is found in the energy-containing, equilibrium range of the spectrum. The time-step dependence of the limiter was introduced in order to ensure that results for initial wave growth become insensitive to the time step. The consequence of making the limiter proportional to the time step is that it will remain active even when the limit $\Delta t \to 0$ is taken. Therefore, the limiter (5.10) is not just a numerical feature; it becomes part of the physics. Usually it is effective only for initial wave growth, in which the energy-containing part of the spectrum extends into the diagnostic high-frequency tail of the model spectrum, beyond the cut-off limiting the prognostic range of the model spectrum. Therefore, for initial growth, the limiter compensates for the lack of physics in the diagnostic part of the spectrum.

The big advantage of the present limiter over an earlier version in the WAM model is that Eq. (5.10) has the correct scaling behaviour with the friction velocity u_* (Hersbach and Janssen, 1999). As a consequence, global implementations of ECWAM on a coarse resolution of 55 km and a fine-mesh version of the ECWAM model, with a resolution of 1 km, give similar growth curves for significant wave height. In addition, for a typical test case good agreement was obtained between an explicit integration with a time step of 1 min and the implicit scheme with only diagonal terms for time steps up to 20 min.

Nevertheless, Hargreaves and Annan (2001) have expressed serious concerns about the introduction of the limiter (5.10), and these authors seem to favour alternative methods such as a variable-time-step method. In their reply, Hersbach and Janssen (2001) show by means of a simple example, provided by Hargreaves and Annan (2001), that the limiter (5.10) is ideally suited for its purpose and that the variable-time-step method will result in such small time steps that it cannot reach beyond a certain finite simulated time and therefore this approach is not very practical in terms of operational implementation. Furthermore, choosing a lower bound to the time step implies a limiter of the type in Eq. (5.10).

5.1.2 Advective terms

The advective terms in the energy balance equation have been written in the flux form. As an illustration we shall consider the one-dimensional advection equation

$$\frac{\partial F}{\partial t} = -\frac{\partial \Phi}{\partial x}, \tag{5.11}$$

with flux $\Phi = v_g F$, since the generalization to four dimensions λ, ϕ, θ and ω is obvious. A number of alternative propagation schemes have been tested by different groups in the past decade. Examples are first-order upwinding schemes, a second-order leap-frog scheme, semi-Lagrangian schemes, third-order schemes, etc.; therefore there exists a considerable amount of experience with discretization of the advection equation. However, none of the schemes give satisfactory results unless special measures are taken. In fact, a propagation scheme with vanishingly small errors would give poor results for sufficiently large propagation times since it would not account for the dispersion associated with the finite resolution of the wave spectrum in frequency and direction (the so-called 'garden-sprinkler effect').

In order to explain the garden-sprinkler effect, let us study the evolution in space and time of one spectral bin having a width Δf, $\Delta \theta$,

$$\frac{\partial F}{\partial t} + v_{\text{g}} \frac{\partial F}{\partial x} = 0, \tag{5.12}$$

where we have taken a group speed which is independent of the spatial coordinate x. For the initial condition

$$F(x, 0) = f(x) \tag{5.13}$$

it is straightforward to solve for the evolution of the wave spectrum. The solution becomes

$$F(x, t) = f(x - v_{\text{g}}t); \tag{5.14}$$

hence the waves with group speed v_{g} propagate over the surface with a spatial distribution that does not change its shape. Since this is a linear problem, the solution for an arbitrary number of spectral bins is obtained by summation of the solution in Eq. (5.13) for different group velocities. Consider the solution for two neighbouring frequency bins and suppose that the two bins have equal spatial distributions of the box type with width Δx, where Δx is the spatial resolution. Clearly, after a finite time τ_{s} there is a separation of the two pulses which is determined by the difference in group velocity and the spatial width. Assuming that the frequency increment Δf is small, one may use a Taylor expansion of the difference in group velocity and the separation time τ_{s} becomes

$$\tau_{\text{s}} = \frac{\Delta x}{v_{\text{g}}} \frac{f}{\Delta f}. \tag{5.15}$$

For longer time periods, droplets are formed on the surface, hence the name 'garden-sprinkler effect'. In a similar vein, it can be shown that a finite directional resolution $\Delta \theta$ will also give rise to the garden-sprinkler effect. Finite directional resolution gives in practice a much shorter separation time since

$$\tau_{\text{s}} = \frac{\Delta x}{v_{\text{g}}} \frac{1}{\Delta \theta} \tag{5.16}$$

and $1/\Delta \theta < f/\Delta f$. It is emphasized that for a continuous spectrum the garden-sprinkler effect will not occur and therefore in a discrete model measures have to be taken to avoid it.

In the past, inaccurate numerical schemes with relatively high numerical diffusion have been used; these schemes are smoothing, to such an extent that the garden-sprinkler effect is prevented from occurring. This explains why a first-order upwinding scheme is so successful, despite the fact that it is only first-order accurate and is highly diffusive having a numerical diffusion coefficient $D \simeq (\Delta x)^2/\Delta t$ (with Δt the time step). However, it is of course preferable to have some control over the amount of horizontal diffusion. This led the WAMDI group (1988) to study the benefits of a second-order, leap-frog scheme. The advection term of the second-order scheme is second-order accurate and has a smaller, inherent, numerical diffusion. The second-order scheme suffers, however, from the drawback that it generates unphysical negative energy in regions of sharp gradients. This can be alleviated by including explicit-diffusion terms. In practice, the explicit diffusion required to remove the negative side lobes in the second-order scheme is of the same order as the implicit numerical diffusion of the first-order scheme so that the effective diffusion is comparable between the two schemes.

As shown in WAMDI (1988) the two schemes have similar propagation and diffusion properties. An advantage of the second-order scheme is that the lateral diffusion is less dependent on the propagation direction than in the first-order scheme, which shows significant differences in the diffusion characteristics for waves travelling along the coordinate axes compared with directions in between. However, this undesirable feature of the first-order scheme may be alleviated by rotating the spectrum by half its angular resolution. In general, the differences between the model results using the first- and second-order propagation schemes were found to be small. Apparently, the additional diffusion introduced in the potentially more accurate second-order scheme to alleviate the problem of negative energy implies loss of accuracy. Thus, there is a preference for the first-order scheme because of its efficiency and simplicity.

Note that the main reason for considering a second-order scheme was not to reduce diffusion, but to be able to control it in such a way as to avoid the garden-sprinkler effect. In other words, one tried to match numerical diffusion to the finite dispersion associated with the finite frequency–direction spectral resolution of the model. But this dispersion increases linearly with respect to propagation time or distance, while standard propagation schemes yield a spreading of the wave groups which increases with the square root of time or space. Linear spreading rates may be achieved by introducing a variable diffusion coefficient proportional to the age of the wave packets (Booij and Holthuijsen, 1987). In its ultimate form this requires the determination and storage of a wave-age parameter for every frequency and direction, which may not be practical. The essence of this idea has been tested with some success in the context of a third-generation wave model by Chi Wai Li (1992) and Tolman (2001) who used a third-order scheme in combination with an average age of the wave packets per ocean basin.

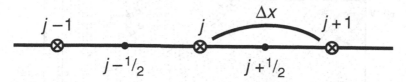

Fig. 5.1. Definition of grid points for first-order upwinding scheme.

To summarize the discussion, we have chosen the first-order upwinding scheme because of its simplicity, because it requires less memory and computer time and because in practice it gives reasonable results. Note that there are other error sources, e.g. the quality of the forcing wind fields, which dominate the error budget of a wave prediction system (for a more detailed discussion of this see Section 5.5).

Applied to the simple advection scheme in flux form (Eq. (5.11)) one obtains the discretization below, where for the definition of grid points we refer to Fig. 5.1. The rate of change of the spectrum ΔF_j at the jth grid point is given by

$$\Delta F_j = -\frac{\Delta t}{\Delta x}(\Phi_{j+1/2} - \Phi_{j-1/2}), \qquad (5.17)$$

where Δx is the grid spacing and Δt the propagation time step, and

$$\Phi_{j+1/2} = \frac{1}{2}(u_j + |u_j|)F_j + \frac{1}{2}(u_j - |u_j|)F_{j+1}, \qquad (5.18)$$

where $u_j = 0.5(v_{g,j} + v_{g,j+1})$ is the mean group velocity and the flux at $j - 1/2$ is obtained from Eq. (5.18) by replacing $j + 1/2$ with $j - 1/2$. The absolute values of the mean speeds arise because of the upwinding scheme. For example, for flow going from the left to the right the speeds are positive and, as a consequence, the evaluation of the gradient of the flux involves the spectra at grid points $j - 1$ and j.

It is important to remark that the first-order upwinding scheme suffers from numerical instabilities when the time step is so large that the so-called 'Courant–Friedrichs–Lévy' (CFL) criterion is violated. The upwinding scheme is stable provided that

$$v_g < \frac{\Delta x}{\Delta t}, \qquad (5.19)$$

in other words, wave groups are only allowed to travel at most one grid length during one time step. The CFL criterion may become critical, particularly near the poles when spherical coordinates are used: when moving towards the poles, the actual distance in the latitudinal direction decreases. Clearly, the closer one moves to the poles the more likely it becomes that the CFL criterion is violated. In ECWAM this problem is solved by choosing an irregular spherical grid in such a way that

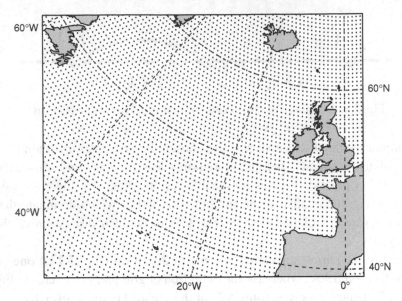

Fig. 5.2. Irregular grid for the North-Atlantic area on a polar stereographic projection.

the latitudinal distance is more or less fixed to its value at the equator. An example of such a grid from the present operational ECWAM model is shown in Fig. 5.2. The advection scheme is still formulated in terms of standard spherical coordinates but the gradient in the longitudinal fluxes is evaluated by linear interpolation of the fluxes from the closest neighbours. The additional advantage of the use of an irregular spherical grid is a reduction in the total number of grid points by 30 per cent, giving a substantial reduction in the number of computations.

Furthermore, it is remarked that there are several alternative advection schemes that avoid the CFL criterion (5.19). One could follow a fully implicit treatment of the advective terms, but the drawback of such an approach is excessive smoothing. Another alternative is a semi-Lagrangian scheme, which has been succesfully implemented in the ECMWF atmospheric model. However, a carefully tuned interpolation scheme is needed to avoid the garden-sprinkler effect.

5.1.3 Software aspects of ECWAM

The WAM model development was finished in the early 1990s. Since then there has been continuous effort at ECMWF to streamline the software in areas such as input–output, archiving, vectorization and to adapt the code to the new massive, parallel (vector) machines. Nevertheless, many of the original features of the WAM model have been retained. These are described in Komen *et al.* (1994). We only give a description here of the additional features that have been introduced at ECMWF.

The ECWAM model software that has been developed over a period of 10 years (1992 to 2002) is fairly general. The software is based on FORTRAN 90 which allows the use of one executable for all the applications that are running at ECMWF because the size of the problem is determined at run time. Thus, maintenance of the model software is reduced to a minimum. A full description of the model software may be found on the ECMWF website (www.ecmwf.int). Here, only a few design choices are discussed.

The model has been developed with an important application in mind, namely operationally predicting waves over the whole globe. Even with a modest spatial resolution of 3° (resulting in approximately 4000 grid points) and 25 frequencies and 12 directions, it follows that about 1.2 million equations have to be solved. Since the most expensive part of the numerical code, the nonlinear source term, cannot be vectorized, vectorization is achieved over the grid points, which are placed in the innermost loop. In order to make this loop as long as possible, a mapping from the two-dimensional spherical grid to a one-dimensional array is performed. If there are no limitations to the amount of internal memory of the computer, the most efficient procedure is to convert the entire global grid to a single one-dimensional array. In the early days of the wave-model development, there were serious restrictions on the amount of memory available. It was, therefore, decided to split up the problem into blocks of a fixed size, and to perform the computations on blocks loaded one block at a time into the memory. The drawback of this approach was that extensive input–output operations were needed, but it was the only approach that allowed the running of large wave prediction systems on computers with only small amounts of memory. The later generation of computers, such as the CRAY-YMP, allowed the whole problem to be loaded into the core of the computer, and nowadays the blocking option as it was originally introduced has become obsolete.

Later generations of computers were based on the concept of massive parallel computing. In this context it is important to distinguish between memory-shared and memory-distributed machines. Machines such as the CRAY-YMP and the CRAY-C90 are examples of shared-memory machines. By using macro-tasking it was relatively straightforward to develop a version of the ECWAM model that utilized more processors in an efficient way. Note that there are limits to the number of processors that can be used, because each processor requires a sufficient amount of work. Therefore, a low-resolution version of the ECWAM model, such as the 1.5° model, could only perform efficiently on about 4 processors, while the high-resolution, 55 km, version still ran efficiently on 16 processors.

The present generation of computers either are memory-distributed machines or have memory distribution over nodes while per node the processors share the memory. In general, a memory-distributed machine requires a different approach; this is described in the next subsection.

Massive parallel computing

Memory-distributed machines such as the Fujitsu VPP series require the introduction of messages passing between processors (known as processing elements
(PEs)). Therefore, one PE can send a message which is received by one or more
other PEs. In its very basic implementation, the message is nothing more than a
one-dimensional array of a given type containing values that are needed by the other
PE(s) plus the necessary information about the sender and receiver. For a succesful
message exchange both send and receive should be completed.

Message passing in the ECWAM model was introduced in 1996 based on the
message-passing interface library MPILIB. This newly developed code can also
run on nondistributed-memory machines and on a single PE.

In the present setting the number of processors is determined at run time. Once
the message-passing program starts simultaneously on all assigned PEs, the parallel environment and the message-passing protocol are initialized. Also, the total
number of PEs is determined as well as the logical PE on which the code is run.

When running in parallel it is important to have an even distribution of work
over the PEs in order to avoid load unbalance. In the case of the ECWAM model it
comes down to splitting the global computation domain into regions of equal size,
bearing in mind that information is only locally known on each PE and can only be
exchanged with the other PEs via message passing (which is a slower process than
computing).

Once the total number of PEs is known, an even decomposition of the total grid
into one subdomain per PE is set up. Since the global grid is mapped onto a one-
dimensional sea-point array following increasing latitude lines, the subdomains
are chosen to be consecutive segments of the full sea-point array (see Fig. 5.3).
The length of each segment is determined by the requirement that the work is
distributed in an even manner over the given number of PEs. Thus, each PE will
only perform the integration of the source functions of the energy balance equation
of one subdomain. However, the upwinding schemes which solves the advection
term, uses neighbouring grid points in the two-dimensional grid that might belong
to another subdomain. The information from the other PEs needed to evaluate the
spatial derivatives of the energy flux is obtained through message passing. Here,
the message is constructed using the geometrical rules displayed in Fig. 5.3, and is
similar in spirit to the method that was developed for the multi-block version of the
WAM model. An important difference is, however, that the domain composition is
done at run time, allowing more flexibility.

The advantage of the above decomposition procedure is that one PE is only
communicating with two other PEs. However, the decomposition requires a relatively large amount of message passing. This is because the domain does not have
an optimal shape so that a large number of grid points inside the domain require

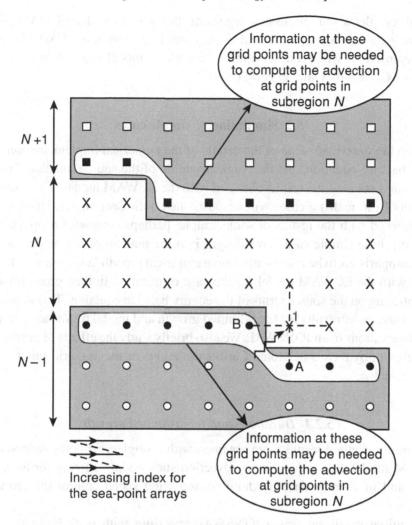

Fig. 5.3. Domain decomposition.

information from neighbouring PEs to evaluate the gradient of the energy flux. A more optimal shape is a rectangle and this has been introduced in the ECWAM model as an alternative way of decomposition. By staggering the rectangles one PE only communicates with six others. No real advantages of this alternative decomposition procedure have been found, however, except for very large applications.

Finally, to allow even more flexibility and to accommodate the modern architecture of having memory distribution over nodes while in each node the processors share the memory, the macro-tasking option used on the Cray machines has been replaced by OpenMP directives.

This completes our discussion regarding the design of the ECWAM model. A more detailed description of this may be found on the ECMWF website (www.ecmwf.int) while a description of the WAM model is given in the manual *Wamodel Cycle* 4 by Günther *et al.* (1991).

5.2 Simulation of simple cases

After having described some of the details of the numerical implementation of the energy balance equation, and the parametrization of the source functions, we will now discuss some of the results obtained with the ECWAM model. The discussion of results from realistic cases will be postponed until Section 5.5. Here, we will be concerned with the results of some simple, perhaps somewhat artificial cases. However, these simple cases give insight into the realism of the model and they allow comparisons to be made with simple empirical growth laws. We first describe results with the ECWAM model for the case of duration-limited growth, thereby concentrating on the source terms of the energy balance equation. This is followed by a discussion of results for fetch-limited growth and model results are compared with observations from JONSWAP. We also briefly study the effects of gustiness on wave growth by means of Monte Carlo simulations of the one-grid-point version of the model.

5.2.1 Duration- and fetch-limited growth

The duration-limited results were obtained with a single-grid-point version of the ECWAM model. Since there is no advection such a simulation is for an infinite ocean, and in deep water only duration and wind speed control the growth of windsea.

The initial condition was a JONSWAP spectrum with peak frequency $f_p = 0.34$ Hz, Phillips' parameter $\alpha_p = 0.025$, overshoot parameter $\gamma = 3$ and spectral width $\sigma = 0.10$. The directional spreading was given by the usual $\cos^2$ distribution. The energy balance equation was integrated for 24 directions and 30 frequencies (on a logarithmic scale with a starting frequency of 0.0418 Hz with a fractional increase of 10 per cent), using the fully implicit scheme. The integration step was 15 min, while wind and waves were coupled every time step. The wind speed was chosen to be 18.45 m/s.

In Fig. 5.4 we show the evolution in time of the one-dimensional frequency spectrum $F(f)$ over the first 24 h of the simulation. The evolution of the simulated frequency spectrum is in accord with the results found during the JONSWAP campaign, as displayed in Fig. 2.5. In particular, a pronounced overshoot of the peak of the spectrum is noted.

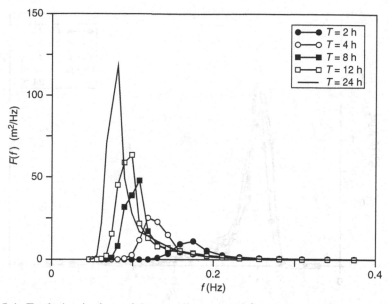

Fig. 5.4. Evolution in time of the one-dimensional frequency spectrum according to the ECWAM model.

It is also of interest to note that the spectral shape is quasi-universal. In order to make this plausible, frequency is normalized with the peak frequency f_p, and the spectrum F is regarded as a function of $x = f/f_p$. Hence, $F(f)\,\mathrm{d}f = E(x)\,\mathrm{d}x \Rightarrow E = f_p F(f)$. Also, in practice, the high-frequency part of the spectrum follows Toba's law closely (Eq. (2.112)), because nonlinear transfer dominates in the equilibrium range. Hence, in agreement with Toba's law, the normalized spectrum $f_p^3 F(f/f_p)/gu_*$ is expected to be quasi-universal when plotted as a function of f/f_p. According to the simulations with the ECWAM model, shown in Fig. 5.5, this is indeed the case. Therefore, if the peak frequency (or the wave variance as we will see in a moment) and the friction velocity are known, it is possible to reconstruct the windsea spectrum. The quasi-universality of the windsea spectrum explains, as already discussed in Section 2.6.1, the relative success of second-generation wave prediction systems (Hasselmann *et al.*, 1976; Sanders, 1976; Janssen *et al.*, 1984).

The energy balance for young windsea (duration 4 h) is shown in Fig. 5.6, by plotting the directional averages of S_{in}, S_{nl}, and S_{ds} as functions of frequency. As expected from the previous discussions the wind input is always positive and the dissipation is always negative, while the nonlinear interactions show a three-lobe structure of different signs. Therefore, the intermediate frequencies receive energy from the airflow, which is transported by the nonlinear interactions towards the high and low frequencies. In the high-frequency range the nonlinear energy flux maintains an f^{-4} spectrum (see the discussion in Section 4.7.2), while in the

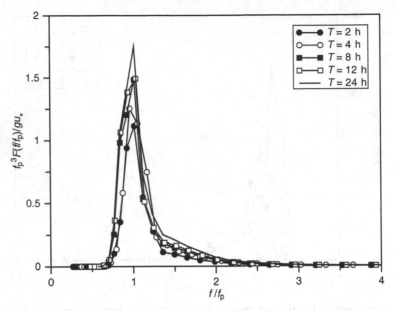

Fig. 5.5. Universality of spectral shape: $f_p^3 F(f/f_p)/gu_*$ versus f/f_p.

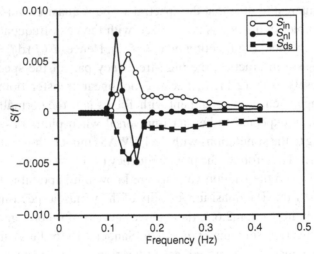

Fig. 5.6. The energy balance for young windsea at a duration of 4 h.

low-frequency range the nonlinear interactions maintain an 'inverse' energy cascade transferring energy towards the region just below the spectral peak, thereby shifting the peak of the spectrum towards lower frequencies. This frequency downshift is to a large extent determined by the shape and the magnitude of the spectral peak itself. For young windsea, having a narrow peak with a considerable peak enhancement,

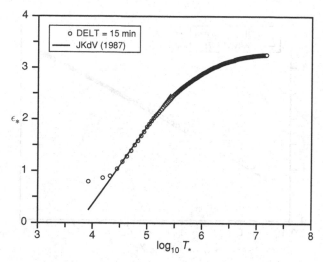

Fig. 5.7. Duration-limited wave growth for time step DELT of 15 min: $\epsilon_* = g^2 m_0/u_*^4$ versus duration gT/u_* plotted on a logarithmic scale. The empirical growth law, labelled 'JKdV', from Sanders (1976) (see also Janssen *et al.*, 1987) is shown for comparison.

the rate of downshifting is considerable, while for gentle, old windsea this is much less so. In the course of time the peak of the spectrum gradually shifts towards lower frequencies (as may be seen from Fig. 5.4) until the peak of the spectrum no longer receives input from the wind because these waves are running faster than the wind. Under these circumstances the waves around the spectral peak are subject to a considerable dissipation so that their steepness reduces. Consequently, because the nonlinear interactions depend on the steepness, the nonlinear transfer is reduced as well, with the result that slowly a quasi-equilibrium spectrum emerges. For old windsea the time scale of downshifting becomes larger than the typical duration of a storm so that for all practical purposes the wind-generated waves evolve towards a steady state.

From the spectrum one may immediately obtain integrated parameters such as the wave variance m_0, the significant wave height H_S, the peak frequency f_p, the wave age $\chi_* = c_p/u_*$, etc. In Fig. 5.7 we have plotted the dimensionless energy $\epsilon_* = g^2 m_0/u_*^4$ versus duration gT/u_*. The comparison with empirical data for duration-limited windsea from the North Sea looks favourable. Note that according to ECWAM the Pierson–Moskowitz limit of the dimensionless energy $\epsilon_* = \mathcal{O}(1500)$ and is therefore considerably larger than the limit values found with the original WAM model. The reason for this difference is that for old windsea the airflow is much smoother in ECWAM than in WAM (because of the different choice for the definition of the prognostic range, see Eq. (5.3)). Hence, for old windsea,

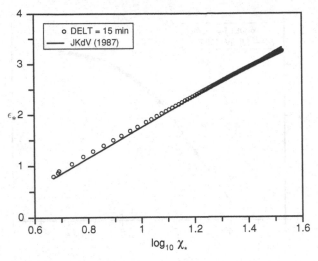

Fig. 5.8. Dimensionless wave variance as function of wave age for time step DELT of 15 min: $\epsilon_* = g^2 m_0/u_*^4$ versus $\chi_* = c_\mathrm{p}/u_*$. The empirical fit from Janssen *et al.* (1987) is also shown.

friction velocities are smaller by about 5 to 10 per cent, and since the dimensionless energy scales with the fourth power of the friction velocity the result is a considerable increase in its limiting value. The actual wave heights of both models are very similar, however.

A consequence of the close agreement of modelled spectra with Toba's law is that the dimensionless wave variance should be proportional to the third power of the wave age χ_* (see Eqs. (2.111) and (2.134)). This follows at once from the results of the numerical simulation with the ECWAM model, shown in Fig. 5.8. For comparison, we also show the empirical fit of Janssen *et al.* (1987) which is based on two North Sea data sets obtained during the summer and winter months.

In order to conclude the discussion of the duration-limited results we show in Fig. 5.9 the dependence of the Phillips parameter α_p on wave age χ_*. According to the numerical simulation the Phillips parameter has an inverse linear dependence on wave age, hence is linearly proportional to the friction velocity. This should not come as a big surprise as the high-frequency part of the ECWAM model follows Toba's law. The inverse linear dependence of the Phillips parameter on wave age is confirmed by comparison with the empirical relation (2.109).

It is emphasized that in the present discussion we have assumed friction-velocity scaling and not 10-m wind-speed scaling. This makes considerable differences, especially for young windseas, because the ratio of friction velocity to 10-m wind speed is not fixed but depends on wave age. For example, while according to the simulation the Phillips parameter is proportional to the inverse of the wave age χ_* it

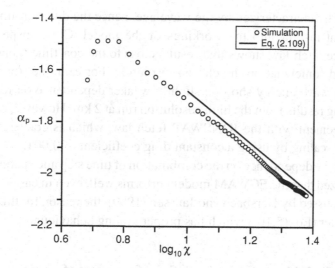

Fig. 5.9. Dependence of the Phillips parameter on wave age.

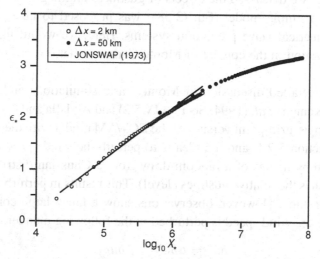

Fig. 5.10. Fetch dependence of wave variance: $\epsilon_* = g^2 m_0 / u_*^4$ versus gX/u_*^2 for two different spatial resolutions, namely $\Delta x = 2$ km, and $\Delta x = 50$ km. The JONSWAP fetch law, (Eq. (2.107)) for a drag coefficient of 0.001 is also shown.

turns out that there is a more sensitive dependence of α_p on the wave age based on the 10-m wind speed. In fact, it is found that $\alpha_p \sim \chi_{10}^{-1.35}$, giving a close agreement with the results found by Battjes *et al.* (1987) (see Eq. (2.125)), but the simulated values are systematically somewhat higher.

We conclude the discussion by showing just one result from the case of fetch-limited wave growth, namely the fetch dependence of the dimensionless energy (see Fig. 5.10). There is no need to discuss other results for windsea because this sea state is fully characterized by the wave age, hence the duration-limited results give sufficient insight into the workings of the model. Close inspection of the wave-variance fetch law shows that results close to the coastline (hence for small fetch) depend somewhat on the chosen resolution. For example, for a resolution of 50 km the wave energy shows a slightly weaker dependence on fetch than the corresponding results from the high-resolution run at 2 km. However, overall, there is a fair agreement with the JONSWAP fetch law, which is converted from U_{10} scaling to u_* scaling by using a constant drag coefficient of 0.001.

Regarding the dependence on the combination of time step and spatial resolution, it is emphasized that the ECWAM model performs well, even in cases of very short fetch. As discussed by Hersbach and Janssen (1999), the reason for this is the more 'liberal' limiter (Eq. (5.10)) which has proper scaling behaviour.

5.2.2 Monte Carlo simulation of the effect of gustiness

In Section 3.3.2 we discussed the effects of gustiness on the growth rate of waves by wind, and a simple model (Eq. (3.58)) was proposed to take gustiness into account in numerical wave prediction systems. Here, we would like to validate this parametrization in the context of Monte Carlo simulations with the ECWAM model.

For a more detailed discussion of Monte Carlo simulations and the effect of gustiness see Komen *et al.* (1994; Section IV.5.2) and Abdalla and Cavaleri (2002). We take the single-grid-point version of the ECWAM model with the same settings as given in Section 5.2.1, and the idea is to perturb the wind speed at every time step of 15 min by means of a random draw from a Gaussian distribution with a width σ (which is the relative gustiness level). This results in perturbations that are uncorrelated in time. However, observations show a fairly large correlation α in time and therefore wind speed is perturbed in the following manner:

$$\delta u^n = \alpha \delta u^{n-1} + \delta u_R, \qquad (5.20)$$

where δu^n is the perturbation at time step n, and δu_R is a random perturbation drawn from a Gaussian, which because of the time correlation has a reduced width $\sigma\sqrt{1-\alpha^2}$ (the need for the reduction in width of the Gaussian can easily be seen by squaring Eq. (5.20), taking the ensemble average, assuming stationarity in time and no correlation between δu_R and δu^{n-1}). In the following simulations the correlation α was taken to be 0.9. Each member of the ensemble was generated by taking a different random draw of the perturbation δu_R. The ensemble consists of 100 members.

In Fig. 5.11 we show, for a mean wind speed of 18.45 m/s, the evolution in time of significant wave height for the case without gustiness. In addition, for a

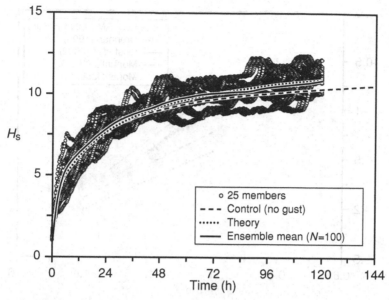

Fig. 5.11. Monte Carlo simulation of duration-limited wave growth with a relative gustiness level of 7 per cent. The figure shows evolution of significant wave height for the case of no gusts, the ensemble-mean evolution (based on 100 members) and the evolution according to Eq. (3.58). Results from the first 25 members of the ensemble are also shown.

quite small gustiness level of 7 per cent the ensemble mean of wave height and the mean evolution according to the parametrization in Eq. (3.58) are shown. The parametrization requires the relative gustiness level in the friction velocity. This was obtained by linearizing the relation between friction velocity and wind speed, while taking a wind-speed-dependent drag coefficient. There is a fair agreement between parametrization and ensemble-mean evolution.

In order to get an idea of the variability in the results, the wave-height evolution according to the first 25 members is also shown in Fig. 5.11. Even for the small level of gustiness chosen here there is a considerable variability in wave height. As in unstable conditions a large variability in wind speed is usually found, one would expect that gustiness should play an important role in the reanalysis of fetch-limited wave-height observations discussed in Section 2.6.1, particularly for large fetches. This might explain some of the differences between wave growth in stable and unstable stratification, but a detailed analysis of this has not been done so far.

Finally, it is of interest to study the effects of wind variability on other quantities of interest such as the roughness of airflow over wind-generated wind waves. In Fig. 5.12 we show the wave-age dependence of the Charnock parameter according to the deterministic version of the ECWAM model. For comparison, some recent

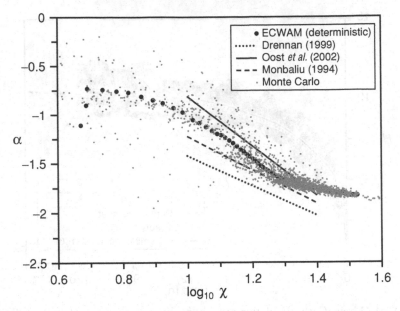

Fig. 5.12. Wave-age dependence of the Charnock parameter α according to the deterministic version of the ECWAM model (shown on a logarithmic scale). For comparison, some recent empirical fits are also shown. Finally, for a gustiness level of 7 per cent results from a number of ensemble members are plotted as dots.

empirical fits are also shown, suggesting that the ECWAM model gives a realistic simulation of surface roughness over the oceans. For a more complete discussion, in particular of the wind-speed dependence of the Charnock parameter, see Bonekamp *et al.* (2002).

Results from Monte Carlo simulations with a gustiness level of 7 per cent are also plotted. Realizing that this is a plot on a logarithmic scale, it is clear that for young windseas small levels of gustiness may give rise to quite a large variability in the results for the Charnock parameter. If this variability is real, then it is understandable why it is so difficult to find in nature a definite relation between quantities such as the Charnock parameter and the wave age.

5.3 Impact of sea state on the atmosphere

In this section a brief description of the impact of sea-state-dependent drag on the atmospheric circulation is given. The basic idea is described in Chapter 3 which also gives a parametrization of the sea-state-dependent roughness. For the orginal papers consult Janssen (1982, 1989, 1991). This parametrization is included in WAMCy4 (Komen *et al.*, 1994) and in ECWAM. A review of the impact on atmospheric circulation is given in Janssen *et al.* (2002). For ease of discussion the basic results for two-way interaction of winds and waves, already discussed in the previous chapters, will be summarized first.

The idea is that momentum transfer from air to sea depends on the sea state because steep waves extract more momentum from the airflow than gentle, smooth waves. Steep waves typically occur in the early stages of wind-wave generation and when a frontal system passes, hence momentum transfer depends on the sea state. In order to account for this effect one needs to calculate the wave-induced stress τ_w which depends on the two-dimensional wave spectrum. This requires the solution of the energy balance equation, which follows from the action balance equation (2.137) by utilizing the relation $N = gF/\sigma$. For deep water and no currents one finds

$$\frac{\partial}{\partial t}F + \frac{\partial}{\partial \mathbf{x}} \cdot (\mathbf{v}_g F) = \hat{S}_{in} + \hat{S}_{nl} + \hat{S}_{ds}, \qquad (5.21)$$

where $F = F(\omega, \theta)$ is the two-dimensional wave spectrum which gives the energy distribution of the ocean waves over angular frequency ω and propagation direction θ. Furthermore, $\mathbf{v}_g$ is the group velocity and on the right-hand side there are the well-known source terms for wind input, nonlinear transfer and dissipation (but note that $\hat{S}_{in} = \sigma S_{in}/g$, etc., because we transformed from action density to wave-variance spectrum).

In the wind–wave interaction problem we only need to know the wave-induced stress τ_w which follows from an integration of the input source function of the energy balance equation (5.21),

$$\tau_w = \rho_w g \int d\omega \, d\theta \, \hat{S}_{in}/c, \qquad (5.22)$$

where c is the phase speed of the gravity waves and ρ_w is the water density. Here, it should be realized that wave momentum P and energy density F of the waves are related by $P = F/c$ and the wave stress is the rate of change of total wave momentum caused by wind input. Because waves grow exponentially fast the source function $\hat{S}_{in}$ is proportional to the wave spectrum itself. The wave-induced stress is mainly determined by the high-frequency part of the wave spectrum because these are the waves that have the largest growth rate due to wind. Since it is known that the high-frequency spectrum depends on the stage of development of the windsea (for example, young wind waves are steeper than old wind waves) it follows that the wave-induced stress depends on the sea state. Therefore, young wind waves represent a rougher surface than gentle, old windsea. The roughness z_0 therefore depends on the sea state and from Eq. (3.164) the roughness length obeys a Charnock relation,

$$z_0 = \alpha u_*^2/g, \qquad (5.23)$$

where the Charnock parameter α depends on the sea state according to

$$\alpha = \frac{\hat{\alpha}}{\sqrt{1 - \tau_w/\tau}}, \quad \hat{\alpha} = 0.01, \tag{5.24}$$

with $\tau = \rho_a u_*^2$ the surface stress and u_* the friction velocity.

In the middle of the 1990s ECMWF developed a coupled ocean-wave–atmospheric model using the parametrization for the roughness in Eqs. (5.23) and (5.24) In order to be able to have two-way interaction at every time step, the wave model is called as a subroutine from the ECMWF's atmospheric model which is called the Integrated Forecasting System (IFS). The coupled model was introduced in operations on 29 June 1998. At present, at every atmospheric time step, wind fields, air density fields and a gustiness factor are passed from the atmospheric model to the wave model. Then the wave model integrates one time step and determines the two-dimensional wave spectrum. The wave-induced stress is obtained from Eq. (5.22) which is followed by a determination of the Charnock parameter field. The loop is closed by passing the Charnock field to the atmospheric model which then continues with the next time step by using the updated Charnock field in the surface drag over the oceans.

Meteorological models

Weather forecasting is basically an initial-value problem and meteorological models integrate numerically a set of equations that describe the evolution of the state of the atmosphere. The equations concern momentum, thermodynamics of the system, conservation of mass and humidity, and the hydrostatic equilibrium (see, among others, Pedlosky (1987) and Holton (1992) for a thorough discussion of the subject). A predictive equation for pressure is obtained by integrating the equation for mass along the vertical. The general formulation is done in three-dimensional space using spherical coordinates.

The classical method for the integration of geophysical differential equations is by finite differences. For the whole globe this requires a proper treatment of the singularity at the poles. An alternative to the classical method is provided by the spectral technique (Orzag, 1970). In a spectral model the horizontal distribution of the physical quantities is represented by means of truncated spherical harmonics. Originally, the linear terms are calculated in spectral form, while the nonlinear terms and the forcing are evaluated in the grid-point domain. This has changed since the introduction of semi-Lagrangian advection schemes; now only the pressure equation and the horizontal diffusion are treated in spectral space. This requires the use of efficient spectral transforms as the model has to switch back and forth between the two representations at each integration step. Aliasing is avoided by use of a special distribution of the grid points which is called the 'linear grid' (Hortal, 1999).

ECMWF runs a spectral version of the atmospheric model. At present the IFS has $M = 60$ layers distributed in a nonuniform manner over the vertical, extending up to 60 to 65 km. The horizontal distribution is represented by $N = 511$ modes corresponding to a spatial resolution of $40\,000/(2 \times 511) \simeq 39.1$ km. In general, a special choice of horizontal and vertical resolution is denoted by $T_l N/LM$, where T_l means truncation on a linear grid and L denotes layer. Hence, the present version of the IFS is called the $T_l 511/L60$ model. When coupling the IFS with the ECWAM model we have to characterize the wave system as well. Hence, we introduce the following notation: $T_l N/LM - D°$ refers to the coupling of an atmospheric model, having a triangular resolution of N modes and M vertical levels, with a wave model which has a horizontal resolution of $D°$.

Because of the sensitive dependence on the initial conditions, a major task in weather forecasting is to construct an analysis of the inital weather from observations of the atmospheric state. This analysis should be optimal in the sense that it is consistent with the evolution of the atmosphere, and with the (known) model errors and observation errors. ECMWF introduced for that purpose a four-dimensional variational approach, called '4DVAR', at the end of 1997. The variational approach is very well suited to dealing with the enormous amount of indirect observations obtained from instruments on board of satellites. As a consequence, a major improvement in weather forecasting has been achieved in the past 5 years (Simmons and Hollingsworth, 2002). Since ocean waves are so sensitive to the forcing wind fields, we will see the consequences for surface wind and wave forecasting when we discuss verification in Section 5.5.

A detailed description of the forecasting system may be found on the ECMWF website (www.ecmwf.int).

With this system we have performed a number of impact studies, the results of which will be briefly described in the following sections. However, before we embark on this study we mention that in weather forecasting it is usually assumed that the Charnock parameter α is a constant. As a result, the drag coefficient, defined as

$$C_D = \left[\kappa / \log(L/z_0) \right]^2, \qquad (5.25)$$

with $L = 10$ the observation height of the wind speed, becomes by virtue of Eq. (5.23) a unique function of surface wind speed. Observations (from the HEXOS campaign and the Lake Ontario data, for example) and the theoretical considerations of Chapter 3 have suggested, however, that the Charnock parameter depends on the sea state. Because of the short fetches, these data sets consist of relatively young windseas which have steep waves and thus a high wave-induced stress. As a consequence relatively high Charnock parameters are observed. On the open ocean, however, windseas have long fetch and/or duration and they interact with swells

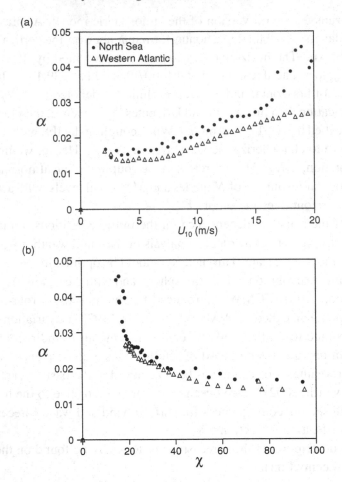

Fig. 5.13. Charnock parameter as function of (a) surface wind and (b) wave age $\chi = c_p/u_*$ for the North Sea and the western Atlantic over a 10-day period in November 2000. The Charnock parameter is averaged over 0.5-m/s wind-speed bins (from Janssen *et al.*, 2002).

generated by winds in remote areas. Under those circumstances, ocean waves are expected to be less steep with relatively low values of the wave-induced stress and therefore low values of the Charnock parameter. This is shown in Fig. 5.13 where for two areas, namely the North Sea and the western-Atlantic area, the relation between the mean Charnock parameter α (averaged over 0.5-m/s wind-speed bins) and 10-m wind speed for the first 10 days of November 2000 is presented. The Charnock parameter was obtained from the coupled ECWAM–ECMWF model. It is clear from Fig. 5.13a that the relation between Charnock parameter and wind speed depends on the geographical location. In the western Atlantic the Charnock parameter is systematically lower compared with the North Sea, presumably because of the relative abundance of swell in the western Atlantic, which through nonlinear transfer and the quasi-linear damping has a calming effect on the wind-generated waves that determine the wave-induced stress.

Figure 5.13a suggests, therefore, that there is no unique relation between Charnock parameter and the surface wind speed. Since we know that in practice the sea state is well characterized by its stage of development parameter, the wave age $\chi = c_p/u_*$, it is expected that the wave age is a more appropriate parameter to characterize the roughness of the ocean. This is indeed confirmed by Fig. 5.13b and by the findings of, for example, Donelan (1982) and HEXOS (Smith *et al.*, 1992).

5.3.1 Impact studies: medium-range forecasting

First we describe results from impact studies on the evolution of a single depression, followed by a discussion of the impact on weather forecasting. The main thread through this discussion is the role of horizontal resolution in representing the delicate interaction between wind and waves which takes place predominantly on scales of 200 km and smaller.

Before we discuss results from the impact studies we first give a simple theoretical picture of the interaction of ocean waves with the large-scale atmosphere. To a first approximation the atmospheric flow over the oceans is determined by the balance between the pressure gradient and the Coriolis force. This is called the 'geostrophic approximation'. Hence from Eq. (2.1) it follows that $\mathbf{f} \times \mathbf{u} = -\nabla p/\rho$, or $\mathbf{u} = \mathbf{f} \times \nabla(p)/\rho f^2$. Thus, isobars (lines of constant pressure) are streamlines and therefore the airflow is along isobars. In the geostrophic approximation a pressure low would not change in time. However, there are important deviations from geostrophy caused by, for example, the inertial terms and the friction terms in the momentum equation (this is called the 'quasi-geostrophic approximation', see Pedlosky (1987)). Thus, close to the surface, friction becomes increasingly important and gives rise to considerable deviations from geostrophy. In fact, friction results in a cross-isobar flow which tends to fill a pressure low. The higher the friction the less deep a low becomes, therefore, in particular for young sea states which are associated with a rough airflow and thus a high friction, an enhanced filling up of a pressure low is to be expected. Following this approach, the time scale of the impact of waves on the atmospheric circulation was estimated to be of the order of 5 days (Janssen and Viterbo, 1996), while due to the two-way interaction the central pressure of the low would change by at most 5 hPa. Because of the long time scale it should be noted that this simple picture of the impact of surface friction on the decay of a depression may be obscured by other effects, such as wave–mean flow interactions in the atmosphere. In addition, it should be realized that enhanced surface roughness will not always lead to the decay of a depression, because heat fluxes are enhanced as well, resulting in vortex stretching and therefore in a deepening of the low. Thus, whether the enhanced roughness, as caused by the ocean waves, results in a filling up or a deepening of the low depends on whether momentum or heat fluxes determine the evolution of that particular low. Nevertheless, this picture of the interaction of ocean waves and large-scale atmosphere cannot be complete, because in the coupled IFS–ECWAM system, in the medium-range time scale, we have observed impacts on depressions of the order of 20 hPa or more.

Impact on a single depression

The impact of two-way interaction on the evolution of a single depression was studied by Doyle (1995). The atmospheric model used in this study was the US Navy's Coupled Ocean Atmospheric Mesoscale Prediction System (COAMPS) and the wave model was the WAM model coupled through Eq. (5.24). The three-dimensional model solves the compressible equations of motion. The model was used in a channel mode with the f-plane approximation. In the vertical, the model has 32 layers with greater resolution in the lower troposphere to enable a good representation of the interaction of the wave-induced stress with the marine boundary layer. For both the atmospheric and the wave model the horizontal resolution was 30 km with periodic boundary conditions in the zonal direction. A time step of 90 s was used in the atmospheric model and 6 min for the ocean-wave model. The simulation was integrated to 96 h which enabled the study of the rapid-development and early-decay phases of an idealized cyclone. Initial conditions were based on mean winter conditions, while the sea surface-temperature pattern resembled the Gulf Stream pattern.

Results of two-way interaction (called 'coupled') were compared with a control run that used a constant Charnock parameter $\alpha = 0.0185$, rather than Eq. (5.24). Doyle's results clearly indicated that at 60 h into the forecast the increased roughness in the coupled run gave, compared with the control run, an increase in central pressure of the low of 6 hPa. The sensible heat flux was increased by 20 per cent, the rainfall maximum increased by 34 per cent, while the kinetic energy at the surface decreased by 20 per cent. These results suggest that frictional effects of wind-generated ocean waves may influence the boundary-layer structure in the vicinity of a marine cyclone.

Hence, Doyle's results suggest that the sea-state-dependent roughness makes a difference to the evolution of a low. On the other hand, Lionello *et al.* (1998) and Lalbeharry *et al.* (2000) found a reduced sensitivity of the development of (synthetic) lows on the sea state, while Bao *et al.* (2000) found, in contrast to the previous works, a deepening of a Gulf of Mexico hurricane by 6 hPa. There may be several reasons for these discrepancies. First, the Lionello *et al.* (1998) and Lalbeharry *et al.* (2000) atmospheric simulations had a coarse resolution, which limits the intensity of the low and therefore reduces the coupling between the ocean waves and the atmosphere. In addition, while all these authors used the same wave prediction system, the atmospheric models were clearly different. A different treatment of the physical processes in the boundary layer will undoubtedly lead to differences in sensitivity to the sea state. Moreover, most atmospheric models have horizontal diffusion in order to ensure numerical stability for feasible integration time steps. A side effect of horizontal diffusion is, however, a reduced level of activity in scales, say of the order of 200 km, that are relevant for the interaction of wind and waves. Finally, in the Bao *et al.* (2000) experiment, and also in that of Lalbeharry *et al.* (2000), sensible and latent heat fluxes have the same sensitivity to the sea state as the momentum flux. Although normally (see Janssen and Viterbo, 1996; Lionello *et al.*, 1998) the dependence of the heat flux on the sea state plays a relatively minor role in the development of a low, under conditions where hurricanes develop the air–sea temperature differences may be large. In that event increased roughness of the waves may result in enhanced heat fluxes giving vortex stretching and therefore a deeper low. However, it should be noted that field observations from HEXOS (DeCosmo, 1991) do not support the sensitive dependence of heat fluxes on the sea state and/or wind speed as chosen by Bao *et al.* (2000) or Lalbeharry *et al.* (2000). Following Beljaars (1995), in the IFS model neutral exchange coefficients for momentum, heat and moisture differ from each other. This means that compared with momentum transfer, heat and moisture transfer have a reduced sensitivity to the sea state because they depend on the square root of the drag coefficient.

In order to illustrate the sensitive dependence of two-way interaction on horizontal resolution, we study one particular case, namely the 4-day forecast from 15 February 1997, shown in Fig. 5.14. The initial data for the atmospheric fields were taken from the operational ECMWF analysis. Initial data for the wave model were generated by using as initial condition a JONSWAP spectrum 10 days before the starting date of the forecast and by running the wave model until the starting date of the experiment with analysed winds from the ECMWF archive. The version of the coupled model was T213/L31 − 0.5°, corresponding to an atmospheric resolution of 95 km while the wave model resolution was only 55 km. The low shown in Fig. 5.14 was observed during FASTEX (Fronts and Atlantic Storm-Track EXperiment) and was called IOP 17. The top-left panel of Fig. 5.14 shows the surface pressure over the North Atlantic for the control run, the top-right panel shows the coupled results, the bottom-left panel shows the operational verifying analysis, while the bottom-right panel shows the difference between coupled and control run. The control day-4 forecast has a good quality, as may be judged from the

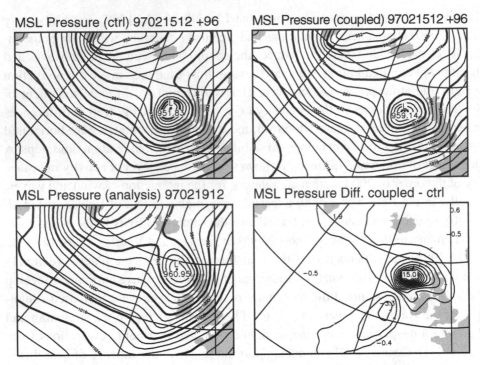

Fig. 5.14. Comparison of 4-day forecast of surface pressure over the North Atlantic, valid for 19 February 1997. Top-left panel, control; top-right panel, coupled; bottom-left panel, operational analysis; bottom-right panel, the difference between coupled and control. Version of coupled model is T213/L31 − 0.5° (from Janssen *et al.*, 2002).

comparison with the analysis, although the fast-moving low to the west of Scotland is misplaced and too deep by about 9 hPa. A measure for the quality of the forecast is the so-called 'anomaly correlation'. It is defined as the correlation between forecast anomaly and the analysis anomaly (both measured with respect to the climate). For this case the anomaly correlation over the North-Atlantic area was greater than 90 per cent. In the coupled run it is seen that the low near Scotland is less deep by 8 hPa so that in this respect there is better agreement between the coupled run and the verifying analysis. However, because the differences are small scale and the scores are evaluated on a coarse grid with a resolution of 2.5°, there is hardly any difference in anomaly correlation between coupled and control run.

It is noted that these results depend in a sensitive manner on the resolution of the coupled system. To appreciate this point we show in Fig. 5.15 results of the T106/L31 − 1.5° model for the same case. While in the high-resolution run we notice differences of up to 15 hPa, the low-resolution results show differences of at most 2 hPa. A similar sensitive dependence was also noticed in other extreme events.

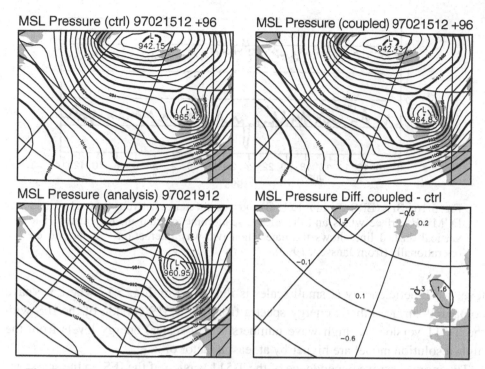

Fig. 5.15. Comparison as in Fig. 5.14 but now for T106/L31 − 1.5° version (from Janssen *et al.*, 2002).

Operational weather forecasting

Trials before the introduction of the coupled ocean-wave–atmosphere prediction system showed a modest impact of sea-state-dependent roughness on atmospheric forecast scores such as the rms error or the anomaly correlation of the 1000- and 500-hPa geopotential for the large domains of the northern or southern hemisphere.

The improvements were more substantial for surface parameters such as the 10-m wind speed or the significant wave height. When the two-way interaction of winds and waves was introduced in operations on 29 June 1998 there was a pronounced improvement of the quality of the surface wind field. Routinely, first-guess (FG) winds are compared with scatterometer winds (from ERS-2 in this case). As shown in Fig. 5.16, which displays time-series of bias (ERS-2-FG) and the rms difference, there is a considerable reduction of 10 per cent in the rms error after the introduction of two-way interaction.

However, currently the impact of two-way interaction of wind and waves is more substantial. The main reason for this is an increase of atmospheric resolution from $T_1 319$ to $T_1 511$ (or from 65 to 40 km) which allows for a more realistic representation of small spatial scales. It is emphasized that these are the scales that matter for air–sea interaction. The fact that the present atmospheric system has more-realistic

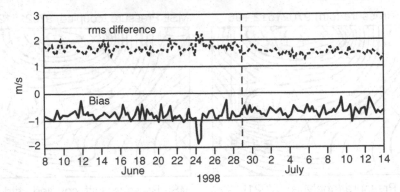

Fig. 5.16. Bias (ERS-2 minus EC FG) and rms difference between the background ECMWF surface winds and the ERS-2 scatterometer wind measurements. The vertical dashed line shows the date when two-way interaction was introduced operationally (from Janssen, 1999).

levels of kinetic energy at small scales is illustrated by Fig. 5.17 where we have compared surface kinetic energy spectra from the $T_l 319$ version of the IFS with the $T_l 511$ version. At high wave numbers (small scales) energy levels from the high-resolution model are higher by at least a factor of 2.

The more sensitive dependence of the $T_l 511$ version of the IFS on the sea-state-dependent drag became evident when we performed experiments with a doubling of angular resolution of the wave spectrum from 12 to 24 directions. Trials with the $T_l 319$ version showed an improvement of forecast skill between 1 and 2 h. However, when experiments were performed with the $T_l 511$ version of the IFS a substantially larger impact of the increase of angular resolution was found. This is illustrated in Fig. 5.18 which compares forecast performance in the northern and southern hemispheres for 24 cases in August 2000. It is worth noting that the sample size is too small to infer general conclusions on the size of the impact, but nevertheless the impact is considerable. Also note that, as a rule of thumb, larger impact is usually found in the summer time, presumably because physical processes near the surface play a more important role in the evolution of the weather. In winter time the atmospheric circulation is dominated by baroclinic activity, and physical processes such as surface friction play a relatively minor role, although there may be a considerable small-scale impact in cases of rapidly developing lows (Doyle, 1995; Janssen *et al.*, 2002).

5.3.2 *Impact studies: seasonal integrations*

Janssen and Viterbo (1996) studied the impact of two-way interaction on the seasonal time scale. In order to obtain reliable information on the impact of waves on

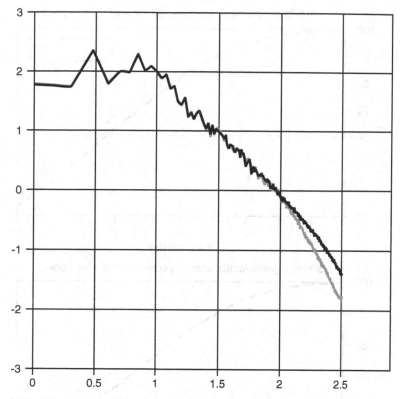

Fig. 5.17. Comparison of surface kinetic energy spectrum as a function of total wave number for T_l511 (dark line) and T_l319 (light line).

the atmospheric circulation there is a need for ensemble forecasting, because the variability of the weather is high, especially over the oceans. Therefore, 15 coupled and control runs were performed for the winter season of 1990 starting from the analysis of 15 consecutive days. The atmospheric resolution was T63, and the wave model had a resolution of 3°, while the length of the runs was 120 days. By taking a time average over the last 90 days, followed by an ensemble average, a reliable estimate of the mean state of one season could be provided. At the same time, information on the variability may be inferred from the scatter around the mean, and thus a Student's *t*-test may be applied to test the statistical significance of the mean difference between coupled and control run. As an illustration of the impact, we have plotted in Fig. 5.19 the ensemble means of the 500-hPa height field and their differences for the northern hemisphere, while for comparison purposes we also display the 90-day mean of the corresponding ECMWF analysis. Contours for the mean are plotted every 60 m, while in the difference plot we have indicated by heavy shading the probability of 95 per cent (or more) that the two fields in question are not equal. Significant differences are noted in the storm track areas

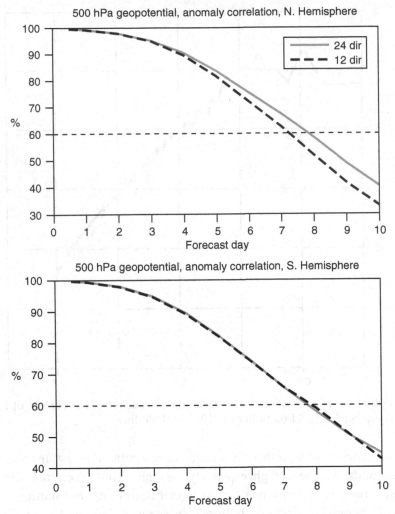

Fig. 5.18. Anomaly correlation of 500-hPa geopotential height for the northern and the southern hemisphere for the last 24 days in August 2000. Here, the impact of increased angular resolution from 12 directions to 24 directions on the forecast performance of the T_l511 IFS forecast system is shown (from Janssen *et al.*, 2002).

of the northern hemisphere (and, not shown, also for the southern hemisphere). We note differences over the northern Pacific, Europe and Siberia. In the last two areas the coupled climate shows, when compared with the analysis, a considerable improvement. There are also improvements in low-frequency variability over the North Atlantic (not shown).

As far as impact of ocean waves on the atmospheric climate is concerned it should be emphasized that here resolution of the atmospheric model also plays a crucial role. Following Weber *et al.* (1993), Janssen and Viterbo (1996) also performed

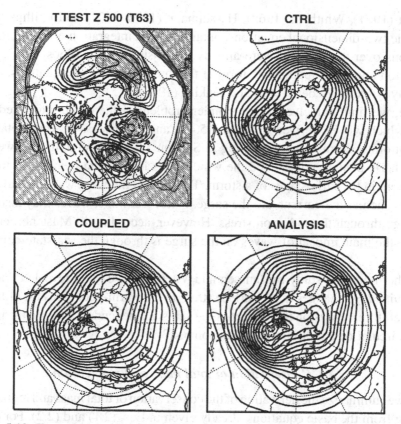

Fig. 5.19. Ensemble means of coupled and control runs and their differences. For comparison the analysed climate is also shown. The period is winter 1990 and the area is the northern hemisphere. The shading indicates the measure of significance; heavy shading means that there is a probability of 95 per cent that the difference is significant (from Janssen and Viterbo, 1996).

seasonal forecasts with the T21 version of the coupled system and particularly in the southern hemisphere a much reduced impact of the sea-state-dependent drag on the atmospheric circulation was found. This should not come as a surprise when it is realized that with T21 the mean wind speeds are reduced by as much as 50 per cent, therefore giving a much weaker coupling between wind and waves.

5.4 Impact of sea state on the ocean circulation

In the literature, relatively little attention has been paid to the study of the impact of ocean waves on the ocean circulation and the mean surface elevation. Only recently has an increased activity been noted, presumably triggered by the thought-provoking work of McWilliams and Restrepo (1999) on the wave-driven ocean circulation. Nevertheless, following the work of Longuet-Higgins and

Stewart (1961), Whitham (1962), Hasselmann (1970, 1971) and Phillips (1977), the basic two-dimensional equations, obtained by an integration of the momentum equations over the vertical, were already established in the 1960s. There is still discussion about the appropriate form of the three-dimensional equations, despite the early work by Weber (1983) and Jenkins (1987).

Here, we limit ourselves to the simpler problem of the depth-integrated equations, which will be derived in Section 5.4.1, and we discuss in some detail their application to storm-surge modelling in Section 5.4.2. The coupling between the atmospheric boundary layer and the wind waves leads to an increase of the drag coefficient in the early stages of a storm. This affects the wind waves but also phenomena such as the current and the surge. Apart from this, waves may also affect the surge through the radiation stress. However, according to Mastenbroek *et al.* (1993), the main impact of waves on the surge is through the sea-state-dependent drag.

Furthermore, the study by Janssen and Viterbo (1996) also revealed that there were quite large changes in the surface stress in the warm-pool area east of Indonesia. Therefore, results on the impact of the sea-state-dependent drag on the tropical ocean circulation are discussed in Section 5.4.3.

5.4.1 Conservation of total mass and momentum

Here, we summarize the derivation of the conservation of total mass and momentum, starting from the basic equations already given in Eqs. (2.1) and (2.2). For ease of reference we repeat them here, but written in a slightly different form.

Consider an incompressible fluid (water) in a constant gravitational field on a rotating Earth. Let the body of water with air above it be of infinite extent in the horizontal while in the vertical it extends from $z = -D$ (with D the water depth) to $z = \eta$, with $\eta(x, y, t)$ the unknown surface elevation. Let us assume that the water motion is governed by the continuity equation

$$\frac{\partial}{\partial t}\rho + \nabla \cdot \rho\mathbf{u} = 0 \tag{5.26}$$

and the momentum equation

$$\frac{\partial}{\partial t}\rho\mathbf{u} + \nabla \cdot \rho\mathbf{u}\mathbf{u} = -\nabla p + \rho\mathbf{g} + \rho\mathbf{u} \times \mathbf{f} + \nabla \cdot \tau. \tag{5.27}$$

These equations apply to the domain $-D < z < \eta$ and the boundary conditions are

$$z = \eta(x, y, t), \frac{\partial}{\partial t}\eta + \mathbf{u} \cdot \nabla_{\mathrm{h}}\eta = w, \ p = p_{\mathrm{a}}, \tag{5.28}$$

where p_{a} is the given air pressure at the sea surface and $\nabla_{\mathrm{h}} = (\partial/\partial x, \partial/\partial y)$ is the horizontal gradient operator. At the flat bottom $D = D_0$ we impose the condition that no fluid penetrates the bottom:

$$z = -D, w = 0. \tag{5.29}$$

According to Longuet-Higgins and Stewart (1961), Whitham (1962), and Phillips (1977) conservation laws for the mean surface elevation ζ and the mean horizontal velocity $\mathbf{U}$ may now be obtained by integration of the continuity equation and the momentum equation over the depth of the water, followed by a suitable ensemble averaging. The ensemble average $\langle\rangle$ is supposed to filter the linear gravity wave motion. Here, the mean surface elevation ζ is defined as

$$\zeta = \langle \eta \rangle, \tag{5.30}$$

while the mean horizontal velocity $\mathbf{U}$ follows from

$$\mathbf{U} = \frac{\mathbf{P}}{\rho h}, \tag{5.31}$$

with $h = D + \zeta$ the slowly varying water depth. Note that $\mathbf{P}$ is the *total* mass flux

$$\mathbf{P} = \left\langle \int_{-D}^{\eta} dz \, \rho \mathbf{u} \right\rangle, \tag{5.32}$$

i.e. it consists of the sum of the water column mean $\mathbf{P}^{\mathrm{m}}$ and the surface-layer mean $\mathbf{P}^{\mathrm{w}}$, defined as (Hasselmann, 1971)

$$\mathbf{P}^{\mathrm{m}} = \left\langle \int_{-D}^{\zeta} dz \, \rho \mathbf{u} \right\rangle, \quad \mathbf{P}^{\mathrm{w}} = \left\langle \int_{\zeta}^{\eta} dz \, \rho \mathbf{u} \right\rangle. \tag{5.33}$$

In the linear approximation the surface-layer mean mass flux may be expressed in terms of the wave momentum

$$\mathbf{P}^{\mathrm{w}} = \rho g \int d\mathbf{k} \, \mathbf{l} \, F/c, \tag{5.34}$$

where c is the phase speed of the gravity waves and $\mathbf{l} = \mathbf{k}/k$ is a unit vector pointing in the direction of the wave propagation. As a consequence, the mean horizontal velocity $\mathbf{U}$ is the sum of the ocean circulation velocity $\mathbf{U}_{\mathrm{c}}$ and the wave-induced drift $\mathbf{U}_{\mathrm{surf}}$,

$$\mathbf{U} = \mathbf{U}_{\mathrm{c}} + \mathbf{U}_{\mathrm{surf}}. \tag{5.35}$$

Note that the momentum in the mean surface drift equals the momentum of the Stokes drift (Phillips, 1977).

The conservation laws become (Mastenbroek *et al.*, 1993)

$$\frac{\partial}{\partial t} \zeta + \nabla_{\mathrm{h}} \cdot (h \mathbf{U}) = 0, \tag{5.36}$$

and

$$\left(\frac{\partial}{\partial t} + \mathbf{U} \cdot \nabla_h\right) \mathbf{U} + g\nabla_h \zeta + \frac{1}{\rho}\nabla_h p_a = \mathbf{U} \times \mathbf{f} + \frac{\tau_a - \tau_b}{\rho h} - \frac{1}{\rho h}\nabla_h \cdot \mathbf{S},$$

$$(5.37)$$

where τ_a and τ_b represent the atmospheric surface stress and the bottom stress. The radiation stress tensor $\mathbf{S}$ represents the contribution of the wave motions to the mean horizontal flux of horizontal momentum. In terms of the wave spectrum it is given by

$$S_{ij} = \rho g \int d\mathbf{k} \left\{ \frac{v_g}{c} l_i l_j + \left(\frac{v_g}{c} - \frac{1}{2}\right) \delta_{ij} \right\} F(\mathbf{k}). \qquad (5.38)$$

Note that the first term corresponds to advection of wave momentum, while the second term consists of a combination of contributions from the wave-induced pressure and the wave-induced stress (Phillips, 1977).

As pointed out by Whitham (1974) the momentum conservation law (5.37) assumes its most simple form when the mass-transport velocity including the wave momentum is used. In this formulation of the conservation laws, ocean waves only appear explicitly through the radiation stress tensor $\mathbf{S}$. Implicitly they also appear through parametrizations of the stress. An example is the case where the bottom stress τ_b is assumed to depend on the current velocity $\mathbf{U}_c$ only rather than on the total velocity $\mathbf{U}$. Mastenbroek *et al.* (1993) assumed that the bottom stress is given in terms of the total velocity and used the above depth-averaged equations in a study of the impact of sea-state-dependent atmospheric stress on a number of storm surges in the North Sea (see Section 5.4.2).

Although the depth-averaged continuity and momentum equations show their simplest form in terms of the total velocity, it is nevertheless instructive to study the evolution equations for the ocean circulation velocity $\mathbf{U}_c$, because they reveal some important physics. Following Hasselmann (1971) one eliminates from Eq. (5.37) the rate of change in time of the wave momentum by means of the energy balance equation (5.21). Equation (5.21) is valid for deep-water gravity waves only, but the extension to shallow water is straightforward by adding an additional source function S_{bot}, representing energy loss of ocean waves by bottom friction. Dividing the energy balance equation by the phase speed and integrating over wavenumber $\mathbf{k}$ gives (dropping the carets over the source functions)

$$\frac{\partial}{\partial t}\mathbf{P}^w = -\rho g \nabla \cdot \int d\mathbf{k} \, \frac{\mathbf{l} v_g}{c} F + \rho g \int \frac{d\mathbf{k}}{c} (S_{in} + S_{nl} + S_{ds} + S_{bot}). \quad (5.39)$$

Substitution of Eq. (5.39) into Eq. (5.37) gives the following evolution equation for the ocean circulation velocity $\mathbf{U}_c$:

$$\left(\frac{\partial}{\partial t} + \mathbf{U}_c \cdot \nabla_h\right)\mathbf{U}_c + g\nabla_h\zeta + \frac{1}{\rho}\nabla_h p_a$$

$$= \mathbf{U}_c \times \mathbf{f} + \mathbf{U}_{surf} \times \mathbf{f} + \frac{\tau_{oc,a} - \tau_{oc,b}}{\rho h} - \frac{1}{\rho h}\nabla_h \cdot \mathbf{T}. \quad (5.40)$$

And it is straightforward to rewrite the continuity equation:

$$\frac{\partial}{\partial t}\zeta + \nabla_h \cdot (h\mathbf{U}_c) = -\nabla_h \cdot (h\mathbf{U}_{surf}). \quad (5.41)$$

The conservation laws for the mean ocean circulation differ in a number of respects from the laws for the total current. First, the continuity equation now shows an explicit dependence on the mass flux related to the ocean waves. Second, in the momentum equation, effects of the advection of wave momentum have been eliminated; therefore, T becomes

$$T_{ij} = \rho g \int d\mathbf{k} \left(\frac{v_g}{c} - \frac{1}{2}\right)\delta_{ij} F(\mathbf{k}). \quad (5.42)$$

Note that for deep-water waves the group speed is half the phase speed of the waves, and therefore T vanishes in deep water. Third, the surface stress and bottom stress are modified accordingly. The surface stress felt by the mean circulation is the total stress minus the net stress going into the waves, or,

$$\tau_{oc,a} = \tau_a - \rho g \int \frac{d\mathbf{k}}{c} (S_{in} + S_{nl} + S_{ds}), \quad (5.43)$$

and the bottom stress becomes

$$\tau_{oc,b} = \tau_b + \rho g \int \frac{d\mathbf{k}}{c} S_{bot}. \quad (5.44)$$

For deep-water waves the bottom stress is unaffected. Fourth, the wave-momentum equation (5.39) does not involve an explicit Coriolis term, and therefore the mean circulation experiences an additional force given by $\rho\mathbf{U}_{surf} \times \mathbf{f}$. It is this additional force that recently has been given considerable attention. The additional contribution of the waves to the Coriolis force gives rise to the excitation of inertial oscillations (Hasselmann, 1970) and it affects the Ekman spiral in the ocean, in particular in the deeper parts of the ocean (Jenkins, 1989).

Jenkins (1987) proposed a three-dimensional equivalent of Eqs. (5.40) and (5.41), by extending the dissipation source function S_{ds} and the wave momentum into the vertical, assuming essentially potential theory. Hence, they decay like $\exp(-2k|z|)$. However, in an Eulerian frame, wave momentum is concentrated at the surface (Phillips, 1977), while most of the wave-energy dissipation is thought to be concentrated in a surface layer with thickness of the order of the significant wave height. The three-dimensional extension of Eq. (5.40) is therefore still a subject of further investigation, but it is clear that this is a complex issue. As a compromise, one might contemplate introducing wave effects into an ocean-circulation model by having a surface layer that is so thick that at the bottom of the layer the effects of ocean waves are vanishingly small. In that event one could apply Eq. (5.40) to the surface layer.

Nevertheless, it is expected that the study of the effect of ocean waves on the ocean circulation will give rise to some interesting new results in the near future. Although we have not discussed the turbulent kinetic energy budget in this context, it is mentioned that breaking waves are also thought to play an important role in determining the thickness of the mixed layer. This may have important consequences for large-scale air–sea interaction processes related to the prediction of El Niño.

Clearly, the study of the impact of ocean waves on ocean circulation is only beginning and at present no complete picture can be given. In the remainder of this section we therefore only discuss the impact of the sea-state-dependent drag on storm surges and tropical ocean circulation.

5.4.2 Impact on storm surges

The wave-age dependence of the surface drag coefficient C_D with higher values in the early stages of a storm, leads to different effects at the different stages of development. It is instructive to study the hindcast of two storms of opposite characteristics, present in the North Sea in February 1989 and December 1990 (Mastenbroek *et al.*, 1993). Both storms resulted in a considerable surge and high waves off the coast of the Netherlands, but while the first was characterized by a fast passage across the North Sea and therefore relatively young wind waves, the second storm moved slowly producing almost fully developed conditions.

The surges were simulated with the barotropic equations (5.36) and (5.37). In the control run the effects of the radiation stress tensor S were ignored, and surface stresses were obtained from 10-m winds U_{10} from a limited-area model using the empirical Smith and Banke (1975) formulation for the drag coefficient,

$$C_D = (0.63 + 0.066 U_{10}) \times 10^{-3}. \tag{5.45}$$

In order to assess the importance of the sea-state-dependent drag coefficient and the radiation stress, two coupled runs were performed, one using the sea-state-dependent drag formulation of Eqs. (5.23)–(5.25) with radiation stresses switched off, and a second one with radiation stresses included.

From the last two experiments it was found that in one case the water levels showed an increase of 10 to 15 cm when the radiation stress was included in the calculation, while in two other cases the impact was less than 5 cm. The effect of the radiation stress, therefore, cannot always be neglected, especially when shallow-water effects are important.

The surge results for the 1989 storm at eight stations along the North Sea coast are shown in Fig. 5.20. There is a systematic underestimation of the surge when the uncoupled model is used, while, with the exception of the Wick and North Shields stations, the coupled model performs rather satisfactorily, its average error on the peak values being less than 0.11 m. Similar results hold for the second storm. The coupling of wind and waves is more important, however, for the first storm since the mean difference in surge between coupled and control surges is 0.45 m while for the second storm the mean difference is only 0.28 m.

The above results have been obtained without additional tuning of the coupled wave-surge model. This therefore gives further indirect evidence for the effect of waves on the momentum flux. Finally, the impact of waves on storm surges in the Irish Sea and the Yellow and East China Seas have been studied by Wu and Flather (1992) and Zhang and Li (1996, 1997) respectively.

5.4.3 Impact on ocean circulation

As already mentioned, the climate study by Janssen and Viterbo (1996) also revealed that there were quite large changes in the surface stress in the warm-pool area east of Indonesia. Because this area plays a prominent role in understanding certain issues in El Niño prediction, it was thought to be of interest to generate stresses over a 1-year period in order to investigate the impact of the sea-state-dependent momentum transfer on ocean circulation. The long period of 1 year was thought to be necessary because of the long response times of the ocean circulation.

The stress fields were supplied to Dave Anderson (then at Oxford University) and Gerrit Burgers (KNMI) who forced their tropical ocean model with the coupled and control fluxes. Both models gave considerable differences in the temperature distribution of the surface layer of the ocean (Burgers *et al.*, 1995). An integration period of 6 months had already given a good idea of the kind of impact, typically of the order of 1 K. However, the difference patterns of the two models were surprisingly dissimilar. One model showed differences with fairly small spatial scale of the order of 2000 km, while the difference pattern in the other model covered the whole tropical Pacific.

Note that such experiments probably exaggerate the size of the impact, because there may be an important feedback from the ocean to the atmosphere. The present ECMWF seasonal forecasting system consists of a coupled atmosphere–ocean-circulation model. The atmospheric model is coupled to the ocean-wave model in two-way interaction mode. Coupling of wind and waves gave a beneficial reduction

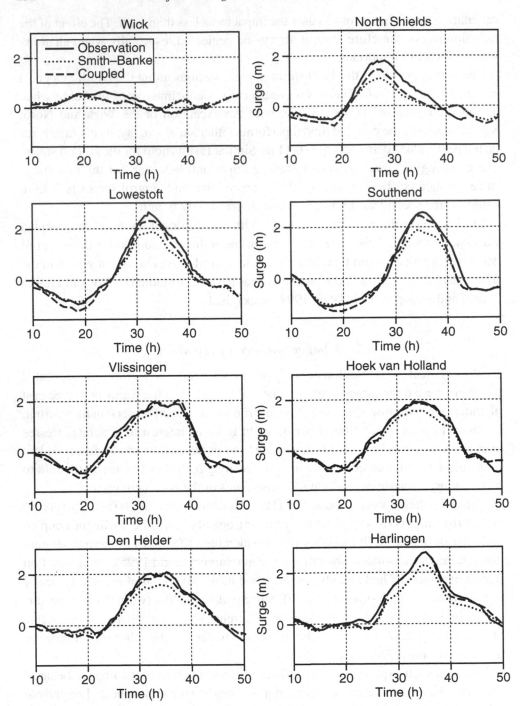

Fig. 5.20. Surge at eight stations along the North Sea coast. Start time is 00 UTC (Universal Time) 11 February 1989. Dotted lines are results using Smith and Banke (1975) and dashed lines are obtained by taking into account the effect of waves on the wind stress (from Mastenbroek *et al.*, 1993).

in the drift in the mean temperature, but the size of the reduction was relatively modest (0.2 K out of a drift of about 1 K in 6 months) (T. Stockdale, private communication, 2003).

5.5 Verification of analysis and forecast

In the penultimate section of this book we shall discuss the performance of the operational ECMWF wave analysis and forecasting system against observations and against the verifying analysis. Currently, the coupled ECWAM–IFS system is used in many applications. Examples are:

(1) *Ten-day deterministic forecasts.* Spatial resolution of the atmospheric model is 40 km, while the wave-model resolution is 55 km. Initial conditions for ocean waves are generated by means of the assimilation of altimeter wave-height data and SAR 2-D spectra from ERS-2. No *in-situ* buoy observations are assimilated.

(2) *Ensemble prediction* needed to estimate forecast uncertainty in wind and waves. Spatial resolution of the atmospheric model is 80 km, while the wave-model resolution is 110 km. Initial conditions for waves are obtained by means of interpolation of the high-resolution wave analysis.

(3) *Monthly and seasonal forecasting.* This is a fairly recent activity at ECMWF. The IFS–ECWAM model is coupled to the Hamburg Ocean Primitive Equation (HOPE) model in order to take advantage of the predictive skill of the ocean over a time scale of a couple of months. The atmospheric component of the monthly forecast has a spatial resolution of 125 km while for the seasonal forecast the resolution is 210 km.

(4) *European reanalysis (ERA) 40: a 40-year reanalysis effort.* This is an extensive effort to obtain the best estimate of the weather over the period September 1957 to August 2002. The analysis is produced with the coupled IFS–ECWAM model, using 3DVAR. Atmospheric resolution is 125 km while the wave model resolution is 1.5°.

An important element of any operational forecasting system is its verification against observations. The main verification activities are concentrated on the deterministic medium-range forecast. Analysed and forecast parameters such as significant wave height and mean period are routinely verified against independent buoy data. A number of operational centres involved in ocean-wave forecasting take part in a project to assess forecast performance against buoy data (Bidlot *et al.*, 2002). However, buoy data are usually only available near coastal areas in the northern hemisphere. In order to assess the global performance of the wave prediction system we compare first-guess wave heights against altimeter wave heights, and we compare forecast wave height against the verifying analysis. Furthermore, the quality of the wave forecast depends to a considerable extent on the quality of the forcing wind fields. For this reason, analysed surface winds are validated against independent altimeter wind-speed observations while forecast wind speed is validated against the verifying analysis. An overview of these activities is given in Janssen *et al.* (1997a) and in Janssen *et al.* (2000).

Although not discussed here, it is mentioned that recently also an extensive validation of products of the wind-wave ensemble prediction system has been performed (Saetra and Bidlot, 2002; Saetra *et al.*, 2002).

In the following subsections, these validation efforts are briefly discussed, while more details may be found in the already referenced literature. Before we present the results of these validation studies we first review what is already known about wave-height errors and their relation to, for example, errors in the forcing wind field. In the past there have been extensive efforts to evaluate the quality of analysed wave results by comparison with buoy observations, particularly in the 1970s and the 1980s when the usefulness of wave prediction first became apparent for the now traditional applications such as ship routing, coastal-defence construction work and offshore operations. The typical performance of early global wave models has been summarized by Clancy *et al.* (1986), Cardone (1987) and Zambresky (1987). Using operationally available winds, these authors found that the scatter index (SI) for the analysed wave height (the ratio of the standard deviation of error to the mean of the observed wave height) ranged from 25 to 40 per cent, while the scatter index for analysed surface wind speed was of the order of 30 per cent or more. Similar results were obtained for limited-area models in shallow water (see, e.g. Janssen *et al.*, 1984).

The first wave model that attempted to give an explicit solution of the energy balance equation was the WAM model. This model has been extensively validated in realistic circumstances against, for example, buoy data (Zambresky, 1989; Wittman *et al.*, 1995; Khandekar and Lalbeharry, 1996; Janssen *et al.*, 1997a) and altimeter wave-height data from Seasat and Geosat (see, for example, Komen *et al.*, 1994) and ERS-1 and-2 (see, for example, Janssen *et al.*, 1997a, b). Although the overall performance of the WAM model was regarded as satisfactory, an underestimation of wave height during extreme events was noted. There may be several reasons for the underestimation of wave height. One reason could be related to problems with the physics of the WAM model, but the major cause of the underestimation of wave height turned out to be the quality of the driving wind field. This was illustrated by the work of Cardone *et al.* (1995) who performed simulations with the WAM model during the Surface Wave Dynamics Experiment (SWADE) using two different wind fields. The first simulation used operational wind products from ECMWF, while the second simulation used manually analysed winds produced by the Ocean Weather Service and the Atmospheric Environment Service (OW/AES). The latter winds are the result of a man–machine mix procedure that takes maximum advantage of all available products from numerical modelling, the know-how of the experienced meteorologist, and all available observations made during the SWADE

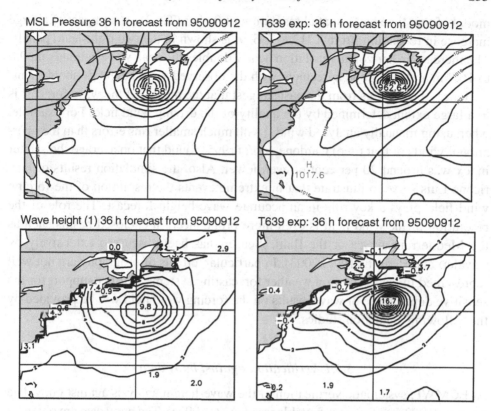

Fig. 5.21. Progress in extreme-sea-state forecasting due to the increase in spatial resolution as illustrated by the 36-h forecast of hurricane Luis. The top-left panel shows the operational forecast mean sea level pressure of 1995090912 UTC with T213 resolution while the top-right panel shows the 36-h forecast with T_l639 resolution. The bottom panels show the corresponding 36-h wave-height forecast forced by low -(left) and high-(right)-resolution winds (from Janssen, 1999).

campaign. Considerable differences between ECMWF and OW/AES winds were found which resulted in large differences in simulated wave height. Compared with buoy data, modelled wave heights based on the ECMWF winds were seriously underestimated while wave heights simulated with the OW/AES winds were in good agreement with the observations.

The SWADE campaign took place from October 1990 to March 1991, but even in 1995 the quality of operational ECMWF wind fields during extreme events was not always optimal. We have illustrated this in Fig. 5.21 which shows the 36-h surface pressure and wave-height forecast for Hurricane Luis in its extra-tropical phase. The two left-hand panels show the operational forecast of that time; compared with observed minimum surface pressure (965 hPa) and observed maximum wave height (17 m) it is clear that the model simulations perform poorly. The operational model of that time had a spatial resolution of 94 km (T213). The impact of a large increase in resolution, from T213 to T_l639, is shown in the two right-hand panels. The increase in horizontal resolution nearly doubles the peak wave height from 9.8 to 16.7 m and is in good agreement with the observed wave height of about 17 m.

Hence, the findings from the past suggest that the quality of the wave forecast is to a large extent determined by the quality of the driving wind field. For example, when using manually analysed winds (with much smaller rms errors than the operational winds of that time) Cardone *et al.* (1995) found that on average the scatter index was around 20 per cent or even lower. Also, the simulation results of hurricane Luis serve to illustrate that in extreme events the resolution of the forcing wind field plays a key role in an accurate wave-height forecast. The role of the resolution of the wind field is even more pronounced in enclosed basins such as the Mediterranean Sea or the Baltic Sea, as has been discussed extensively by Cavaleri and Bertotti (1997, 2003). In particular, effects of orography are not well represented in low-resolution weather forecasting systems. Further support for the sensitive dependence of wave results on the forcing wind field will be provided by the following validation studies.

5.5.1 Verification against buoy data

At ECMWF, systematic verification of the wave-height analysis against buoy data began in 1993 (see for more detail Janssen *et al.*, 1997a). The buoy data are obtained through the Global Telecommunication System (GTS) and are purposefully not used in the wave analysis, so that the comparison of analysis and buoy data provides an independent test of the quality of the analysed wave height. The modelled result is obtained by linear interpolation in space towards the buoy location and is compared with the adequate observed value at one of the synoptic times (00, 06, 12, 18 UTC). Buoy observations and the model represent different scales. Buoys exhibit high-frequency variability on a time scale of 1 h; this is absent in the model because the model value does represent a mean value over a box of size 55 km × 55 km. Averaging of the observed wave height is therefore preferable where the averaging period should match the scales still represented by the model. With a mean group velocity of 5 m/s an averaging time of the order of 4 h thus seems appropriate to represent a spatial scale of 55 km.

Most buoy data are reported every hour via the GTS and are archived in the ECMWF data base, provided they arrive in time. It is a simple matter to use monthly time series to perform a quality check on the data. This quality check will only keep values that are within an acceptable physical range. It will try to detect faulty instruments by removing all constant records of over 1-day long, and it will remove outliers by looking at the deviations from the mean of each monthly data record

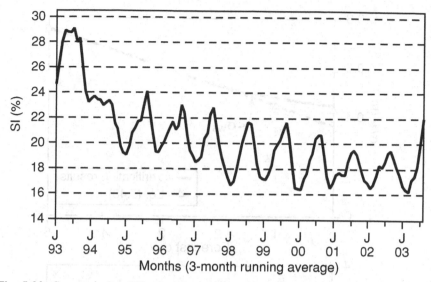

Fig. 5.22. Scatter index (SI) of analysed wave-height error against buoy observations over the period January 1993 to July 2003. The SI averaged over 3 months is shown.

and from the deviation from one hourly value to the next. Furthermore, only buoys are selected that have fairly continuous data records, that are sufficiently far away from the coast and that are located in deep water.

In order to have an indication of the progress that has been made over the past 10 years we show in Fig. 5.22 a monthly time series of the SI of significant wave height. The SI is defined as the ratio of the standard deviation error of the difference between model and observation, normalized with the mean observed value. The plot shows clearly that significant progress has been made over the past decade. Since 2000 the SI is below 20 per cent in the northern-hemisphere summer time, while in the winter time it is even as low as 15 to 16 per cent. This suggests that the present operational ECMWF winds probably have a similar, or perhaps even better quality than the manually analysed winds from OW/AES. Note that, quite remarkably, the SI shows a significant increase in the summer of 2003. This was caused by the failing temporary archiving system on board of the ERS-2 satellite. Data can only be disseminated now when the satellite is in sight of a ground station. This has resulted in a dramatic reduction in the amount of altimeter wave-height data available for use in the wave analysis. Hence, Fig. 5.22 also illustrates the value of the assimilation of altimeter data, but it may also indicate that there are problems with the wave model. We will discuss this issue more thoroughly in Section 5.5.4.

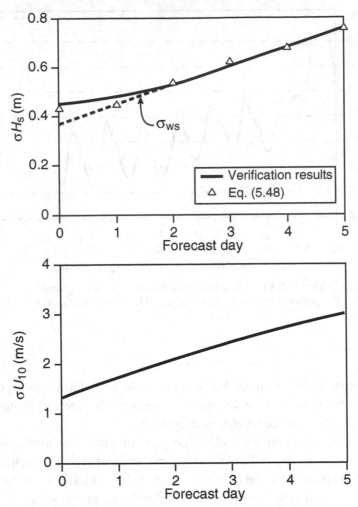

Fig. 5.23. The rms error growth in wave and wind forecast during the period June
to September 1995 (from Janssen *et al.*, 1997a).

As an example of forecast verification we have taken results from the second
half of the year 1995. The evolution of rms error of significant wave height and
surface wind speed with forecast time is shown in Fig. 5.23. It is seen that while
the rms wind-speed error grows almost linearly with time up to day 4, error growth
in significant wave height is slower, at least for the first 2 days of the forecast.
Beyond day 2 of the forecast, wave-height error also grows linearly with time.
This different behaviour of error growth in wave height and wind speed requires an
explanation.

An attempt to explain the relation between wave-height and wind-speed errors
starts from the following empirical relation for the significant wave height H_S of

equilibrium wind waves (see Eq. (2.93)):

$$H_S = \beta U_{10}^2/g, \quad \beta = 0.22, \tag{5.46}$$

where U_{10} is the surface wind speed at 10 m height and g is the acceleration of gravity. First, let us assume that wave-height errors are just caused by local wind-speed errors. The simplicity of this assumption should be emphasized, because there may be many other causes for wave-height error, e.g. wave-model errors and errors in the nonlocal part of the sea state, namely swell. Nevertheless, making this simple assumption and using Eq. (5.46), the rms wave-height error σ_{ws} follows at once,

$$\sigma_{ws} = \sqrt{\langle \delta H_S^2 \rangle} \approx 2\beta U_{10}/g \sqrt{\langle \delta U_{10}^2 \rangle}. \tag{5.47}$$

Then, using as wind speed the average wind speed over the validation period in question and the rms errors in wind speed, the rms error in wave height may be obtained and is plotted in Fig. 5.23. By comparing with the actual rms errors in wave height it is seen that the simple model of Eq. (5.47) explains a considerable part of the error in wave height except for the analysis and the day-1 forecast.

This discrepancy may be attributed to: (1) nonlocal errors (e.g. swell is an important component of the sea state, certainly in the open ocean) and (2) wave-model errors. Let us denote this second error by σ_{sw}, then the total wave-height rms error becomes

$$\sigma_{hs} = \sqrt{\sigma_{ws}^2 + \sigma_{sw}^2}, \tag{5.48}$$

and for the choice of $\sigma_{sw} = 20$ cm we have also plotted the total wave-height error in Fig. 5.23. Compared with the actual error growth curve a good agreement is obtained. Since at day 0 (the analysis) $\sigma_{hs} = 45$ cm while $\sigma_{sw} = 20$ cm, the wind-speed errors are seen to dominate. This just supports the common belief in the ocean-wave community that a considerable part of the wave-height error is caused by errors in the wind field.

5.5.2 Verification against analysis

A problem with the verification of model products against buoy data is the limited coverage over the globe. Typically, most buoys are located in the northern-hemisphere storm tracks near the coasts so that, from the forecast verification against buoys, no information is available on the quality of the wave forecast in the tropics and in the southern hemisphere, and even on the open oceans of the northern hemisphere. A way out of the problem posed by the limited coverage of the buoy observations is to validate the wave forecast against the analysis. However, this is only meaningful when the wave analysis is of sufficient quality.

The wave-height analysis is obtained through the method of optimum interpolation (Lionello *et al.*, 1992), in which equal weight is given to the first-guess model wave height and the wave-height observations from the ERS-1/ERS-2 altimeters. Thus, (systematic) errors in the altimeter wave height will induce an analysed wave-height error but with a weight of 50 per cent. The quality of the altimeter wave heights from ERS-1 and ERS-2 has been studied extensively. Janssen *et al.* (1997b) compared ERS-1 and ERS-2 altimeter wave heights to buoy data over the period of June 1995 to May 1996. The verification was restricted to cases with wave heights larger than 1.5 m, because of known altimeter problems at low wave height. The standard deviation of error (the random error for short) was 35 and 30 cm for respectively ERS-1 and ERS-2 while there was a systematic underestimation of wave height by respectively 15 and 8 per cent (Janssen *et al.*, 1997b). Later, the ERS-2 altimeter wave-height verification was extended to a period of 4 years and an even lower systematic error of only 5 per cent was found. We will return to the issue of the quality of altimeter observations in Section 5.5.4.

Therefore, the quality of the wave analysis is expected to be good, even in the tropical and southern oceans. However, up to May 1996 (before that time ERS-1 data were used in the wave analysis) it seems plausible that the analysis underestimates wave height by about 8 per cent (which implies with an average global wave height of 2.5 m, a systematic error of about 20 cm), while after May 1996, when the switch was made from ERS-1 data to ERS-2 data, the underestimation of wave height is about 4 per cent.

In Figs. 5.24, 5.25 and 5.26 we show, for the northern hemisphere, the tropics and the southern hemisphere respectively, the monthly mean of the random error of significant wave height H_S and surface wind speed U_{10} for different forecast times. The period is August 1994 to August 2003. Over this 9-year period considerable improvements in the skill of the ECMWF wave forecasting system may be noted. For example, in the northern hemisphere the day-1 wave-height random error is reduced by 35 per cent, while the day-1 wind-speed error is reduced by 40 per cent. Similar reductions in random error are also noted in the southern hemisphere. However, since 2001, there is a saturation in the time series for the 1-day forecast error in the southern hemisphere, although remarkably the 3-day forecast error continues to decrease. This improvement in the short-term forecast scores for the northern and southern hemispheres seems to be related to a reduction of the seasonal cycle in the random error, suggesting that most of the improvements have been achieved in northern and southern winter time. The most prominent change in skill scores occurred, by the way, in the tropics when in May 1997 a new formulation of the background cost function J_b was introduced in the atmospheric analysis (Derber and Bouttier, 1999).

Wind speed at 10-m surface level

– – – T+ 24	▬ ▬ ▬ T+ 24 MA	⋯⋯ T+ 72	●●●● T+ 72 MA
—— T+120	▬▬▬ T+120 MA	– ⋅ – T+168	▬ ⋅ ▬ T+168 MA

MA = 12-month moving average

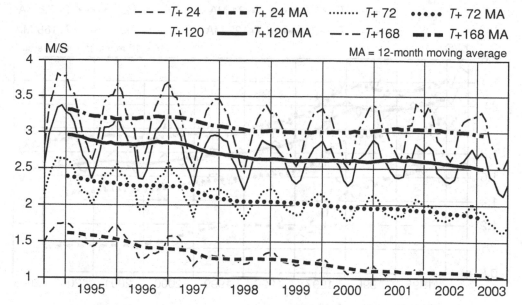

Height of waves at surface level

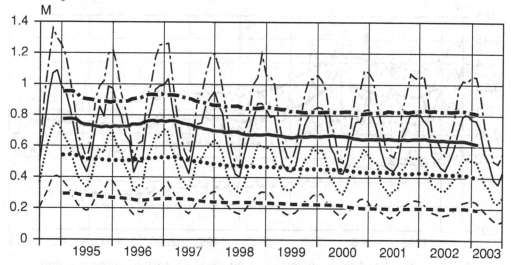

Fig. 5.24. Standard deviation of forecast wind-speed and wave-height error against analysis for the northern hemisphere over the period of August 1994 to August 2003. Forecast ranges are $T + 24$, $T + 72$, $T + 120$ and $T + 168$. The 12-month moving average is also shown.

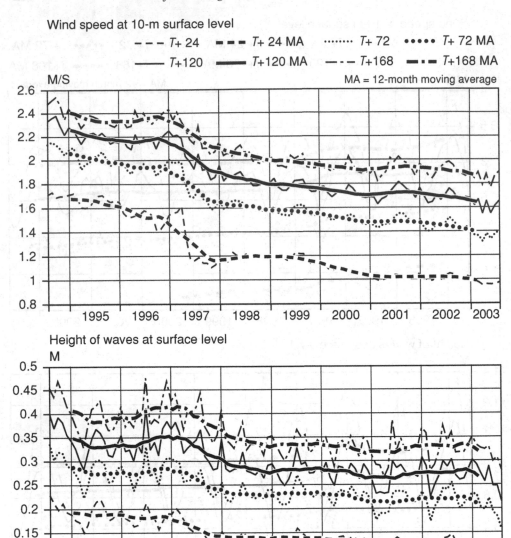

Fig. 5.25. Standard deviation of forecast wind-speed and wave-height error against analysis for the tropics over the period of August 1994 to August 2003. Forecast ranges are $T + 24$, $T + 72$, $T + 120$ and $T + 168$. The 12-month moving average is also shown.

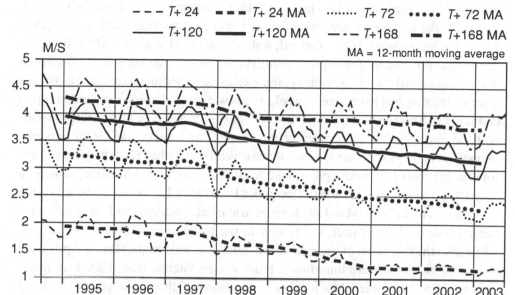

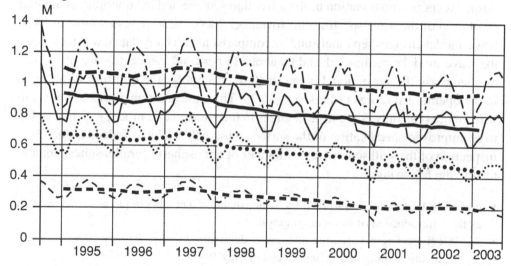

Fig. 5.26. Standard deviation of forecast wind-speed and wave-height error against analysis for the southern hemisphere over the period of August 1994 to August 2003. Forecast ranges are $T + 24$, $T + 72$, $T + 120$ and $T + 168$. The 12-month moving average is also shown.

Thus, the time series of random error in wave height and surface wind are a useful tool to diagnose changes in the ECMWF wind and wave forecasting system. However, the interesting question of what has caused the improved skill requires additional information. In this context, it should be noted that at ECMWF there is a continuous programme for improvement. Three times a year changes are introduced in the operational ECMWF system after extensive experimentation and a parallel suite which may last from several weeks to a few months. During the parallel runs a comparison between wave and wind scores of the old and new forecasting systems is made and therefore the impact of the changes on forecast skill is in principle known. Over the past 9 years we have made several major changes to the ocean-wave forecasting model. For example, spatial resolution was increased twice, namely in July 1994 (from $3°$ to $1.5°$) and in December 1996 (from $1.5°$ to $0.5°$), the advection scheme was modified in May 1997 in order to alleviate problems with shadow effects behind islands. Furthermore, the limiter in the integration scheme was changed, while also the definition of the prognostic range was modified. Also, in April 2002 effects of gustiness were introduced. Finally, we switched from ERS-1 altimeter data to ERS-2 altimeter data in May 1996. None of these changes in the wave model and wave analysis resulted in large reductions in the random wave-height error, except the modification of the advection scheme and the doubling of angular resolution of the wave spectrum in November 2000. We have tested most of these wave-model changes separately and by comparison with buoy data it was found that the wave-model error has reduced by about 15 per cent. Hence, the overall impact of changes in the wave model is too small to account for the improvements found in the operational scores of Figs. 5.24–5.26. Thus, the additional improvements in wave scores can only come from changes in the atmospheric model which have led to an improved specification of the surface wind fields. This is indeed found upon inspection of the skill scores from a number of atmospheric parallel suites, notably testing the following:

- cycle 13R4, introduced in April 1995 (which included a number of physics changes such as the reintroduction of mean orography);
- 3DVAR (including the use of scatterometer data) introduced in January 1996;
- the formulation of the new J_b introduced in May 1997;
- 4DVAR introduced in December 1997;
- the coupling between wind and waves introduced in June 1998;
- the introduction of the high-resolution $T_l 511$ atmospheric model, combined with the doubling of angular resolution in the wave model, in November 2000.

All these changes combined have led to the considerable reduction of the day-1 random wave-height error of about 35 per cent. This once more supports the contention that the main contribution to the wave-height error comes from errors in the driving wind fields.

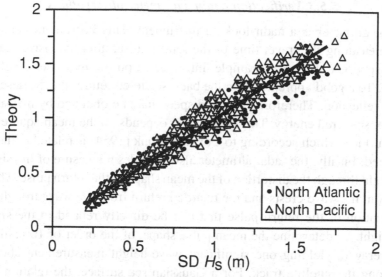

Fig. 5.27. Comparison of random wave-height error obtained from Eq. (5.48) with the actual forecast error from the ECMWF archives for the North Atlantic and the North Pacific.

Using the scores given in Figs. 5.24, 5.25 and 5.26 the validity of the model for H_S error growth given in Eq. (5.48) has been tested. The result is presented in Fig. 5.27 where on the y-axis we have plotted the random wave-height error according to Eq. (5.48) using the monthly mean wind and random wind error from the 12-h forecast until day 10 of the forecast; on the x-axis we have plotted the corresponding random wave-height error from the verification of the wave forecast against the analysis. We have chosen two areas, namely the North Atlantic and the North Pacific. It is evident that there is a close relation between random wave-height and wind-speed error.

A more detailed discussion of the relation between wave-height and wind-speed error is given by Janssen (1998). According to the ECWAM model the waves are driven by the friction velocity rather than the wind speed. Therefore, a growth relation for wave height based on friction-velocity scaling would be more appropriate than Eq. (5.48). Since in the later stages of the forecast the relative wave-height errors become large (as is also evident from Fig. 5.27), even higher moments of the wind-speed error distribution should be taken into account. Nevertheless, as seen from Fig. 5.27, a simple relation such as given in Eq. (5.48) seems to work relatively well in practice.

As the wave-height error is dominated by the wind-speed error, one may turn things around so that wave-model results may be used as a diagnostic tool for the detection of possible errors or problems in the atmospheric model. A number of examples of this approach have been discussed by Janssen *et al.* (2000).

5.5.3 Verification against altimeter observations

The radar altimeter is a nadir-looking instrument. This instrument emits pulses, and by measuring the travel time of the return pulse, after extensive corrections for atmospheric delays, for example, information on the mean sea level may be obtained. To a good approximation the back-scattered return may be described by specular reflection. Therefore, the wind speed may be obtained by measuring the total back-scattered energy. The back-scatter depends on the mean-square slope of the sea surface, which according to Cox and Munk (1954) is related to the surface wind speed. Finally, the radar altimeter also provides a measure of the significant wave height through the distortion of the mean shape of the return pulse. The earlier return from the wave crests and the retarded return from the wave troughs lead to a deformation of the return pulse that can be directly related to the significant wave height. To determine the mean pulse shape, of the order of 100 pulses need to be averaged, yielding one significant wave-height measurement about every 7 km along the satellite track. For a Gaussian sea surface, the relation between pulse shape and the rms sea-surface displacement can be determined theoretically (although corrections are needed due to deviations from normality). This model has been confirmed by numerous comparisons with *in-situ* measurements. The typical accuracy of radar altimeter wave-height measurements is of the order of 10 per cent in the range of 1 to 20 m.

The validation of wave-model results against altimeter data started relatively recently, at least compared with the verification against buoy data. The first validation studies used altimeter wave-height data obtained during the Seasat mission in 1978 (Janssen *et al.*, 1989; Francis and Stratton, 1990; Bauer *et al.*, 1992). The scatter index for significant wave height was typically between 30 and 40 per cent. The poor performance during the Seasat period may be attributed to the poor quality of the driving wind field. This follows from the study by Romeiser (1993) who compared WAM-modelled wave height with Geosat data during the year 1988. The mean scatter index was found to be around 25 per cent, a number that is consistent with the findings of Zambresky (1989) who verified modelled WAM wave height against buoy data.

In July 1991, ESA launched the ERS-1 satellite. ECMWF was involved in the validation of a number of instruments on board of ERS-1, for example, the radar altimeter, the scatterometer and the synthetic-aperture radar (SAR). The comparison of altimeter wave height and wind speed with the corresponding modelled ECMWF parameters showed the high quality of the altimeter measurements. As a consequence, since August 1993, ERS-1 altimeter wave-height data have been used to give an improved specification of the initial conditions for the wave forecast, while altimeter wind speeds have been used to provide an independent validation of the

ECMWF analysed surface wind speed. Therefore, after August 1993 it is only meaningful to compare first-guess wave heights with the altimeter data, which are formally independent, because they have not been used in the analysis yet. In 1995, the mean scatter index obtained from the comparison of first-guess wave height and ERS-1 altimeter wave-height data was found to be of the order of 18 per cent (Janssen *et al.*, 1997a).

Just as for the buoy data, the ERS altimeter data are received in real time from the ESA through the GTS. The along-track resolution is 7 km, corresponding to approximately one measurement per second. In order to obtain observations that represent similar spatial and temporal scales as resolved by the wave model, the altimeter time series are smoothed to a resolution of 200 km. Unrealistic, rapid changes in the signal were filtered out by applying quality control in a similar manner to that of Janssen *et al.* (1989) and Bauer *et al.* (1992). Altimeter data and model data are routinely compared in this fashion. For ERS-2 data, results of this comparison are presented for wave height in Fig. 5.28, while those for wind speed are presented in Fig. 5.29. These plots show time series of the standard deviation of error of first-guess wave height and analysed wind speed over the period May 1995 (the beginning of the ERS-2 mission) to October 2000 for the whole globe. Note that

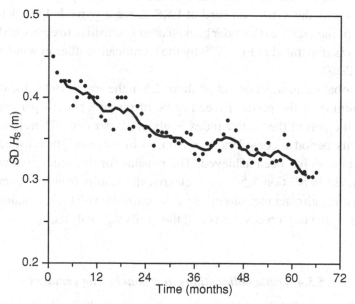

Fig. 5.28. Standard deviation (SD) of first-guess wave-height error as obtained from the comparison between first-guess wave-height field and ERS-2 altimeter wave height over the period May 1995 to October 2000. Area is the whole globe. The filled circles are monthly data while the continuous line is a 6-month running average.

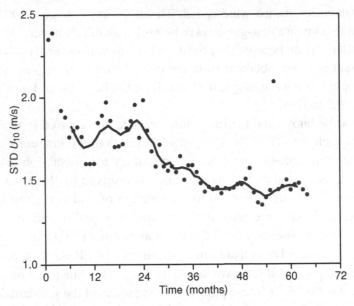

Fig. 5.29. Same comparison as in Fig. 5.28 but now for analysed wind speed.

at the end of the period the quality of the altimeter wind speed worsened because of problems with the gyros on board of ERS-2; these gyros help to determine the orientation of the satellite (the radar back-scatter is sensitive to errors in the azimuth angle). This is illustrated in Fig. 5.29 by the significant outlier in wind-speed error in January 2000.

With a global mean wave height of about 2.5 m the scatter index for wave height in the beginning of the period is seen to be of the order of 18 per cent while at the end of the period the scatter index is about 12 per cent. Therefore, it is seen that over this period a considerable reduction in wave-height error (and also in wind-speed error) has been achieved. The reasons for this reduction have already been discussed in Section 5.5.2, and, clearly, the results from the comparison of model data and altimeter measurements are consistent with those obtained from the comparison with buoy observations and the verifying analysis.

5.5.4 Triple collocations and concluding remarks

In the validation studies discussed so far we have made the implicit assumption that the observations or the analysis represent the truth. This is not necessarily the case and therefore this assumption needs to be checked.

The observation error consists of several components. The instrumental, measurement error usually only gives a small contribution to the total error. More significant are representativeness errors and errors caused by the finite distance and time between two observations.

When comparing several types of data it is desirable to have an idea about the size of the errors. For example, when calibrating one instrument against another it is important to know their error because the calibration constants depend on them. The example of linear regression is discussed by Marsden (1999), see also Tolman (1998). Furthermore, data assimilation requires knowledge of the weights given to the data and to the first-guess field. These weights depend on the ratio of the first-guess error and the observation error. In wave forecasting these errors are usually not known, and one assumes, as is done in the optimum interpolation (OI) scheme of the ECMWF wave forecasting system, that the errors are equal. Hence first guess and observations get equal weight during the analysis.

The need for estimates of errors of different data sources was realized by Stoffelen (1998). He proposed to use a triple collocation method to calibrate observations of winds from a scatterometer using winds from buoys, a model analysis and the ERS-1 scatterometer. In his approach it was assumed that error and truth were not correlated. In a similar vein, Caires and Sterl (2003) applied a triple collocation method to estimate and calibrate analysed winds and wave heights from the ERA-40 analysis effort. Quilfen *et al.* (2001) followed a different approach proposed by Freilich and Vanhoff (1999) to estimate and calibrate ERS scatterometer wind measurements over the period 1992 to 1998. However, in this methodology the true wind speed was assumed to be Weibull distributed and the data sets were not independent because, through the data assimilation, the analysed wind depends on both buoy winds and scatterometer winds. In a somewhat different context, Tokmakian and Challenor (1999) estimated errors in the mean-sea-level anomalies of the model and the ERS-2 and Topex-Poseidon satellites; they used a method that only assumes that there is no correlation between the respective errors. However, a calibration is then not possible.

The method suggested by Stoffelen (1998) is quite elegant and it is straightforward to show that, if one has at least three data sets which have uncorrelated errors, then the error of each data type can be estimated from the variances and covariances of the data sets. Suppose we have three estimates of the truth, denoted by X, Y and Z, obtained from observations or from simulations of the truth by means of a forecasting system. In the following all these estimates of the truth will be referred to as measurements. Furthermore, it is assumed that the measurements depend on the truth T in a linear fashion:

$$X = \beta_X T + e_X,$$
$$Y = \beta_Y T + e_Y, \qquad (5.49)$$
$$Z = \beta_Z T + e_Z,$$

where e_X, e_Y and e_Z denote the errors in the measurements X, Y and Z, while β_X, β_Y and β_Z are the calibration constants. Since we are estimating wave height, which is a quantity that is positive definite, no intercept is included in the model for the measurements. A finite intercept (such as used by Caires and Sterl (2003)) gives rise to negative values of the mean value of either the truth or the measurement, which physically does not make sense.

It is emphasized that the linear dependence of the measurement on the truth is an assumption which needs not to be true and, therefore, one cannot assume that the errors are random. For example, if actually there is a nonlinear relation between measurement and truth but one takes the linear model of Eq. (5.49) instead, the error will have a random and a systematic component. Furthermore, if two types of measurements have a similar nonlinear relation with the truth, then in the context of the linear model of Eq. (5.49) there is now the possibility of correlated errors. This may be the case when comparing two altimeters that share the same measurement principle.

Let us now assume that the linear model of Eq. (5.49) is valid and that the measurement results X, Y and Z have uncorrelated errors,

$$\langle e_X e_Y \rangle = \langle e_X e_Z \rangle = \langle e_Y e_Z \rangle = 0, \tag{5.50}$$

where the angle brackets denote the average over a sufficiently large sample. In order to eliminate the calibration constants we introduce the new variables $X' = X/\beta_X$, $e'_X = e_X/\beta_X$, etc. so that

$$X' = T + e_{X'},$$
$$Y' = T + e_{Y'}, \tag{5.51}$$
$$Z' = T + e_{Z'},$$

and the primed observations also have uncorrelated errors. We now eliminate the truth to obtain

$$X' - Y' = e_{X'} - e_{Y'},$$
$$X' - Z' = e_{X'} - e_{Z'}, \tag{5.52}$$
$$Y' - Z' = e_{Y'} - e_{Z'}.$$

Then, multiplying the first with the second equation of Eq. (5.52) and utilizing the assumption of independent errors in Eq. (5.50) one immediately obtains the variance of error in X' in terms of the variance of X' and the covariances of X' and Y', X' and Z', and Y' and Z'. In a similar manner, by multiplying the first with the third equation of Eq. (5.52) one obtains the variance of error in Y', while the variance of error in Z' is obtained by multiplying the second and the third equation.

Hence,

$$\langle e_{X'}^2 \rangle = \langle (X' - Y')(X' - Z') \rangle,$$
$$\langle e_{Y'}^2 \rangle = \langle (Y' - X')(Y' - Z') \rangle, \qquad (5.53)$$
$$\langle e_{Z'}^2 \rangle = \langle (Z' - X')(Z' - Y') \rangle.$$

Therefore, if errors are uncorrelated only three collocated data sets are needed to estimate the variance of the error in each of them.

The next step is to perform a calibration of the measurements. Since the truth is not known, only two of the three calibration constants can be obtained. Therefore, we arbitrarily choose X as the reference. Since the errors in the measurements are now known the calibration constants for Y and Z may be obtained using neutral regression (Marsden, 1999).

Having performed the calibration of Y and Z it is clear that the work is not finished yet because this calibration will affect the estimation of the errors in X, Y and Z and hence the calibration constants, etc. An iteration procedure is then started until convergence is achieved (Janssen *et al.*, 2003). This approach can be extended to more than three data sets in a straightforward manner.

Janssen *et al.* (2003) have applied this approach to a collocated data set over a 2-year period starting in January 2000, which consisted of collocated buoy, ERS-2 altimeter, first-guess and analysed wave-height data. Formally, buoy, altimeter and first-guess data are independent, but it was found that there are small but significant correlations between altimeter and first-guess wave-height data during the northern-hemisphere summer time. Most probably the reason for this is that the ERS-2 altimeter has problems at low wave heights because it cannot measure significant wave height below 60 cm. This systematic error is then passed on to the analysed wave field and hence to the first-guess field, resulting in a correlation between first-guess and altimeter data. Therefore, a fifth, independent data set was generated by running the ECWAM model in stand-alone mode forced by operational analysed winds but without using ERS-2 altimeter data in the analysis step.

Results for first-guess, analysed, ERS-2 altimeter and buoy wave-height scatter index over the period January 2000 to December 2001 are displayed in Fig. 5.30. It is striking that these errors are relatively small, with the buoy errors the largest and the analysis errors the smallest. The high-quality analysis is a consequence of the properties of the Optimum Interpolation (OI) scheme used to produce the wave analysis. This is explained in more detail in Janssen *et al.* (2003). The reason for the relatively high errors in buoy wave height is probably that, despite the averaging over a 4-h period, the buoy data do not represent the scales of the model field. Note that Stoffelen (1998), who studied errors in buoy, scatterometer and first-guess model winds, obtained a similar ordering of the errors as found in the present study.

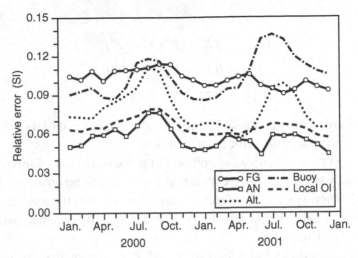

Fig. 5.30. Monthly scatter index of first-guess (FG), analysed (AN), ERS-2 altimeter (Alt.) and buoy wave height. Maximum relative collocation difference is 5 per cent. For comparison the analysis error according to a local optimum interpolation (OI) scheme is also shown.

The statistical analysis also allowed an estimation of the regression constant β. As a reference the *in-situ* buoy wave-height data were chosen, and it was found that on average, altimeter data were too low by 3 per cent, while the analyzed wave height was too low by 4 to 5 per cent.

Combining now the results of the verification studies, it is concluded that the quality of present-day wave forecasting systems is high. From Fig. 5.30 it is inferred that the scatter index for analysed wave height is only of the order of 6 per cent. Recalling that two decades ago the typical value of the wave-height scatter index was 25 per cent or more, it is evident that considerable progress has been achieved. Because wave results are so sensitive to the quality of the forcing wind fields, a considerable part of the progress may be attributed to the increased quality of the atmospheric analysis and forecast. However, wave-model improvements have, no doubt, contributed as well.

We close the discussion on verification by presenting a new diagnostic tool which enables one to study problems in the modelled spectral shape. This is now opportune because a consequence of the large improvements in the forcing wind fields is that it is nowadays more straightforward to identify (systematic) errors in the wave model. This tool was first introduced by Voorrips *et al.* (2002) in a study to validate SAR- and WAM-model frequency spectra against buoy spectra. One simply determines for each period the wave variance from the modelled and observed frequency spectra in a period bin of, say, 2 s and one obtains the 'equivalent' wave height by the usual definition. The resulting period-dependent bias is then plotted as a function of time.

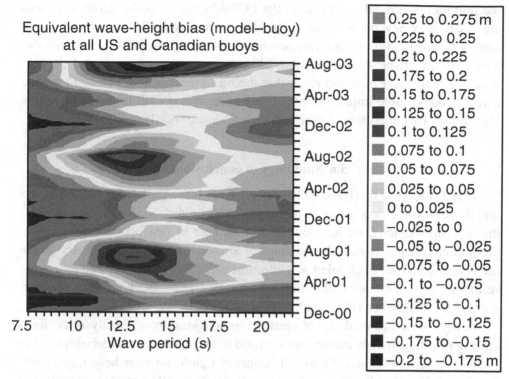

Fig. 5.31. Period spectral bias (model–buoy) at all US and Canadian buoy locations for the period December 2000 to August 2003.

In Fig. 5.31 (obtained from Jean Bidlot, private communication, 2003) an example is given involving all US and Canadian one-dimensional frequency spectra over the period of December 2000 to August 2003. In the range of 10 to 15 s there is a clear seasonal dependence of the 'equivalent' wave-height bias, it being large in the summer time and vanishingly small in the winter time.

It turns out that these large positive biases are related to swell events generated by the storms in the southern-hemisphere winter time. It would be tempting to speculate on the causes of the overestimate by the ECWAM model. An obvious candidate would be the dissipation source function, because this source term is the least well understood. However, a closer inspection of these results reveals that the main problem occurs in the Pacific Ocean and not in the Atlantic (not shown). Increasing the dissipation source term would therefore have a detrimental impact on results in the Atlantic. Hence, it was decided to look for an explanation that takes properties of a particular ocean basin into account. An important difference between the Pacific and the Atlantic Ocean is that in the equatorial region of the Pacific there are a vast number of small islands and atolls that are not resolved by the present operational resolution of the ECWAM model. Although these islands are small, they nevertheless block considerable amounts of low-frequency wave energy (Tolman, 2003). Therefore, using a high-resolution global topography of 2 min, Bidlot (private communication, 2003) determined a wavenumber-dependent blocking factor and he could show that the bias problem disappeared to a large extent. Considerable improvements in wave forecasting skill in the tropics were also found. This important change will be introduced shortly at ECMWF.

5.6 Summary of conclusions

In the last chapter of this book we have applied our basic knowledge on the physics and dynamics of ocean waves to numerical wave prediction. Apart from the 'traditional' applications, we have discussed a number of new applications of wave prediction systems, namely in the areas of air–sea interaction and impact on the atmospheric and ocean circulation. Unfortunately, many interesting and promising results in areas such as satellite remote sensing or ensemble prediction of ocean waves have hardly been discussed.

We have presented evidence of great progress in atmospheric analysis and forecasting resulting in high-quality surface wind fields. Although two decades ago the wind-speed error dominated the error budget of significant wave height, this problem is vanishing rapidly. The implication is that now it will be easier to identify and resolve wave-model errors. Considerable progress in wave modelling is therefore expected in the near future.

Epilogue

In this book we have given an overview of the role that ocean waves play in the problem of the interaction of atmosphere and ocean. However, in order to appreciate this role we had to elaborate on how ocean waves evolve in space and time. It was found that ocean waves evolve according to the well-known energy balance equation which states that the wave spectrum changes due to advection with the group velocity, and due to physical processes such as generation by wind, nonlinear transfer by four-wave interactions and dissipation by, for example, white capping. A detailed exposition of the derivation of the physical source functions was given, followed by a study of the impact of ocean waves on the atmospheric circulation and one aspect of the ocean circulation, namely storm surges. It was also pointed out that the study of the effects of ocean waves on ocean circulation is only beginning, but that promising improvements on its wind-driven component are expected in the near future.

This book was concluded by an extensive discussion of the verification of the ECMWF forecast of wave parameters, such as significant wave height. The impression from this verification study is that the quality of the ECMWF wave analysis and wave forecast is high, certainly if the results are put in a historical perspective.

This impressive improvement in wave forecasting is to a large extent related to the significant improvements seen in the last decade in forecasting of the forcing wind fields, caused by improvements in the numerical treatment of the atmospheric equations, the better representation of physical processes, the introduction of the 4DVAR assimilation and the consequent more efficient utilization of satellite observations. An important contribution to the improved forecasting skill is also given by a better numerical treatment of the energy balance equation, the introduction of processes such as the two-way interaction of wind and waves, and processes such as gustiness in wind-wave growth and the improved quality of the altimeter wave-height data.

Despite these impressive improvements, it cannot be concluded that the physics of waves is entirely understood. Far from it! Regarding wind input and dissipation one may argue that perhaps we know what the most important parameters are that govern these source functions, but, to be really convincing, more work is needed in understanding the role of air turbulence and of airflow separation in wind-wave growth. The dissipation of ocean waves is attributed to a number of processes such as white capping and the generation of Langmuir circulation, for example. Work is needed to be able to assess the relative importance of these two processes in order to estimate the amount of wave energy dissipation in a convincing way. The recent observational evidence of the spectral distribution of dissipation caused by white capping will most certainly play an important role in resolving this issue.

Furthermore, more research is needed into a better numerical treatment of four-wave interactions, advection of wave energy, and refraction by bottom topography and currents. Although recent attempts to improve on this have not been very successful, it cannot be denied that these improvements have been masked by the relatively large uncertainties in the forcing wind field. With the rapid improvements in the quality of the winds, seen in the past 5 to 10 years and ongoing, it seems now opportune to pay more attention to these issues.

Finally, it is justified to ask the question whether there is still a need for further wave model development. There are most certainly definite reasons for further development and they derive, as they should, from a number of applications in which the wave spectrum plays an important role. Just recently we have seen rapid progress in the understanding of the mechanisms behind the generation of extreme sea states such as freak waves. Prediction of enhanced probabilities of extreme events would be of tremendous benefit to the marine world, but clearly an accurate prediction of the detailed low-frequency part of the wave spectrum is of utmost importance. This capability of wind-wave forecasting systems is, as yet, unproven. Furthermore, more research into the relation between spectral shape and the occurrence of extreme states is highly desirable.

On the other hand, knowledge of the high-frequency part of the wave spectrum is important in all remote-sensing applications that depend to some extent on properties of the sea surface. We mention instruments such as the altimeter, the scatterometer, the special sensor microwave/imager (SSM/I), in short, any instrument that involves aspects of specular reflection. In fact, the ocean-surface albedo depends in a straightforward manner on properties of the slope spectrum. Also, we have seen that knowledge of the high-frequency spectrum is important in order to determine the air–sea momentum exchange, and as a consequence it is also important for the exchange of 'passive' scalars such as CO_2. In this book we have introduced a parametrization of the high-frequency spectrum that may be regarded as a good first guess. Nevertheless, the actual spectral shape is not well understood and a significant amount of experimental and theoretical work is still needed to obtain a convincing and working model for the high frequencies.

Combined with the prospects of the impact of ocean waves on ocean circulation, very exciting times in the field of ocean waves are indeed expected to lie ahead of us.

References

Abdalla, S. and L. Cavaleri, 2002. Effect of wind variability and variable air density on wave modeling. *J. Geophys. Res.* **107** (C7), **17**-1.

Abramowitz, M. and I. A. Stegun, 1965. *Handbook of Mathematical Functions*. New York: Dover Publications.

Alber, I. E., 1978. The effects of randomness on the stability of two dimensional surface wave trains. *Proc. R. Soc. Lond.* **A363**, 525–546.

Alber, I. E. and P. G. Saffman, 1978. *Stability of Random Nonlinear Deepwater Waves with Finite Bandwidth Spectra*. TRW Defense and Space Systems Group, 31326-6035-RU-00. Los Angeles, California: TRW Defense and Space System Group.

Alves J. H. G. M. and M. L. Banner, 2003. Performance of a saturation-based dissipation-rate source term in modeling the fetch-limited evolution of wind waves. *J. Phys. Oceanogr.* **33**, 1274–1298.

Al-Zanaidi, M. A. and W. H. Hui, 1984. Turbulent air flow water waves. *J. Fluid Mech.* **148**, 225–246.

Annenkov, S. Yu. and V. I. Shrira, 1999. Sporadic wind wave horse-shoe patterns. *Nonlin. Proc. Geophys.* **6**, 27–50.

2001. On the predictability of evolution of surface gravity and gravity-capillary waves. *Physica D*, **152–153**, 665–675.

Banner, M. L., 1990. Equilibrium spectra of wind waves. *J. Phys. Oceanogr.* **20**, 966–984.

Banner, M. L. and W. K. Melville, 1976. On the separation of air flow over water waves. *J. Fluid Mech.* **77**, 825–842.

Banner, M. L. and J. B. Song, 2002. On determining the onset and strength of breaking for deep water waves. Part II: influence of wind forcing and surface shear. *J. Phys. Oceanogr.* **32**, 2559–2570.

Banner, M. L. and I. R. Young, 1994. Modeling spectral dissipation in the evolution of wind waves. Part I: assessment of existing model performance. *J. Phys. Oceanogr.* **24**, 1550–1571.

Bao, J.-W., J. M. Wilczak, J.-K. Choi and L. H. Kantha, 2000. Numerical simulations of air–sea interaction under high wind conditions using a coupled model: a study of hurricane development. *Mon. Weath. Rev.* **128**, 2190–2210.

Barnett, T. P., 1968. On the generation, dissipation and prediction of ocean wind waves. *J. Geophys. Res.* **73**, 513–530.

Barrick, D. E. and B. L. Weber, 1977. On the nonlinear theory for gravity waves on the ocean's surface. Part II: interpretation and applications. *J. Phys. Oceanogr.* **7**, 11–21.

Battjes, J. A., T. J. Zitman and L. H. Holthuijsen, 1987. A reanalysis of the spectra observed in JONSWAP. *J. Phys. Oceanogr.* **17**, 1288–1295.

Bauer, E., S. Hasselmann, K. Hasselmann and H. C. Graber, 1992. Validation and assimilation of Seasat altimeter wave heights using the WAM wave model. *J. Geophys. Res.* **C97**, 12 671–12 682.

Belcher, S. E. and J. C. R. Hunt, 1993. Turbulent shear flow over slowly moving waves. *J. Fluid Mech.* **251**, 109–148.

   1998. Turbulent flow over hills and waves. *Annu. Rev. Fluid Mech.* **30**, 507–538.

Belcher, S. E., J. A. Harris and R. L. Street, 1994. Linear dynamics of wind waves in a coupled turbulent air–water flow. Part 1: theory. *J. Fluid Mech.* **271**, 119–151.

Beljaars, A. C. M., 1995. The parametrization of surface fluxes in large scale models under free convection. *Q. J. R. Meteorol. Soc.* **121**, 255–270.

Beljaars, A. C. M. and P. A. Taylor, 1989. On the inner-layer scale height of boundary-layer flow over low hills. *Boundary Layer Meteorol.* **49**, 433–438.

Benjamin, T. B.; 1959. Shearing flow over a wavy boundary. *J. Fluid Mech.* **6**, 161–205.

   1967. Instability of periodic wavetrains in nonlinear dispersive systems. *Proc. R. Soc. Lond.* **A299**, 59–75.

Benjamin, T. B. and J. E. Feir, 1967. The disintegration of wave trains on deep water. Part I: theory. *J. Fluid Mech.* **27**, 417–430.

Bernstein, I. B. and F. Engelmann, 1966. Quasi-linear theory of plasma waves. *Phys. Fluids* **9**, 937.

Bidlot, J.-R., D. J. Holmes, P. A. Wittmann, R. Lalbeharry and H. S. Chen, 2002. Intercomparison of the performance of operational ocean wave forecasting systems with buoy data. *Weath. Forecasting* **17**, 287–310.

Birch, K. G. and J. A. Ewing, 1986. *Observations of Wind Waves on a Reservoir*. IOS report 234. Wormley, UK: IOS.

Blake, R. A., 1991. The dependence of wind stress on wave height and wind speed. *J. Geophys. Res.* **96**, 20 531–20 545.

Bonekamp, H., G. J. Komen, A. Sterl, P. A. E. M. Janssen, P. K. Taylor and M. J. Yelland, 2002. Statistical comparisons of observed and ECMWF modeled open ocean surface drag. *J. Phys. Oceanogr.*, **32**, 1010–1027.

Booij, N. and L. H. Holthuijsen, 1987. Propagation of ocean waves in discrete spectral wave models. *J. Comp. Phys.* **68**, 307–326.

Broer, L. J. F., 1974. On the Hamiltonian theory of surface waves. *Appl. Sci. Res.* **30**, 430–446.

Brüning, C., W. Alpers, L. F. Zambresky and D. G. Tilley, 1988. Validation of a SAR ocean wave imaging theory by the shuttle imaging radar-B experiment over the North Sea. *J. Geophys. Res.* **93**, 15 403–15 425.

Burgers, G. J. H., P. A. E. M. Janssen and D. L. T. Anderson, 1995. Impact of sea-state dependent fluxes on the tropical ocean circulation. *International Scientific Conference on the Tropical Ocean's Global Atmosphere (TOGAS), 2–7 April 1995, Melbourne*, pp. 295–297.

Caires, S. and A. Sterl, 2003. Validation of the ERA-40 ocean wave dataset using triple collocation. *J. Geophys. Res.* **108** (C3), 3098.

Caponi, E. A., P. G. Saffman and H. C. Yuen, 1982. Instability and confined chaos in a nonlinear dispersive wave system. *Phys. Fluids* **25**, 2159–2166.

Cardone, V., 1987. The present status of operational wave forecasting. *Johns Hopkins APL Tech. Dig.* **8**, 24–32.

Cardone, V., H. C. Graber, R. Jensen, S. Hasselmann and M. Caruso, 1995. In search of the true surface wind field in SWADE IOP-1: ocean wave modelling perspective. *Global Atmos. Ocean Syst.* **3**, 108–150.

Cavaleri, L. and L. Bertotti, 1997. In search of the correct wind and wave fields in a minor basin. *Mon. Weath. Rev.* **125**, 1964–1975.

2003. The characteristics of wind and wave fields modelled with different resolutions. *Q. J. R. Meteorol. Soc.* **129**, 1647–1662.

Chalikov, D. V., 1976. A mathematical model of wind-induced waves. *Dokl. Akad. Nauk SSR* **229**, 1083–1086.

Chalikov, D. V. and M. Y. Belevich, 1993. One-dimensional theory of the wave boundary layer. *Boundary Layer Meteorol.* **63**, 65–96.

Chalikov, D. V. and V. K. Makin, 1991. Models of the wave boundary layer. *Boundary Layer Meteorol.* **56**, 83–99.

Charnock, H., 1955. Wind stress on a water surface. *Q. J. R. Meteorol. Soc.* **81**, 639–640.

Chi Wai Li, 1992. A split operator scheme for ocean wave simulation. *Int. J. Numer. Meth. Fluids* **15**, 579–593.

Clancy, R. M., J. E. Kaitala and L. Zambresky, 1986. The Fleet Numerical Oceanography Center Global Spectral Ocean Wave model. *Bull. Am. Meteorol. Soc.* **67**, 498–512.

Conte, S. D. and J. W. Miles, 1959. On the integration of the Orr–Sommerfeld equation. *J. Soc. Ind. Appl. Math.* **7**, 361–369.

Cook, I, 1974. Advanced nonlinear theory of plasmas. In *Plasma Physics*, ed. B. E. Keen. London: The Institute of Physics, pp. 225–242.

Cornish, V., 1934. *Ocean Waves and Kindred Geophysical Phenomena*. Cambridge: Cambridge University Press.

Cox, C. S. and W. H. Munk, 1954. Statistics of the sea surface derived from sun glitter. *J. Marine Res.* **13**, 198–227.

Crawford, D. R., P. G. Saffman and H. C. Yuen, 1980. Evolution of a random inhomogeneous field of nonlinear deep-water gravity waves. *Wave Motion* **2**, 1–16.

Crawford, D. R., B. M. Lake, P. G. Saffman and H. C. Yuen, 1981. Stability of weakly nonlinear deep-water waves in two and three dimensions. *J. Fluid Mech.* **105**, 177–191.

Creamer, D. B., F. Heney, R. Schult and J. Wright, 1989. Improved linear representation of ocean surface waves. *J. Fluid Mech.* **205**, 135–161.

Davidson, K. L. and A. K. Frank, 1973. Wave-related fluctuations in the airflow above natural waves. *J. Phys. Oceanogr.* **3**, 102–119.

Davidson, R. C., 1972. *Methods in Nonlinear Plasma Theory*. New York and London: Academic Press.

DeCosmo, J., 1991. Air–sea exchange of momentum, heat and water vapor over whitecap sea states. Ph.D. thesis, University of Washington.

Derber, J. and F. Bouttier, 1999. A reformulation of the background error covariance in the ECMWF global data assimilation system. *Tellus*, **51A**, 195–221.

Dobson, F. W., 1971. Measurements of atmospheric pressure on wind-generated sea waves. *J. Fluid Mech.* **48**, 91.

Dold, J. W. and D. H. Peregrine, 1986. Water-wave modulation. In *Proceedings of the 20th International Conference on Coastal Engineering, Taipeh, Taiwan*. Reston, Virginia: ASCE, pp. 163–175.

Donelan, M. A. 1982. The dependence of the aerodynamic drag coefficient on wave parameters. In *Proceedings of the First International Conference on Meteorological and Air/Sea Interaction of the Coastal Zone, 10–14 May 1982, The Hague, the Netherlands*. Boston, Massachusetts: American Meteorological Society, pp. 381–387.

1983. *Attenuation of Laboratory Swell in an Adverse Wind*. Canada Centre for Inland Waters, Burlington, Ontario: National Water Research Institute.

1990. Air–sea interaction. In *The Sea*, vol. 9, *Ocean Engineering Science*. New York: Wiley, pp. 239–292.

2001. A nonlinear dissipation function due to wave breaking. *ECMWF Workshop on Ocean Wave Forecasting*. Reading, UK: ECMWF, pp. 87–94.

Donelan, M. A. and W. J. Pierson, 1987. Radar scattering and equilibrium ranges in wind-generated waves with application to scatterometry. *J. Geophys. Res.* **92**, 4971–5029.

Donelan, M. A., J. Hamilton and W. H. Hui, 1985. Directional spectra of wind generated waves. *Philos. Trans. R. Soc. Lond.* **A315**, 509–562.

Donelan, M. A., F. W. Dobson, S. D. Smith and R. J. Andersson, 1993. On the dependence of the sea surface roughness on wave development. *J. Phys. Oceanogr.* **23**, 2143–2149.

Donelan, M. A., W. M. Drennan and E. A. Terray, 1999. Wavenumber spectra of wind waves in the range of 1–50 m. In *The Wind-Driven Air–Sea Interface*, ed. M. L. Banner. Sydney: The University of New South Wales, School of Mathematics, pp. 35–42.

Dorrestein, R. 1960. Simplified method of determining refraction coefficients for sea waves. *J. Geophys. Res.* **65**, 637–642.

Doyle, J. D. 1995. Coupled ocean wave/atmosphere mesoscale simulations of cyclogenesis. *Tellus* **47A**, 766–778.

Drazin, P. G. and W. H. Reid, 1981. *Hydrodynamic Stability*. Cambridge: Cambridge University Press.

Drennan, W. M., K. K. Kahma and M. A. Donelan, 1999. On momentum flux and velocity spectra over waves. *Boundary Layer Meteorol.* **92**, 489–515.

Drummond, W. E. and D. Pines, 1962. Nonlinear stability of plasma oscillations. *Nucl. Fusion Suppl.* **3**, 1049–1052.

Dyachenko, A. I. and V. E. Zakharov, 1994. Is free surface hydrodynamics an integrable system? *Phys. Lett. A* **190**, 144.

Dysthe, K. B., K. Trulsen, H. E. Krogstad and H. Socquet-Juglard, 2003. Evolution of a narrow-band spectrum of random surface gravity waves. *J. Fluid Mech.* **478**, 1–10.

Ewans, K. C., 1998. Observations of the directional spectrum of fetch-limited waves. *J. Phys. Oceanogr.* **28**, 495–512.

Ewing, J. A. 1971. A numerical wave prediction method for the North Atlantic Ocean. *Dtsch. Hydrogr. Z.* **24**, 241–261.

Ewing, J. A. and A. K. Laing, 1987. Directional spectra of seas near full development. *J. Phys. Oceanogr.* **17**, 1696–1706.

Eymard, L., G. Caniaux, H. Dupuis, L. Prieur, H. Giordani, R. Troadec P. Bessemoulin, G. Lachaud, G. Bouhours, D. Bourras, C. Guerin, P. Le Borgne, A. Brisson and A. Marsouin 1999. Surface fluxes in the North Atlantic current during CATCH/FASTEX. *Q. J. R. Meteorol. Soc.* **125**, 3562–3599.

Fabrikant, A. L. 1976. Quasilinear theory of wind-wave generation. *Izv. Acad. Sci. USSR, Atmos. Ocean. Phys.* **12**, 524–526.

Felizardo, F. C. and W. K. Melville, 1995. Correlations between ambient noise and the ocean surface wave field. *J. Phys. Oceanogr.* **25**, 513–532.

Fermi, E., J. Pasta and S. Ulam, 1955. Studies of nonlinear problems. In *Collected Papers of Enrico Fermi*, vol. 2. Chicago: University of Chicago Press, pp. 978–988.

Freilich, M. H. and B. A. Vanhoff, 1999. QuikScat vector wind accuracy: initial estimates. In *Proceedings of the QuikScat Cal/Val Early Science Meeting, 2–5 November 1999, Pasadena, California*. Pasadena, California: Jet Propulsion Laboratory.

Forristall, G. Z. 1981. Measurements of saturated range in ocean wave spectra. *J. Geophys. Res.* **86**, 8075–8089.

Francis, P. E. and R. A. Stratton, 1990. Some experiments to investigate the assimilation of SEASAT altimeter wave height data into a global wave model. *Q. J. R. Meteorol. Soc.* **116**, 1225–1251.

Garratt, J. R. 1977. Review of drag coefficients over oceans and continents. *Mon. Weath. Rev.* **105**, 915–929.

Gelci, R., H. Cazalé and J. Vassal, 1957. Prévision de la houle. La méthode des densités spectroangulaires. *Bull. Inform. Comité Central Océanogr. Etude Côtes* **9**, 416–435.

Gent, P. R. and P. A. Taylor, 1976. A numerical model of the air flow above water waves. *J. Fluid Mech.* **77**, 105–128.

Groen, P. and R. Dorrestein, 1976. *Zeegolven*. 's-Gravenhage, the Netherlands: Staatsdrukkerij.

Groves, G. W. and J. Melcer, 1961. On the propagation of ocean waves on a sphere. *Geof. Int.* **8**, 77–93.

Günther, H. 1981. A parametric surface wave model and the statistics of the prediction parameters. Ph.D. thesis, University of Hamburg.

Günther, H., S. Hasselmann and P. A. E. M. Janssen, 1991. *Wamodel Cycle 4*. DKRZ report 4. Hamburg, Germany: Deutsches Klima Rechen Zentrum.

Hara, T. and S. E. Belcher, 2002. Wind forcing in the equilibrium range of wind-wave spectra. *J. Fluid Mech.* **470**, 223–245.

Hara, T. and A. V. Karachintsev, 2003. Observation of nonlinear effects in ocean surface wave frequency spectra. *J. Phys. Oceanogr.* **33**, 422–430.

Hargreaves, J. C. and J. D. Annan, 2001. Comments on 'Improvement of short-fetch behavior in the Wave Ocean Model (WAM).' *J. Atmos. Oceanic Technol.*, **18** 711–715.

Harris, D. L. 1966. The wave-driven wind. *J. Atmos. Sci.* **23**, 688–693.

Hashimoto, N. and K. Kawagushi, 2001. Extension and modification of the Discrete Interaction Approximation (DIA) for computing nonlinear energy transfer of gravity wave spectra. In *Proceedings of the Fourth International Symposium on Ocean Waves, Measurement and Analysis, WAVES-2001, 2–6 September 2001, San Francisco, California*. Reston, Virginia: ASCE.

Hasimoto, H. and H. Ono, 1972. Nonlinear modulation of gravity waves. *J. Phys. Soc. Japan* **33**, 805–811.

Hasselmann, D. E. and J. Bösenberg, 1991. Field measurements of wave-induced pressure over wind sea and swell. *J. Fluid Mech.* **230**, 391–428.

Hasselmann, D. E., M. Dunckel and J. A. Ewing, 1980. Directional wave spectra observed during JONSWAP 1973. *J. Phys. Oceanogr.* **10**, 1264–1280.

Hasselmann, K. 1962. On the non-linear energy transfer in a gravity-wave spectrum. Part 1: general theory. *J. Fluid Mech.* **12**, 481.

  1967. Nonlinear interactions treated by the methods of theoretical physics (with application to the generation of waves by wind). *Proc. R. Soc. Lond.* **A299**, 77–100.

  1970. Wave-driven inertial oscillations. *Geophys. Fluid Dyn.* **1**, 463–502.

  1971. On the mass and momentum transfer between short gravity waves and larger-scale motions. *J. Fluid Mech.* **50**, 189–205.

  1974. On the spectral dissipation of ocean waves due to whitecapping. *Boundary Layer Meteorol.* **6**, 107–127.

Hasselmann, K. and S. Hasselmann, 1991. On the nonlinear mapping of an ocean wave spectrum into a SAR image spectrum and its inversion. *J. Geophys. Res.* **C96**, 10 713–10 729.

Hasselmann, K., T. P. Barnett, E. Bouws, H. Carlson, D. E. Cartwright, K. Enke, J. A. Ewing, H. Gienapp, D. E. Hasselmann, P. Kruseman, A. Meerburg, P. Müller, D. J. Olbers, K. Richter, W. Sell and H. Walden, 1973. Measurements of wind-wave growth and swell decay during the Joint North Sea Wave Project (JONSWAP). *Dtsch. Hydrogr. Z. Suppl. A* **8**(12).

Hasselmann, K., D. B. Ross, P. Müller and W. Sell, 1976. A parametric wave prediction model, *J. Phys. Oceanogr.* **6**, 200–228.

Hasselmann, S. and K. Hasselmann, 1981. A symmetrical method of computing the non-linear transfer in a gravity-wave spectrum. *Hamb. Geophys. Einzelschr.* Serie A. **52**.

Hasselmann, S. and K. Hasselmann, 1985. Computations and parameterizations of the nonlinear energy transfer in a gravity-wave spectrum. Part 1: a new method for efficient computations of the exact nonlinear transfer integral. *J. Phys. Oceanogr.* **15**, 1369–1377.

Hasselmann, S., K. Hasselmann, J. H. Allender and T. P. Barnett, 1985. Computations and parameterizations of the nonlinear energy transfer in a gravity wave spectrum. Part 2: parameterizations of the nonlinear energy transfer for application in wave models. *J. Phys. Oceanogr.* **15**, 1378–1391.

Hersbach, H. and P. A. E. M. Janssen, 1999. Improvement of the short-fetch behavior in the Wave Ocean Model (WAM). *J. Atmos. Oceanic Technol.* **16** 884–892.

2001. Reply. *J. Atmos. Oceanic Technol.* **18**, 716–721.

Holland, J. Z. 1981. Atmospheric boundary layer. In *IFYGL: The International Field Year for the Great Lakes*, eds. E. J. Aubert and T. L. Richards. Ann Arbor, Michigan: NOAA.

Holton, J. E. 1992. *An Introduction to Dynamic Meteorology*, 3rd edn. New York: Academic Press.

Hortal, M. 1999. *The Development and Testing of a New Two-Time Level Semi-Lagrangian Scheme (SETTLS) in the ECMWF Forecast Model*. ECMWF Technical Memorandum 292. Reading, UK: ECMWF.

Hristov, T. S., S. D. Miller and C. A. Friehe, 2003. Dynamical coupling of wind and ocean waves through wave-induced air flow. *Nature*, **442**, 55–58.

Hwang, P. A. 1997. A study of the wavenumber spectra of short water waves in the ocean. Part II: spectral model and mean square slope. *J. Atmos. Oceanic Technol.* **14**, 1174–1186.

Hwang, P. A., S. Atatürk, M. A. Sletten and D. B. Trizna, 1996. A study of the wavenumber spectra of short water waves in the ocean. *J. Phys. Oceanogr.* **26**, 1266–1285.

Hwang, P. A., D. W. Wang, E. J. Walsh, W. B. Krabill and R. N. Swift, 2000a. Airborne measurements of the wavenumber spectra of ocean surface waves. Part I: spectral slope and dimensionless spectral coefficient. *J. Phys. Oceanogr.* **30**, 2753–2767.

2000b. Airborne measurements of the wavenumber spectra of ocean surface waves. Part II: directional distribution. *J. Phys. Oceanogr.* **30**, 2768–2787.

Jackson, F. C., W. T. Walton and C. Y. Peng, 1985. A comparison of *in situ* and airborne radar observations of ocean wave directionality. *J. Geophys. Res.* **90**, 1005–1018.

Jackson, F. C., W. T. Walton, B. A. Walter, D. E. Hines and C. Y. Peng, 1992. Sea surface mean squared slope from Ku-band radar backscatter data. *J. Geophys. Res.* **C97**, 11 411–11 427.

Jackson, P. S. and J. C. R. Hunt, 1975. Turbulent wind flow over a low hill. *Q. J. R. Meteorol. Soc.* **101**, 929–955.

Jacobs, S. J. 1987. An asymptotic theory for the turbulent flow over a progressive water wave. *J. Fluid Mech.* **174**, 69–80.

Jähne, B. and K. S. Riemer, 1990. Two-dimensional wave number spectra of small-scale water surface waves. *J. Geophys. Res.* **95**, 11 531–11 546.

Janssen, P. A. E. M. 1981. Modulational instability and the Fermi–Pasta–Ulam recurrence. *Phys. Fluids* **24**, 23–26.

    1982. Quasilinear approximation for the spectrum of wind-generated water waves. *J. Fluid Mech.* **117**, 493–506.

    1983. Long-time behaviour of a random inhomogeneous field of weakly nonlinear surface gravity waves. *J. Fluid Mech.* **133**, 113–132.

    1986. *On the Effect of Gustiness on Wave Growth.* KNMI Afdeling Oceanografisch Onderzoek memo, 00-86-18. De Bilt, the Netherlands: KNMI.

    1989. Wave-induced stress and the drag of air flow over sea waves. *J. Phys. Oceanogr.* **19**, 745–754.

    1991. Quasi-linear theory of wind wave generation applied to wave forecasting. *J. Phys. Oceanogr.* **21**, 1631–1642.

    1992. Experimental evidence of the effect of surface waves on the airflow. *J. Phys. Oceanogr.* **22**, 1600–1604.

    1998. *On Error Growth in Wave Models.* ECMWF Technical Memorandum 249, Reading, UK: ECMWF.

    1999. On the effect of ocean waves on the kinetic energy balance and consequences for the inertial dissipation technique. *J. Phys. Oceanogr.* **29**, 530–534.

    2000. ECMWF wave modeling and satellite altimeter wave data. In *Satellites, Oceanography and Society*, ed. D. Halpern. Amsterdam: Elsevier, pp. 35–56.

    2001. Reply. *J. Phys. Oceanogr.* **31**, 2537–2544.

    2003. Nonlinear four-wave interactions and freak waves. *J. Phys. Oceanogr.* **33**, 863–884.

Janssen, P. A. E. M. and G. J. Komen, 1985. Effect of the atmospheric stability on the growth of surface gravity waves. *Boundary Layer Meteorol.* **32**, 85–96.

Janssen, P. A. E. M. and P. Viterbo, 1996. Ocean waves and the atmospheric climate. *J. Climate* **9**, 1269–1287.

Janssen, P. A. E. M., G.J. Komen and W. J. P. de Voogt, 1984. An operational coupled hybrid wave prediction model. *J. Geophys. Res.* **C89**, 3635–3654.

    1987. Friction velocity scaling in wind-wave generation. *Boundary Layer Meteorol.* **38**, 29–35.

Janssen, P. A. E. M., P. Lionello, M. Reistad and A. Hollingsworth, 1989. Hindcasts and data assimilation studies with the WAM model during the Seasat period. *J. Geophys. Res.* **C94**, 973–993.

Janssen, P. A. E. M., A. C. M. Beljaars, A. Simmons and P. Viterbo, 1992. On the determination of surface stresses in an atmospheric model. *Mon. Weath. Rev.* **120**, 2977–2985.

Janssen, P. A. E. M., B. Hansen and J.-R. Bidlot, 1997a. Verification of the ECMWF wave forecasting system against buoy and altimeter data. *Weath. Forecasting*, **12**, 763–784.

    1997b. Validation of ERS satellite wave products with the WAM model. In *CEOS Wind and Wave Validation Workshop, 3–5 June 1997, ESTEC, The Netherlands*, ESA WPP-147 Noordwijk, the Netherlands: ESA, pp. 101–108.

Janssen, P. A. E. M., H. Wallbrink, C. J. Calkoen, D. van Halsema, W. A. Oost and P. Snoeij, 1998. VIERS-1 scatterometer model. *J. Geophys. Res.* **103**, 7807–7831.

Janssen, P. A. E. M., J. Bidlot and B. Hansen, 2000. *Diagnosis of the ECMWF Ocean-Wave Forecasting System.* ECMWF Technical Memorandum, 318. Reading, UK: ECMWF.

Janssen, P. A. E. M., J. D. Doyle, J. Bidlot, B. Hansen, L. Isaksen and P. Viterbo, 2002. Impact and feedback of ocean waves on the atmosphere. In *Atmosphere–Ocean Interactions*, vol. I, ed. W. Perrie, *Advances in Fluid Mechanics* **33**, Boston, Massachusetts: WIT Press, pp. 155–197.

Janssen, P. A. E. M., S. Abdalla and H. Hersbach, 2003. *Error Estimation of Buoy, Satellite and Model Wave Height Data*. ECMWF Technical Memorandum, 402. Reading, UK: ECMWF.

Jeffreys, H. 1924. On the formation of waves by wind. *Proc. R. Soc.* **A107**, 189–206.
1925. On the formation of waves by wind. II. *Proc. R. Soc.* **A110**, 341–347.

Jenkins, A. D., 1987. A Lagrangian model for wind- and wave-induced flux of near-surface currents. *Coastal Engng* **11**, 513–526.
1989. The use of a wave prediction model for driving a near-surface current model. *Dtsch. Hydrogr. Z.* 3–6, 133–149.
1992. A quasi-linear eddy-viscosity model for the flux of energy and momentum to waves, using conservation law equations in a curvilinear system. *J. Phys. Oceanogr.* **22**, 843–858.

Kahma, K. K. 1981. A study of the growth of the wave spectrum with fetch. *J. Phys. Oceanogr.* **11**, 1503–1515.

Kahma, K. K. and C. J. Calkoen, 1992. Reconciling discrepancies in the observed growth of wind-generated waves. *J. Phys. Oceanogr.* **22**, 1389–1405.

Kawai, S. 1979. Generation of initial wavelets by instability of a coupled shear flow and their evolution to wind waves. *J. Fluid Mech.* **93**, 661–703.

Kawai, S., K. Okuda and Y. Toba, 1977. Field data support of three-seconds power law and $gu_*\sigma^{-4}$ spectral form for growing wind waves. *J. Oceanogr. Soc. Japan* **33**, 137–150.

Katsaros, K. B., S. D. Smith and W. A. Oost, 1987. HEXOS: Humidity Exchange Over the Sea. A program for research on water-vapor and droplet fluxes from sea to air at moderate to high wind speeds. *Bull. Am. Meteorol. Soc.* **68**, 466–476.

Khandekar, M. L. and R. Lalbeharry, 1996. An evaluation of Environment Canada's operational wave model based on moored buoy data. *Weath. Forecasting* **11**, 139–152.

Kitaigorodskii, S. A. 1962. Application of the theory of similarity to the analysis of wind-generated water waves as a stochastic process. *Bull. Acad. Sci. USSR Geophys. Ser.* **1**.
1970. *The Physics of Air–Sea Interaction*. Jerusalem: Israel Program for Scientific Translations.
1983. On the theory of the equilibrium range in the spectrum of wind-generated gravity waves. *J. Phys. Oceanogr.* **13**, 816–827.

Komen, G. J., K. Hasselmann and S. Hasselmann, 1984. On the existence of a fully developed windsea spectrum. *J. Phys. Oceanogr.* **14**, 1271–1285.

Komen, G. J., L. Cavaleri, M. Donelan, K. Hasselmann, S. Hasselmann and P. A. E. M. Janssen, 1994. *Dynamics and Modelling of Ocean Waves*. Cambridge: Cambridge University Press.

Komen, G. J., P. A. E. M. Janssen, V. Makin and W. Oost, 1998. On the sea state dependence of the Charnock parameter. *Global Atmos. Ocean Syst.* **5**, 367–388.

Krasitskii, V. P. 1990. Canonical transformation in a theory of weakly nonlinear waves with a nondecay dispersion law. *Sov. Phys. JETP* **71**, 921–927.
1994. On reduced equations in the Hamiltonian theory of weakly nonlinear surface waves. *J. Fluid Mech.* **272**, 1–20.

Krasitskii, V. P. and V. A. Kalmykov, 1993. Four-wave reduced equations for surface gravity waves. *Izv. Acad. Sci. USSR, Atmos. Ocean. Phys.* **29**, 222–228.

Krogstad, H. E., R. L. Gordon and M. C. Miller, 1988. High-resolution directional wave spectra from horizontally mounted acoustic Doppler current meters. *J. Atmos. Oceanic Technol.* **5**, 340–352.

Kuik, A. J., G. Ph. van Vledder and L. H. Holthuijsen, 1988. A method for the routine analysis of pitch-and-roll buoy wave data. *J. Phys. Oceanogr.* **18**, 1020–1034.

Lake, B. M., H. C. Yuen, H. Rungaldier and W. E. Ferguson, Jr, 1977. Nonlinear deep-water waves: theory and experiment. Part 2, Evolution of a continuous wave train. *J. Fluid Mech.* **83**, 49–74.

Lalbeharry, R., J. Mailhot, S. Desjardins and L. Wilson, 2000. Examination of the impact of a coupled atmospheric and ocean wave system. Part II: ocean wave aspects. *J. Phys. Oceanogr.* **30**, 402–415.

Lamb, H. 1932. *Hydrodynamics*, 6th edn. New York: Dover.

Landau, L. 1946. On the vibrations of the electronic plasma. *J. Phys. (USSR)* **10**, 25.

Landau, L. D. and E. M. Lifshitz, 1960. *Mechanics*. Reading, Massachusetts: Pergamon Press/Addison-Wesley.

Large, W. G. and S. Pond, 1982. Sensible and latent heat flux measurements over the ocean. *J. Phys. Oceanogr.* **12**, 464–482.

Larisch-Mönnich, Fz, 1925. *Sturmsee und Brandung*. Bielefeld/Leipzig, Germany: Velhagen und Klasing.

Levi-Città, T. 1925. Determination rigoureuse des ondes permanentes d'ampleur finie. *Math. Ann.* **93**, 264–314.

Lighthill, M. J. 1962. Physical interpretation of the mathematical theory of wave generation by wind. *J. Fluid Mech.* **14**, 385–398.

   1965. Contributions to the theory of waves in nonlinear dispersive systems. *J. Inst. Math. Appl.* **1**, 269–306.

Lin, R. Q. and W. Perry, 1999. Wave–wave interactions in finite depth water. *J. Geophys. Res.* **104**, 11 193–11 213.

Lionello, P., H. Günther and P. A. E. M. Janssen, 1992. Assimilation of altimeter data in a global third generation wave model. *J. Geophys. Res.* **C97**, 14 453–14 474.

Lionello, P., P. Malguzzi and A. Buzzi, 1998. Coupling between the atmospheric circulation and the ocean wave field: an idealized case. *J. Phys. Oceanogr.* **28**, 161–177.

Longuet-Higgins, M. S. 1978. The instabilities of gravity waves of finite amplitude in deep-water. II. Subharmonics. *Proc. R. Soc. Lond.* **A360**, 489–505.

Longuet-Higgins, M. S. and R. W. Stewart, 1961. Radiation stress and mass transport in gravity waves, with application to 'surf-beats'. *J. Fluid Mech.* **10**, 529–549.

Longuet-Higgins, M. S., D. E. Cartwright and N. D. Smith, 1963. Observations of the directional spectrum of sea waves using the motions of a floating buoy. In *Ocean Wave Spectra*. Englewood Cliffs, New Jersey: Prentice Hall, pp. 111–136.

Luke, J. C. 1967. A variational principle for a fluid with a free surface. *J. Fluid Mech.* **27**, 395–397.

Lundgren, T. S. 1967. Distribution function in the statistical theory of turbulence. *Phys. Fluids* **10**, 969–975.

Maat, N., C. Kraan and W. A. Oost, 1991. The roughness of wind waves. *Boundary Layer Meteorol.* **54**, 89–103.

Makin, V. K. and D. V. Chalikov, 1979. Numerical modeling of air structure above waves. *Izv. Acad. Sci. USSR, Atmos. Ocean. Phys.* **15**, 292–299 (English translation: *Izv. Atmos. Ocean. Phys.* **15**, 199–204).

Makin, V. K. and V. N. Kudryavstev, 2002. Impact of dominant waves on sea drag. *Boundary Layer Meteorol.* **103**, 83–99.

Makin, V. K., V. N. Kudryavstev and C. Mastenbroek, 1995. Drag of the sea surface. *Boundary Layer Meteorol.* **73**, 159–182.

Marsden, R. E. 1999. A proposal for a neutral regression. *J. Atmos. Oceanic Technol.* **16**, 876–883.

Mastenbroek, C. 1996. Wind–wave interaction. Ph.D. thesis, Delft Technical University.

Mastenbroek, C., G. J. H. Burgers and P. A. E. M. Janssen, 1993. The dynamical coupling of a wave model and a storm surge model through the atmospheric boundary layer. *J. Phys. Oceanogr.* **23**, 1856–1866.

Mastenbroek, C., V. K. Makin, M. H. Garat and J. P. Giovanangeli, 1996. Experimental evidence of the rapid distortion of the turbulence in the air flow over water waves. *J. Fluid Mech.* **318**, 273–302.

McLean, J. W. 1982. Instabilities of finite amplitude water waves. *J. Fluid Mech.* **114**, 331–341.

McLean, J. W., Y. C. Ma, D. U. Martin, P. G. Saffman and H. C. Yuen, 1982. Three-dimensional instability of finite amplitude gravity waves. *Phys. Rev. Lett.* **46**, 817–820.

McWilliams, J. C. and J. M. Restrepo, 1999. The wave-driven ocean circulation. *J. Phys. Oceanogr.* **29**, 2523–2540.

Melville, W. K. 1982. The instability and breaking of deep-water waves. *J. Fluid Mech.* **115**, 165–185.

Melville, W. K. and P. Matusov, 2002. Distribution of breaking waves at the ocean surface. *Nature* **417**, 58.

Miles, J. W. 1957. On the generation of surface waves by shear flows. *J. Fluid Mech.* **3**, 185–204.

1959. On the generation of surface waves by shear flows. Part 2. *J. Fluid Mech.* **6**, 568–582.

1961. On the stability of heterogeneous shear flows. *J. Fluid Mech.* **10**, 496.

1965. A note on the interaction between surface waves and wind profiles. *J. Fluid Mech.* **22**, 823–827.

1977. On Hamilton's principle for surface waves. *J. Fluid Mech.* **83**, 153.

1993. Surface wave generation revisited. *J. Fluid Mech.* **256**, 427–441.

1997. On Janssen's model for surface wave generation by gusty wind. *J. Phys. Oceanogr.* **27**, 592–593.

Mitsuyasu, H. and T. Honda, 1982. Wind-induced growth of water waves. *J. Fluid Mech.* **123**, 440.

Mitsuyasu, H., F. Tasai, T. Suhara, S. Mizuno, M. Ohkusu, T. Honda and K. R. Kiishi, 1975. Observations of the directional spectrum of ocean waves using a cloverleaf buoy. *J. Phys. Oceanogr.* **5**, 750–760.

1980. Observations of the power spectrum of waves using a cloverleaf buoy. *J. Phys. Oceanogr.* **10**, 286–296.

Mizuno, S. 1976. Pressure measurements above mechanically generated water waves. 1. *Rep. Res. Inst. Appl. Mech., Kyushu Univ.* **23**, 113–129.

Monbaliu, J. 1994. On the use of the Donelan wave spectral parameter as a measure for the roughness of wind waves. *Boundary Layer Meteorol.* **67**, 277–291.

Mori, N. and T. Yasuda, 2002. A weakly non-gaussian model of wave distribution for random wave train. *Ocean Engng* **29**, 1219–1231.

Morland, L. C. 1996. Oblique wind waves generated by the instability of wind blowing over water. *J. Fluid Mech.* **316**, 163–172.

Nikolayeva, Y. I. and L. S. Tsimring, 1986. Kinetic model of the wind generation of waves by a turbulent wind. *Izv. Acad. Sci. USSR, Atmos. Ocean Phys.* **22**, 102–107.

Nordeng, T. E. 1991. On the wave age dependent drag coefficient and roughness length at sea. *J. Geophys. Res.* **C96**, 7167–7174.

Onorato, M., A. R. Osborne, M. Serio and S. Bertone, 2001. Freak waves in random oceanic sea states. *Phys. Rev. Lett.* **86**, 5831–5834.

Onorato, M., A. R. Osborne, M. Serio, D. Resio, A. Pushkarev, V. Zakharov and C. Brandini, 2002. Freely decaying weak turbulence for sea surface gravity waves. *Phys. Rev. Lett.* **89**.

Oost, W. A., G. J. Komen, C. M. J. Jacobs and C. van Oort, 2002. New evidence for a relation between wind stress and wave age from measurements during ASGAMAGE. *Boundary Layer Meteorol.* **103**, 409–438.

Orzag, S. A. 1970. Transform method for calculation of vector coupled sums: application to the spectral form of the vorticity equations. *J. Atmos. Sci.* **27**, 890–895.

Osborne, A. R., M. Onorato and M. Serio, 2000. The nonlinear dynamics of rogue waves and holes in deep water gravity wave trains. *Phys. Lett. A* **275**, 386–393.

Pedlosky, J. 1987 *Geophysical Fluid Dynamics*. New York: Springer Verlag.

Phillips, O. M., 1957. On the generation of waves by turbulent wind. *J. Fluid Mech.* **2**, 417–445.

  1958. The equilibrium range in the spectrum of wind-generated water waves. *J. Fluid Mech.* **4**, 426–434.

  1960. The dynamics of unsteady gravity waves of finite amplitude. Part 1. *J. Fluid Mech.* **9**, 193–217.

  1977. *The Dynamics of the Upper Ocean*. Cambridge: Cambridge University Press.

  1985. Spectral and statistical properties of the equilibrium range in wind-generated gravity waves. *J. Fluid Mech.* **156**, 505–531.

Pierson, W. J. and L. Moskowitz, 1964. A proposed spectral form for fully developed wind seas based on the similarity theory of S. A. Kitaigorodskii. *J. Geophys. Res.* **69**, 5181.

Pierson, W. J., G. Neumann and R. W. James, 1955. *Practical Methods for Observing and Forecasting Ocean Waves by Means of Wave Spectra and Statistics*. Hydrographic Office Publication 603. Washington, DC: US Navy Hydrographic Office.

Pierson, W. J., M. A. Donelan and W. H. Hui, 1992. Linear and nonlinear propagation of water wave groups. *J. Geophys. Res.* **C97**, 5607–5621.

Plant, W. J. 1982. A relation between wind stress and wave slope. *J. Geophys. Res.* **C87**, 1961–1967.

Plant, W. J. and J. W. Wright, 1977. Growth and equilibrium of short gravity waves in a wind-wave tank. *J. Fluid Mech.* **82**, 767–793.

Polnikov, V. G. 1991. A third generation spectral model for wind waves. *Izv. Acad. Sci. USSR, Atmos. Ocean. Phys.* **27**, 615–623.

Polnikov, V. G. and L. Farina, 2002. On the problem of optimal approximation of the four-wave kinetic integral. *Nonlinear Proc. Geophys.* **9**, 497–512.

Quilfen, Y., B. Chapron and D. Vandemark, 2001. The ERS scatterometer wind measurement accuracy: evidence of seasonal and regional biases. *J. Atmos. Oceanic Technol.* **18**, 1684–1697.

Rapp, R. J. and W. K. Melville, 1990. Laboratory measurements of deep water breaking waves. *Philos. Trans. R. Soc. Lond.*, **A331**, 735–780.

Reutov, V. P. 1980. The plasma-hydrodynamic analogy and the nonlinear stage of instability of wind waves. *Izv. Acad. Sci. USSR, Atmos. Ocean. Phys.* **16**, 938–943.

Riley, D. S., M. A. Donelan and W. H. Hui, 1982. An extended Miles' theory for wave generation by wind. *Boundary Layer Meteorol.* **22**, 209–225.

Romeiser, R., 1993. Global validation of the wave model WAM over a one year period using GEOSAT wave height data. *J. Geophys. Res.* **C98**, 4713–4726.

Rossby, C. G. and Montgomery, 1935. *Papers in Physical Oceanography* 3, No. 3.

Saetra, Ø. and J.-R. Bidlot, 2002. *Assessment of the ECMWF Ensemble Prediction System for Waves and Marine winds*. ECMWF Technical Memorandum. Reading, UK: ECMWF.

Saetra, Ø., J.-R. Bidlot, H. Hersbach and D. Richardson, 2002. *Effect of Observation Errors on the Statistics for Ensemble Spread and Reliability*. ECMWF Technical Memorandum 393. Reading, UK: ECMWF.

Saffman, P. G. and D. C. Wilcox, 1974. Turbulence-model predictions for turbulent boundary layers. *AIAA J.* **12**, 541–546.

Sanders, J. 1976. A growth stage scaling model for the wind-driven sea. *Dtsch. Hydrogr. Z.* **29**, 136–161.

Simmons, A. J. and A. Hollingsworth, 2002. Some aspects of the improvement in skill of numerical weather prediction. *Q. J. R. Meteorol. Soc.* **128**, 647–677.

Smith, S. A. and E. G. Banke, 1975. Variation of the sea surface drag coefficient with wind speed. *Q. J. R. Meteorol. Soc.* **101**, 665–673.

Smith, S. D. 1980. Wind stress and heat flux over the ocean in gale force winds. *J. Phys. Oceanogr.* **10**, 709–726.

Smith, S. D., K. B. Katsaros, W. A. Oost and P. G. Mestayer, 1990. Two major experiments in the Humidity Exchange Over the Sea (HEXOS) program. *Bull. Am. Meteorol. Soc.* **71**, 161–172.

Smith, S. D., R. J. Anderson, W. A. Oost, C. Kraan, N. Maat, J. DeCosmo, K. B. Katsaros, K. L. Davidson, K. Bumke, L. Hasse and H. M. Chadwick, 1992. Sea surface wind stress and drag coefficients: the HEXOS results. *Boundary Layer Meteorol.* **60**, 109–142.

Snyder, R. L. 1974. A field study of wave-induced pressure fluctuation above surface gravity waves. *J. Mar. Res.* **32**, 497–531.

Snyder, R. L., F. W. Dobson, J. A. Elliott and R. B. Long, 1981. Array measurements of atmospheric pressure fluctuations above surface gravity waves. *J. Fluid Mech.* **102**, 1–59.

Snyder, R. L., W. C. Thacker, K. Hasselmann, S. Hasselmann and G. Barzel, 1993. Implementation of an efficient scheme for calculating nonlinear transfer from wave–wave interactions. *J. Geophys. Res.* **C98**, 14 507–14 525.

Song, J.-B. and M. L. Banner, 2002. On determining the onset and strength of breaking for deep water waves. Part I: unforced irrotational wave groups. *J. Phys. Oceanogr.* **32**, 2541–2558.

Stevensen, Th. 1874. *Design and Construction of Harbours*, 2nd edn.

Stewart, R. H. and C. Teague, 1980. Dekameter radar observations of ocean wave growth and decay. *J. Phys. Oceanogr.* **10**, 128–143.

Stewart, R. W. 1967. Mechanics of the air–sea interface. *Phys. Fluids Suppl.* **10**, S47–55. 1974. The air–sea momentum exchange. *Boundary Layer Meteorol.* **6**, 151–167.

Stiassnie, M. and L. Shemer, 1984. On modifications of the Zakharov equation for surface gravity waves. *J. Fluid Mech.* **143**, 47–67.

Stoffelen, A. 1998. Error modeling and calibration: towards the true surface wind speed. *J. Geophys. Res.* **103**, 7755–7766.

Stoffelen, A. and D. Anderson, 1995. The ECMWF contribution to the characterisation, interpretation, calibration and validation of ERS-1 scatterometer backscatter measurements and winds and their use in numerical weather prediction models. ESA Contract Report. Reading, UK: ECMWF.

Stokes, G. G. 1847. On the theory of oscillatory waves. *Trans. Camb. Philos. Soc.* **8**, 441–455.

Sverdrup, H. U. and W. H. Munk, 1947. *Wind Sea and Swell: Theory of Relations for Forecasting*. Hydrographic Office Publication 601. Washington, DC: US Navy Hydrographic Office.

Sverdrup, H. U., M. W. Johnson and R. H. Fleming, 1946. *The Oceans*. New York: Prentice Hall.

Su, M. Y. 1982. Three-dimensional deep-water waves. Part 1: experimental measurement of skew and symmetric wave patterns. *J. Fluid. Mech.* **124**, 73–108.

Sullivan, P., J. McWilliams and C. -H. Moeng, 2000. Simulation of turbulent flow over idealized water waves. *J. Fluid. Mech.* **404**, 47–85.

Teixeira, M. A. and S. E. Belcher, 2002. On the distortion of turbulence by a progressive surface wave. *J. Fluid Mech.* **458**, 229–267.

Tennekes, H. and J. Lumley, 1974. *A First Course in Turbulence*. Cambridge, Massachusetts: Addison-Wesley.

Terray, E. A., W. M. Drennan, and M. A. Donelan, 1999. The vertical structure of shear and dissipation in the ocean surface layer. In *The Wind-Driven Air–Sea Interface*, ed. M. L. Banner Sydney: The University of New South Wales, School of Mathematics, pp. 239–245.

Toba, Y. 1972. Local balance in the air–sea boundary process, 1. On the growth process of wind waves. *J. Oceanogr. Soc. Japan* **28**, 109–121.

1973. Local balance in the air–sea boundary process, 3. On the spectrum of wind waves. *J. Oceanogr. Soc. Japan* **29**, 209–220.

Tokmakian, R. and P. G. Challenor, 1999. On the joint estimation of model and satellite sea surface height anomaly errors. *Ocean Modelling* **1**, 39–52.

Tolman, H. L. 1998. Effect of observation errors in linear regression and bin-average analyses. *Q. J. R. Meteorol. Soc.* **124**, 897–917.

2001. Numerics in wind wave models. In *ECMWF Workshop on Ocean Wave Forecasting, 2–4 July 2001, Reading*. Reading, UK: ECMWF, pp. 5–16.

2003. Treatment of unresolved islands and ice in wind wave models. *Ocean Modelling* **5**, 219–231.

Trulsen, K. and K. Dysthe, 1997. Freak waves: a three-dimensional wave simulation. In *Proceedings of the 21st Symposium on Naval Hydrodynamics*. Washington, DC: National Academy Press, pp. 550–558.

Trulsen, K. and C. T. Stansberg, 2001. Spatial evolution of water surface waves: numerical simulation and experiment of bichromatic waves. In *Proceedings of the 11th International Offshore and Polar Engineering Conference, 17–22 June 2001, Stavanger, Norway*. Danvers, Massachusetts: ISOPE, pp. 71–77.

Tsimring, L. Sh. 1983. Induced scattering of surface wind waves. *Izv. Acad. Sci. USSR, Atmos. Ocean. Phys.* **19**, 47–50.

Tsytovich, V. N. 1970. *Nonlinear Effects in Plasmas*. Translated by M. Hamburger. New York and London: Plenum Press.

Tulin, M. P. and T. Waseda, 1999. Laboratory observations of wave group evolution, including breaking effects. *J. Fluid Mech.* **378**, 197–232.

Valenzuela, G. R. 1976. The growth of gravity-capillary waves in the coupled shear flow. *J. Fluid Mech.* **76**, 229–250.

van Driest, E. R. 1951. Turbulent boundary layer in compressible fluids. *J. Aeronaut. Sci.* **18**, 145–160.

van Duin, C. A. and P. A. E. M. Janssen, 1992. An analytic model of the generation of surface gravity waves by turbulent air flow. *J. Fluid Mech.* **236**, 197–215.

van Gastel, K., P. A. E. M. Janssen and G. J. Komen, 1985. On phase velocity and growth rate of wind-induced gravity-capillary waves. *J. Fluid Mech.* **161**, 199–216.

van Kampen, N. G. 1955. On the theory of stationary waves in plasmas. *Physica* **21**, 949–963.

Van Vledder, G. Ph., 2001. Extension of the discrete interaction approximation for computing nonlinear quadruplet wave–wave interactions in operational wave prediction models. In *Proceedings of the Fourth International Symposium on Ocean Waves, Measurement and Analysis, WAVES-2001, 2–6 September 2001, San Francisco, California*. Reston, Virginia: ASCE.

Van Vledder, G. Ph., T. H. C. Herbers, R. E. Jensen, D. T. Resio and B. Tracy, 2000. Modelling of the nonlinear quadruplet wave–wave interactions in operational wave models. In *Proceedings of the 27th International Conference on Coastal Engineering, Sydney, Australia*. Reston, Virginia: ASCE.

Vedenov, A. A., E. P. Velikhov and R. Z. Sagdeev, 1961. Nonlinear oscillations of a rarefied plasma. *Nucl. Fusion* **1**, 182.

Voorrips, A. C., C. Mastenbroek, and B. Hansen, 2001. Validation of two algorithms to retrieve ocean wave spectra from ERS synthetic aperture radar. *J. Geophys. Res.* **106** (C8) 16 825–16 840.

Walmsley, J. L. and P. A. Taylor, 1996. Boundary-layer flow over topography: impacts of the Askervein study. *Boundary Layer Meteorol.* **78**, 291–320.

WAMDI group: S. Hasselmann, K. Hasselmann, E. Bauer, P. A. E. M. Janssen, G. J. Komen, L. Bertotti, P. Lionello, A. Guillaume, V. C. Cardone, J. A. Greenwood, M. Reistad, L. Zambresky and J. A. Ewing, 1988. The WAM model: a third generation ocean wave prediction model. *J. Phys. Oceanogr.* **18**, 1775–1810.

Weber, J. E. 1983. Steady wind- and wave-induced currents in the open ocean. *J. Phys. Oceanogr.* **13**, 524–530.

Weber, S. L., H. von Storch, P. Viterbo and L. Zambresky, 1993. Coupling an ocean wave model to an atmospheric general circulation model. *Climate Dynamics* **9**, 63-69.

Whitham, G. B. 1962. Mass, momentum and energy flux in water waves. *J. Fluid Mech.* **12**, 135–147.

 1974. *Linear and Nonlinear Waves*. New York: Wiley.

Willebrand, J. 1975. Energy transport in a nonlinear and inhomogeneous random gravity wave field. *J. Fluid Mech.* **70**, 113–126.

Wittmann, P. A., R. M. Clancy and T. Mettlach, 1995. Operational wave forecasting at Fleet Numerical Meteorology and Oceanography Center, Monterey, CA. In *Fourth International Workshop on Wave Hindcasting and Forecasting, Banff, Alberta, Canada*. Atmospheric Environment Service, pp. 335–342.

Wu, J. 1982. Wind-stress coefficients over sea surface from breeze to hurricane. *J. Geophy. Res.* **C87**, 9704–9706.

Wu, X. and R. A. Flather, 1992. Hindcasting waves using a coupled wave-tide-surge model. In *Third International Workshop on Wave Hindcasting and Forecasting, 19–22 May 1992, Montreal, Quebec*. Ontario: Environment Canada, pp. 159–170.

Wyatt, L. R. 1995. The effect of fetch on the directional spectrum of Celtic Sea storm waves. *J. Phys. Oceanogr.* **25**, 1550–1559.

 2000. Limits to the inversion of HF radar backscatter for ocean wave measurement. *J. Atmos. Oceanic Technol.* **17**, 1651–1665.

Yelland, M. J. and P. K. Taylor, 1996. Wind stress measurements from the open ocean. *J. Phys. Oceanogr.* **26**, 541–558.

Young, I. R. and R. H. Sobey, 1985. Measurements of the wind-wave energy flux in an opposing wind. *J. Fluid Mech.* **151**, 427–442.

Young, I. R., L. A. Verhagen and M. L. Banner, 1995. A note on the bimodal directional spreading of fetch-limited wind waves. *J. Geophy. Res.* **100**, 773–778.

Yuen, H. C. and W. E. Ferguson, Jr, 1978. Relationship between Benjamin–Feir instability and recurrence in the nonlinear Schrödinger equation. *Phys. Fluids* **21**, 1275–1278.

Yuen, H. C. and B. M. Lake, 1982. Nonlinear dynamics of deep water gravity waves. *Adv. Appl. Mech.* **22**, 67–229.

Zakharov, V. E. 1968. Stability of periodic waves of finite amplitude on the surface of a deep fluid. *J. Appl. Mech. Tech. Phys.* **9**, 190–194.

1992. Inverse and direct cascade in the wind-driven surface wave turbulence and wave-breaking. In *Breaking Waves*, eds. M. L. Banner and R. H. J. Grimshaw. Berlin and Heidelberg: Springer Verlag, pp. 69–91.

Zakharov, V. E. and N. N. Filonenko, 1967. Energy spectrum for stochastic oscillations of the surface of a liquid. *Sov. Phys. Doklady* **11**, 881.

Zakharov, V. E. and A. Pushkarev, 1999. Diffusion model of interacting gravity waves on the surface of a deep fluid. *Nonlinear Proc. Geophys.* **6**, 1–10.

Zakharov, V. E. and A. B. Shabat, 1972. Exact theory of two-dimensional self-focussing and one-dimensional self-modulating waves in nonlinear media. *Sov. Phys. JETP* (English translation) **34**, 62.

Zakharov, V. E. and M. M. Zaslavskii, 1982. The kinetic equation and Kolmogorov spectra in the weak turbulence theory of wind waves. *Izv. Acad. Sci. USSR, Atmos. Ocean. Phys.* **18**, 747–753.

Zambresky, L. 1987. The operational performance of the Fleet Numerical Oceanography Center Global Spectral Ocean-Wave Model. *Johns Hopkins APL Tech. Dig.* **8**, 33–36.

1989. *A Verification Study of the Global WAM Model. December 1987–November 1988.* ECMWF Technical Report 63. Reading, UK: ECMWF.

1991. An evaluation of two WAM hindcasts for LEWEX. In *Directional Ocean Wave Spectra*, ed. R. C. Beal. Baltimore, Maryland: The Johns Hopkins University Press, pp. 167–172.

Zhang, M. Y. and Y. S. Li, 1996. The synchronous coupling of a third-generation wave model and a two-dimensional storm surge model. *Ocean Engng* **6**, 533–543.

1997. The dynamic coupling of a third-generation wave model and a 3d hydrodynamic model through boundary layers. *Continental Shelf Res.* **17**, 1141–1170.

Index